ADVANCES IN
COMPUTATIONAL BIOLOGY

Volume 2 • 1996

ADVANCES IN COMPUTATIONAL BIOLOGY

Editor: HUGO O. VILLAR
*Terrapin Technologies Inc.
South San Francisco, California*

VOLUME 2 • 1996

JAI PRESS INC.

Greenwich, Connecticut   London, England

Copyright © 1996 by JAI PRESS INC.
55 Old Post Road, No. 2
Greenwich, Connecticut 06836

JAI PRESS LTD.
The Courtyard
28 High Street
Hampton Hill, Middlesex TW12 1PD
England

All rights reserved. No part of this publication may be reproduced, stored on a retrieval system, or transmitted in any form or by any means, electronic, mechanical, photocopying, filming, recording, or otherwise, without prior permission in writing from the publisher.

ISBN: 1-55938-979-6

Manufactured in the United States of America

CONTENTS

LIST OF CONTRIBUTORS   vii

PREFACE
   Hugo O. Villar   ix

COMPUTER-ASSISTED IDENTIFICATION OF PROTEIN SORTING SIGNALS AND PREDICTION OF MEMBRANE PROTEIN TOPOLOGY AND STRUCTURE
   Gunnar von Heijne   1

COMPUTATIONAL APPROACH TO LIPID-PROTEIN INTERACTIONS IN MEMBRANES
   Ole G. Mouritsen, Maria M. Sperotto, Jens Risbo, Zhengping Zhang, and Martin J. Zuckermann   15

PROTEIN MODEL BUILDING USING STRUCTURAL SIMILARITY
   Raúl E. Cachau   65

STATISTICAL ANALYSIS OF PROTEIN SEQUENCES
   Volker Brendel   121

PROGRESS IN LARGE-SCALE SEQUENCE ANALYSIS
   Jean-Michel Claverie   161

COMPUTATIONAL GENE IDENTIFICATION: UNDER THE HOOD
   James W. Fickett   209

THE ARCHITECTURE OF LOOPS IN PROTEINS
   Anna Tramontano   239

INDEX   261

LIST OF CONTRIBUTORS

Volker Brendel　　　　　　　　Department of Mathematics
　　　　　　　　　　　　　　　　Stanford University
　　　　　　　　　　　　　　　　Stanford, California

Raúl E. Cachau　　　　　　　　Structural Biochemistry Laboratory
　　　　　　　　　　　　　　　　Frederick Biomedical Supercomputing Center
　　　　　　　　　　　　　　　　National Cancer Institute
　　　　　　　　　　　　　　　　Frederick Cancer Research and Development Center
　　　　　　　　　　　　　　　　Frederick, Maryland

Jean-Michel Claverie　　　　　Structure and Genetic Information
　　　　　　　　　　　　　　　　CNRS
　　　　　　　　　　　　　　　　Marseille, France

James W. Fickett　　　　　　　Theoretical Biology and Biophysics Group
　　　　　　　　　　　　　　　　Los Alamos National Laboratory
　　　　　　　　　　　　　　　　Los Alamos, New Mexico

Ole G. Mouritsen　　　　　　　Department of Physical Chemistry
　　　　　　　　　　　　　　　　The Technical University of Denmark
　　　　　　　　　　　　　　　　Lyngby, Denmark

Gunnar von Heijne　　　　　　Department of Biochemistry
　　　　　　　　　　　　　　　　Stockholm University
　　　　　　　　　　　　　　　　Stockholm, Sweden

Jens Risbo　　　　　　　　　　Department of Physical Chemistry
　　　　　　　　　　　　　　　　The Technical University of Denmark
　　　　　　　　　　　　　　　　Lyngby, Denmark

Maria M. Sperotto　　　　　　 Department of Physical Chemistry
　　　　　　　　　　　　　　　　The Technical University of Denmark
　　　　　　　　　　　　　　　　Lyngby, Denmark

Anna Tramontano	Biocomputing Unit Instituto di Ricerche di Biologia Moleculare P. Angeletti Pomezia, Italy
Zhengping Zhang	Centre for the Physics of Materials Department of Physics McGill University Montreal, Quebec, Canada
Martin J. Zuckermann	Centre for the Physics of Materials Department of Physics McGill University Montreal, Quebec, Canada

PREFACE

The second volume of this collection finds us at a very exciting time. The human genome project continues to generate results that require the efficient use of computational methods for storage and manipulation of large volumes of data. From the side of computational biology, the excitement translates in the development of modern techniques and algorithms. The data stored are extremely heterogeneous in nature and certainly not error free, which possess unique problems that require the use of the most modern database technologies and statistical analysis of the information. A significant portion of this volume is devoted to the developing areas of research and to the needs created by the accumulation of genomic data.

Dr. Volker Brendel presents in this series a review of the methods that are used in the analysis of protein sequences. His chapter covers many of the aspects that are involved in the statistical analysis of single, pairs and multiple sequences. Dr. James Fickett reviews the methodologies for gene identification. The work shows how today methodologies can be useful in projects of medical and biological importance. Dr. Jean-Michel Claverie also shows another side to the problem, such as the large-scale sequence analysis. This chapter addresses some of the methodologies that deal with the problems presented by artifactual matches, as well as low-complexity sequences. The methods are applied to the identification of exons in mammalian genomic sequences. Dr. Gunnar von Heijne describes the methods useful to sort out signal peptides which may allow their use to design ligands with different subcellular compartments specificities.

Preface

In parallel, other areas of biology continued to benefit from the use of computers. Computational tools in structural biology continue to be essential to model both the structure and the molecular processes that occur in a biological environment, an area where computational biology has traditionally been of use. This volume also covers some of those areas of research.

Because the number of sequences available is growing at a faster pace than the number of three dimensional structures available for them, methods that allow structural predictions based on sequence are increasingly important. Homology modeling is one of the methods aimed to reducing that gap. Dr. Raúl Cachau introduces some new ideas that can be readily applied to the generation of structures based on a closely related protein. A related issue is the difficulties that exist in predicting the conformation of loops in proteins. Dr. Anna Tramontano reviews our current understanding of the architecture of loops in proteins, an essential step to improve our ability to predict structure. Finally, there are a series of biological systems that are also of interest beyond proteins. We covered the treatment of some of them in the previous volume. In this volume we present the work by Dr. Ole Mouritsen and coworkers on the study of lipid protein interactions.

I am thankful to the all contributors to the series for their remarkable efforts in producing such an interesting volume.

Hugo O. Villar
Editor

COMPUTER-ASSISTED IDENTIFICATION OF PROTEIN SORTING SIGNALS AND PREDICTION OF MEMBRANE PROTEIN TOPOLOGY AND STRUCTURE

Gunnar von Heijne

	Abstract	2
I.	Introduction	2
II.	Sorting Signals in Proteins	2
	A. Signal Peptides	3
	B. Mitochondrial Targeting Peptides	5
	C. Chloroplast Transit Peptides	6
	D. Nuclear Localization Signals	6
	E. Other Organelles	7
III.	Topology and Structure of Membrane Proteins	7
	A. Helix-Bundle Proteins	8
	B. β-Barrel Proteins	9
IV.	Conclusions	9
	Acknowledgment	10
	References	11

ABSTRACT

Many proteins contain sorting signals that serve to target them for import into different subcellular compartments. Similar signals also guide the insertion of integral membrane proteins into lipid bilayers. This chapter describes current methods for detecting such signals in amino acid sequences.

I. INTRODUCTION

Many functional characteristics of proteins can be correlated with more or less well-defined linear motifs in their amino acid sequences. The best example is provided by the PROSITE database which currently holds some 800 different motifs (Bairoch, 1993). To be of any use, however, a motif must be sufficiently narrowly defined to avoid picking up too many false positive hits, and at the same time sufficiently tolerant to match most known true positives. This is certainly not always the case, a fact that attests to the importance of long-range interactions and three-dimensional structure in proteins.

Nevertheless, many simple motifs work surprisingly well and have considerable predictive power. This chapter deals with motifs known to be important for intracellular protein sorting, i.e., with sequence elements that serve as "address labels" and allow the cell to transport a given class of proteins to a given subcellular organelle. As it turns out, related motifs also guide membrane proteins into the lipid bilayer. In most cases, these motifs are encoded within a simple, contiguous stretch of the chain, and are thus quite easy to identify in the primary sequence.

II. SORTING SIGNALS IN PROTEINS

The general problem of sorting proteins between different subcellular compartments is faced by all cells. In a eukaryotic cell one can distinguish many tens of distinct organelles or suborganellar compartments, whereas a simple Gram-negative bacterium like *Escherichia coli* has only five: the cytoplasm, the inner membrane, the periplasmic space, the outer membrane, and the surrounding medium. Secretion of proteins through both membranes to the medium is a complicated process requiring a battery of gene products, some of which are parts of the general inner membrane translocation machinery, others which are specific for the particular protein being exported (Pugsley, 1993). The latter components act on signals that are not well understood and that cannot yet be identified in the sequence a priori; therefore, our discussion will deal only with the signals specifying translocation across the inner membrane and integration into the outer membrane. Similar signals specify targeting to the secretory pathway in eukaryotic cells, whereas other kinds of signals serve to route proteins into the nucleus, the mitochondrion, or the chloroplast.

Obviously, the cell can recognize and act on these different signals with high specificity—missorting is believed to be a rare event *in vivo*—and it is tempting for the theoretical biologist to try to reproduce this feat on the computer. Indeed, the best expert systems can already correctly place around 80% of the proteins in a sample of bacterial sequences in one of four locations (Nakai and Kanehisa, 1991), and around 60% of the proteins in a sample of eukaryotic proteins in one of 14 locations (Nakai and Kanehisa, 1992).

A. Signal Peptides

Signal peptides are transient N-terminal extensions on the mature polypeptide chain that target proteins for the secretory pathway, i.e., for translocation across the inner membrane of bacteria or the endoplasmic reticulum (ER) membrane of eukaryotic cells. Reviews describing the genetics (Schatz and Beckwith, 1990; Sanders and Schekman, 1992) and biochemistry (Wickner et al., 1991; Rapoport, 1992) of these processes are available; for our purposes, only the design of the signal peptide is important.

The basic design of signal peptides has been defined both by statistical and experimental studies (Gierasch, 1989; von Heijne, 1990). Signal peptides generally have three distinct regions: a short positively charged, N-terminal domain (n-region), a central hydrophobic stretch of 7–15 residues (h-region), and a more polar C-terminal part of 3–7 residues that defines a processing site for the signal peptidase enzyme (c-region).

All three regions can be quite efficiently recognized by simple algorithms based on hydrophobicity calculations and patterns encoded in weight matrices. McGeoch (McGeoch, 1985) devised a method where two parameters are used in the decision function: the hydrophobicity according to the Kyte–Doolittle scale (Kyte and Doolittle, 1982) of the most hydrophobic 8-residue stretch in the first 30 residues of the protein (= "peak hydrophobicity") and the total length of the uncharged stretch encompassing the peak hydrophobicity segment (= "UR length"). On a two-dimensional plot of "peak hydrophobicity" vs. "UR length", signal peptides are well separated from N-terminal regions of cytoplasmic proteins, with the number of false negatives and false positives being no more than ~5%.

Signal peptidase I cleavage sites can be rather well-predicted by a simple weight-matrix approach, where the position-specific amino acid frequencies in positions −13 to +2 (the cleavage site is between positions −1 and +1) are used to calculate an overall score (von Heijne, 1986c; Folz and Gordon, 1987; Popowicz and Dash, 1988; Daugherty et al., 1990). With this method, the correct site is identified ~75% of the time.

Neural network methods have also been applied to detect signal peptides, and it has been reported that the number of false positives is around ~5% if one first screens with the weight matrix described above, and then allows the neural net to further prune the set of acceptable sequences (Ladunga et al., 1991). In another

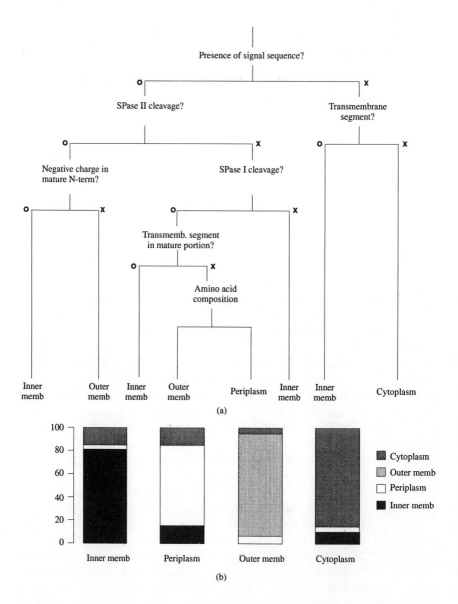

Figure 1. a) Reasoning tree used by Nakai & Kanehisa (1991) for predicting localization sites in prokaryotic proteins. b) Prediction results on a sample of 106 bacterial proteins (with permission from the publisher).

study, a Perceptron-type neural network was trained to predict signal peptidase I cleavage sites in *E. coli* proteins with apparent success (Schneider et al., 1993).

An interesting extension of these approaches has been pioneered by Nakai and Kanehisa (1991). They have developed an expert system that not only detects the presence or absence of an N-terminal signal peptide but also tries to decide whether a protein having a signal peptide will end up in the inner membrane, the periplasm, or the outer membrane. The basic strategy is outlined in Figure 1(a). First, a decision function based on McGeoch's method (see above) is used to decide whether or not a signal peptide is present within the first 50 residues of the protein. Second, a simple amino acid motif based on known sequences (von Heijne, 1989) is used to decide whether the signal peptide will be cleaved by the signal peptidase II enzyme (Lsp) which cleaves only a subset of all exported proteins (the lipoproteins). If this test is passed, the protein is allocated either to the inner or outer membrane, depending on the identity of the residue in positions +2 and +3 (Yamaguchi et al., 1988).

If no signal peptidase II site is found, the weight-matrix method described above is used to decide whether or not the signal peptide will be cleaved by signal peptidase I (Lep). If the answer is no, it is assumed that the protein remains anchored to the inner membrane by its uncleaved signal peptide; if yes, putative transmembrane segments (i.e., long stretches of hydrophobic amino acids; see below) are sought using a decision function derived by Klein et al. (Klein et al., 1985). If such a segment is found, the protein is assumed to be in the inner membrane; if not, it can either be periplasmic or in the outer membrane. This is decided on the basis of the overall amino acid composition of the chain, using an optimized decision function derived by principal component analysis. Finally, if no signal sequence is found among the first 50 residues, there may still be transmembrane segments in other parts of the chain. Again, if such a segment is found, the protein is in the inner membrane; if not, it is predicted to be cytoplasmic. As can be seen in Figure 1(b), this method does a good job of allocating proteins between the four different locations. Overall, 88 out of 106 proteins (83%) were correctly localized in the original study.

B. Mitochondrial Targeting Peptides

It appears from both theoretical (von Heijne, 1986b) and experimental (Roise et al., 1988; Bedwell et al., 1989; Endo et al., 1989; Lemire et al., 1989) studies that mitochondrial targeting peptides form amphiphilic α-helices with one hydrophobic and one positively charged face. In addition, they are characterized by at least three different cleavage site motifs (Hendrick et al., 1989; Gavel and von Heijne, 1990a), one of which has a high predictive value (the Arg-X-Tyr↓(Ser/Ala) motif found in ~25% of all targeting peptides).

Different discriminant functions based on these features were tried by Nakai and Kanehisa (1992), but they concluded that the best discrimination (~90% correct predictions on the training set) was achieved by a simple function based on overall amino acid composition of the N-terminal 20 residues of the chain (mitochondrial targeting peptides tend to be rich in arginines and lack acidic residues). Amphiphilicity is thus a rather weak predictor, possibly because it will be displayed by a large proportion of random sequences with the overall amino acid composition of mitochondrial targeting peptides (Flinta et al., 1983; Gavel et al., 1988).

The amphiphilic targeting peptides deliver proteins to the mitochondrial matrix compartment. Some proteins, however, are destined for the intermembrane space between the outer and inner mitochondrial membranes. In such cases, the matrix-targeting signal is often attached to a second signal that seems to be composed of a cluster of positively charged residues followed by a stretch of uncharged residues (von Heijne et al., 1989; Beasley et al., 1993; Schwarz et al., 1993), a structure reminiscent of the signal peptides.

C. Chloroplast Transit Peptides

Chloroplast transit peptides are highly variable in length and very rich in hydroxylated amino acids, serine in particular. They are largely devoid of acidic residues. Their cleavage sites are not well conserved, though the motif (Val/Ile)-X-(Ala/Cys)↓Ala is found in about 30% of the sequences (Gavel and von Heijne, 1990b). It is not known whether they form any specific secondary structure (von Heijne and Nishikawa, 1991), and it has been suggested that part of their targeting function may be through interactions with chloroplast-specific lipids (van't Hof et al., 1993).

As with the mitochondrial targeting peptides, discrimination based on overall amino acid composition seems to work well (Nakai and Kanehisa, 1992). In fact, the mitochondrial and chloroplast sorting signals can be efficiently discriminated from each other on the simple basis of whether the fraction of serine, f_{Ser}, is smaller or greater than $0.07 + 1.4 \times f_{Arg}$ (von Heijne et al., 1989).

Transit peptides target proteins to the stromal compartment, and further transport into the thylakoids normally depends on a second sorting signal located immediately downstream of the transit peptide. Such thylakoid transfer domains look very much like bacterial signal peptides with n-, h-, and c-regions, and even have similar cleavage sites (Halpin et al., 1989; von Heijne et al., 1989; Shackleton and Robinson, 1991). Thylakoid transfer domains can be identified by the same methods used for signal peptides (Nakai and Kanehisa, 1992), and a weight matrix for predicting their cleavage sites has also been published (Howe and Wallace, 1990).

D. Nuclear Localization Signals

Nuclear localization signals are not N-terminally located and are not removed from the mature protein (Silver, 1991). It has been suggested that a bipartite consensus motif is responsible for targeting: two basic residues, a spacer of any ten residues, and a second basic cluster with at least three out of five basic residues (Dingwall and Laskey, 1991). About 50% of all known nuclear proteins have this motif, whereas it is much rarer in nonnuclear proteins (~5%). In addition, nuclear proteins can often be identified on the basis of their very high overall content of basic residues. A cutoff level of 20% or more arginine+lysine apparently works well (Nakai and Kanehisa, 1992).

E. Other Organelles

A number of other organelles in eukaryotic cells also import or retain proteins. Certain motifs have been implicated in such sorting events and can easily be sought in new sequences.

Many, but not all, peroxisomal proteins end with the tripeptide Ser-Lys-Leu-COOH or a variant thereof (Gould et al., 1989; Swinkels et al., 1992), and this turns out to be an efficient discriminator (Nakai and Kanehisa, 1992).

Many organelles within the secretory pathway collect their specific complement of proteins from proteins in transit toward the plasma membrane by virtue of specific retention signals. Such proteins thus have an N-terminal signal peptide that targets to the endoplasmic reticulum (ER), and also contain a retention or retrieval signal that prevents them from progressing all the way to the plasma membrane. Lumenal ER proteins normally end with the tetrapeptide Lys-Asp-Glu-Leu-COOH (Pelham, 1990); again, this is an efficient discriminator (Nakai and Kanehisa, 1992). Golgi retention signals generally map to transmembrane segments, but no specific motifs have been detected so far (Hurtley, 1992; Machamer et al., 1993; Nothwehr et al., 1993; Townsley et al., 1993). Some plasma membrane proteins have internalization signals in their cytoplasmic tails that usually contain a critical tyrosine residues located in a turn structure (Bansal and Gierasch, 1991). Finally, lysosomal proteins need to acquire a special sugar group for proper sorting; this modification depends on a patch of residues on the surface of the molecule and thus provides an example of a nonlocal signal (Baranski et al., 1991) that cannot be predicted unless the whole three-dimensional structure can be calculated.

III. TOPOLOGY AND STRUCTURE OF MEMBRANE PROTEINS

Integral membrane proteins span a lipid bilayer one or more times. Two structurally distinct classes are known to exist: the helix-bundle proteins where the transmembrane segments are long hydrophobic α-helices (Rees et al., 1989; Henderson et al., 1990), and β-barrel proteins where the transmembrane segments are in an extended β-conformation, together forming a continuous β-barrel structure (Weiss et al., 1991a; Cowan et al., 1992).

To predict the structure of a protein a priori, one probably needs to be able to simulate the entire folding pathway. For an integral membrane protein, the early steps on the folding pathway are dominated by the membrane insertion process during which the basic topology of the molecule is defined, i.e., the locations and orientations of the transmembrane segments. This section will briefly introduce the current thinking on how certain features of the amino acid sequence dictate the insertion process, and how these ideas can be incorporated into prediction schemes. For a more thorough review of the field, see (von Heijne, 1994).

A. Helix-Bundle Proteins

Helix-bundle proteins are found in the inner membrane of bacteria and in most membrane systems in eukaryotic cells. Their hydrophobic transmembrane segments are reminiscent of signal peptides and can indeed trigger the secretory machinery of both prokaryotic and eukaryotic cells. When placed downstream of a signal peptide, a long hydrophobic segment can also function as a stop-transfer signal, i.e., it will block further translocation of the chain and impart a transmembrane topology to the molecule.

The first step in a topology prediction is thus to pick out candidate transmembrane segments by looking for long stretches of hydrophobic residues (von Heijne, 1987). The most straightforward way is to scan the amino acid sequence with a (possibly nonrectangular) window while calculating some kind of overall hydrophobicity for the residues in the window, and then to use the resulting profile together with a cutoff to predict the candidate segments. Simple schemes use a rectangular window, a hydrophobicity scale based on experimental or statistical data, and a cutoff set by trial and error, though more sophisticated, optimized versions of this basic approach also exist (Klein et al., 1985; Cornette et al., 1987; Edelman and White, 1989; Edelman, 1993).

Even if successful, hydrophobicity analysis does not allow predicting the orientation of the transmembrane segments, i.e., whether their N-terminus points toward the cytoplasmic or the extracytoplasmic compartment. Fortunately, segments of the chain located on opposite sides of the membrane have markedly different amino acid compositions, in particular as regards their content of positively charged amino acids. Thus, segments rich in arginines and lysines are generally not translocated across membranes—the "positive inside" rule (von Heijne, 1986a; von Heijne and Gavel, 1988). This allows predicting the orientation of transmembrane segments quite efficiently and can even be used to choose between different possible topologies when hydrophobicity analysis cannot unambiguously assign all of the transmembrane segments (von Heijne, 1992; Sipos and von Heijne, 1993). For bacterial inner membrane proteins, this prediction strategy works exceedingly well; at our latest count, we could correctly predict the topology of 34 out of 35 bacterial inner membrane proteins with experimentally determined topologies, and correctly identify 222 transmembrane segments with no false negatives and one false positive prediction (our unpublished data).

To arrive at a full three-dimensional structure prediction, one needs to be able to calculate the optimal packing of the transmembrane helices identified in the topology prediction step. This problem is far from solved, although some progress has been reported recently. A full-blown molecular dynamics simulation starting from a number of different helix–helix orientations is one possibility and seems to give reasonable results (Treutlein et al., 1992). A computationally simpler method has been proposed by Efremov et al. (Efremov et al., 1992; Efremov et al., 1992; Efremov et al., 1993), which is based on a calculation of the "hydrophobic interaction energy" between two helices in different relative orientations. The main obstacle in this area is not so much computational ingenuity but rather the very limited number of solved structures; to date, high-resolution models are available for only three helix-bundle proteins (Henderson et al., 1990; Deisenhofer and Michel, 1991; Kühlbrandt and Wang, 1991).

B. β-Barrel Proteins

In the β-barrel proteins, the need to satisfy all of the backbone hydrogen bonds in the membrane-embedded domain is solved by the formation of a closed β-barrel. The X-ray structures of porins from the outer membrane of bacteria provide beautiful illustrations of how hydrophobic amino acids positioned at every second position of the transmembrane β-strands together form a continuous hydrophobic lipid-facing surface, while the inner lining of the barrel has a more polar character (Weiss et al., 1991b; Cowan et al., 1992).

Predictions of the β-strands can be based on a number of features of the amino acid sequence: predicted turn-forming segments that may join adjacent strands (Paul and Rosenbusch, 1985), putative transmembrane strands characterized by hydrophobic amino acids in every second position (van der Ley and Tommassen, 1987), and the frequent occurrence of aromatic amino acids (Phe, Trp, and Tyr) at the ends of the transmembrane strands (Schirmer and Cowan, 1993). Because many of the known porins are quite homologous, it is of course also possible in many cases to predict the structure of one based on the known structure of another (Welte et al., 1991).

IV. CONCLUSIONS

How well can we predict the subcellular location of a protein from its amino acid sequence? Or, put differently, how well can we recognize the various sorting signals? The complexity of the problem is well-illustrated in Figure 2, which shows the reasoning tree used by Nakai and Kanehisa (1992) in their recent expert system implementation of a collection of sorting signal algorithms. Some signals are quite well-understood and can be reliably recognized (e.g., signal peptides), while very little is known about others (e.g., Golgi retention signals). Nevertheless, it is quite encouraging that ~60% of a test set of 106 proteins could be assigned to the correct

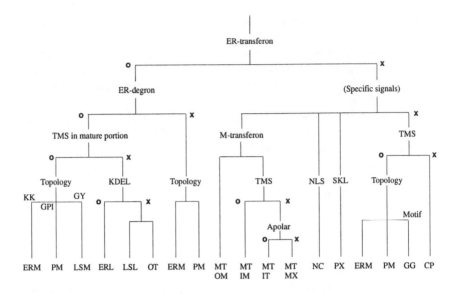

Figure 2. Reasoning tree used by Nakai & Kanehisa (1992) for predicting localization sites in eukaryotic proteins (with permission from the publisher). ERM: ER membrane; PM: plasma membrane; LSM: lysosomal membrane; ERL: ER lumen; LSL: lysosomal lumen; OT: extracellular space; MT-OM: mitochondrial outer membrane; MT-IM: mitochondrial inner membrane; MI-IT: mitochondrial intermembrane space; MT-MX: mitochondrial matrix; NC: nucleus; PX: peroxisome; GG: Golgi complex; CP: cytoplasm.

subcellular location by the expert system. Equally encouraging, it appears that our understanding of the signals that guide the assembly of membrane proteins is sufficiently advanced to allow quite reliable predictions of topology from sequence, at least for bacterial inner membrane proteins (von Heijne, 1992).

For those of us working in the field, a possibly more discouraging possibility is that all of the "easy" problems have now been solved—those where the sorting information is mainly contained within a short, linear stretch of polypeptide—and that further progress will have to rely on much more complex three-dimensional structure prediction algorithms. Incidentally, this will also mean that the prediction field will by necessity become more "professionalized" in the sense that better programming skills will be needed; the day of the polymath who does experiments one day and develops new computer methods the next may well be over before we know it.

ACKNOWLEDGMENT

This work was supported by grants from the Swedish Natural Sciences Research Council (NFR).

REFERENCES

Bairoch, A. (1993). The PROSITE dictionary of sites and patterns in proteins, its current status. Nucleic Acids Res. 21, 3097–3103.

Bansal, A. & Gierasch, L. M. (1991). The NPXY internalization signal of the ldl receptor adopts a reverse-turn conformation. Cell 67, 1195–1201.

Baranski, T. J., Koelsch, G., Hartsuck, J. A., & Kornfeld, S. (1991). Mapping and molecular modeling of a recognition domain for lysosomal enzyme targeting. J. Biol. Chem. 266, 23365–23372.

Beasley, E. M., Muller, S., & Schatz, G. (1993). The signal that sorts yeast cytochrome-b2 to the mitochondrial intermembrane space contains 3 distinct functional regions. EMBO J. 12, 2303–2311.

Bedwell, D. M., Strobel, S. A., Yun, K., Jongeward, G. D., & Emr, S. D. (1989). Sequence and structural Requirements of a mitochondrial protein import signal defined by saturation cassette mutagenesis. Mol. Cell Biol. 9, 1014–1025.

Cornette, J. L., Cease, K. B., Margalit, H., Spouge, J. L., Berzofsky, J. A., & De Lisi, C. (1987). Hydrophobicity scales and computational techniques for detecting amphipathic structures in Proteins. J. Mol. Biol. 195, 659–685.

Cowan, S. W., Schirmer, T., Rummel, G., Steiert, M., Ghosh, R., Pauptit, R. A., Joansonius, J. N., & Rosenbusch, J. P. (1992). Crystal structures explain functional properties of two *E. coli* porins. Nature 358, 727–733.

Daugherty, B. L., Zavodny, S. M., Lenny, A. B., Jacobson, M. A., Ellis, R. W., Law, S. W., & Mark, G. E. (1990). The uses of computer-aided signal peptide selection and polymerase chain reaction in gene construction and expression of secreted proteins. DNA Cell Biol. 9, 453–459.

Deisenhofer, J. & Michel, H. (1991). Structures of bacterial photosynthetic reaction centers. Annu. Rev. Cell Biol. 7, 1–23.

Dingwall, C. & Laskey, R. A. (1991). Nuclear targeting sequences—A consensus? Trends Biochem. Sci. 16, 478–481.

Edelman, J. (1993). Quadratic minimization of predictors for protein secondary structure—Application to transmembrane alpha-helices. J. Mol. Biol. 232, 165–191.

Edelman, J. & White, S. H. (1989). Linear optimization of predictors for secondary structure. Application to transbilayer segments of membrane proteins. J. Mol. Biol. 210, 195–209.

Efremov, R. G., Gulyaev, D. I., & Modyanov, N. N. (1992). Application of 3-dimensional molecular hydrophobicity potential to the analysis of spatial organization of membrane protein domains. 2. Optimization of hydrophobic contacts in transmembrane hairpin structures of Na+, K+-ATPase. J. Protein Chem. 11, 699–708.

Efremov, R. G., Gulyaev, D. I., & Modyanov, N. N. (1993). Application of 3-dimensional molecular hydrophobicity potential to the analysis of spatial organization of membrane domains in proteins. 3. Modeling of intramembrane moiety of Na+, K+-ATPase. J. Protein Chem. 12, 143–152.

Efremov, R. G., Gulyaev, D. I., Vergoten, G., & Modyanov, N. N. (1992). Application of 3-dimensional molecular hydrophobicity potential to the analysis of spatial organization of membrane domains in proteins. 1. Hydrophobic properties of transmembrane segments of Na+, K+-ATPase. J. Protein Chem. 11, 665–675.

Endo, T., Shimada, I., Roise, D., & Inagaki, F. (1989). N-Terminal half of a mitochondrial presequence Peptide takes a helical conformation when bound to dodecylphosphocholine micelles—A proton nuclear magnetic resonance study. J. Biochem. 106, 396–400.

Flinta, C., von Heijne, G., & Johansson, J. (1983). Helical sidedness and the distribution of polar residues in transmembrane helices. J. Mol. Biol. 168, 193–196.

Folz, R. J. & Gordon, J. I. (1987). Computer-assisted predictions of signal peptidase processing sites. Biochem. Biophys. Res. Comm. 146, 870–877.

Gavel, Y., Nilsson, L., & von Heijne, G. (1988). Mitochondrial targeting sequences. Why 'non-amphiphilic' peptides may still be amphiphilic. FEBS Lett. 235, 173–177.

Gavel, Y. & von Heijne, G. (1990a). Cleavage-site motifs in mitochondrial targeting peptides. Protein Eng. 4, 33–37.
Gavel, Y. & von Heijne, G. (1990b). A conserved cleavage-site motif in chloroplast transit peptides. FEBS Lett. 261, 455–458.
Gierasch, L. M. (1989). Signal sequences. Biochemistry 28, 923–930.
Gould, S. J., Keller, G.-A., Hosken, N., Wilkinson, J., & Subramani, S. (1989). A conserved tripeptide sorts proteins to peroxisomes. J. Cell Biol. 108, 1657–1664.
Halpin, C., Elderfield, P. D., James, H. E., Zimmermann, R., Dunbar, B., & Robinson, C. (1989). The reaction specificities of the thylakoidal processing peptidase and *Escherichia coli* leader peptidase are identical. EMBO J. 8, 3917–3921.
Henderson, R., Baldwin, J. M., Ceska, T. A., Zemlin, F., Beckmann, E., & Downing, K. H. (1990). A model for the structure of bacteriorhodopsin based on high resolution electron cryomicroscopy. J. Mol. Biol. 213, 899–929.
Hendrick, J. P., Hodges, P. E., & Rosenberg, L. E. (1989). Survey of amino-terminal proteolytic cleavage sites in mitochondrial precursor proteins: leader peptides cleaved by two matrix proteases share a three-amino acid motif. Proc. Natl. Acad. Sci. USA 86, 4056–4060.
Howe, C. J. & Wallace, T. P. (1990). Prediction of leader peptide cleavage sites for polypeptides of the thylakoid lumen. Nucl. Acid Res. 18, 3417.
Hurtley, S. M. (1992). Golgi localization signals. Trends Biochem. Sci. 17, 2–3.
Klein, P., Kanehisa, M., & DeLisi, C. (1985). The detection and classification of membrane-spanning Proteins. Biochim. Biophys. Acta 815, 468–476.
Kühlbrandt, W. & Wang, D. N. (1991). Three-dimensional structure of plant light-harvesting complex determined by electron crystallography. Nature 350, 130–134.
Kyte, J. & Doolittle, R. F. (1982). A simple method for displaying the hydropathic character of a protein. J. Mol. Biol. 157, 105–132.
Ladunga, I., Czakó, F., Csabai, I., & Geszti, T. (1991). Improving signal peptide prediction accuracy by simulated neural network. CABIOS 7, 485–487.
Lemire, B. D., Fankhauser, C., Baker, A., & Schatz, G. (1989). The mitochondrial targeting function of randomly generated peptide sequences correlates with predicted helical amphiphilicity. J. Biol. Chem. 264, 20206–20215.
Machamer, C. E., Grim, M. G., Esquela, A., Chung, S. W., Rolls, M., Ryan, K., & Swift, A. M. (1993). Retention of a cis-golgi protein requires polar residues on one face of a predicted α-helix in the transmembrane domain. Mol. Biol. Cell 4, 695–704.
McGeoch, D. J. (1985). On the predictive recognition of signal peptide sequences. Virus Res. 3, 271–286.
Nakai, K. & Kanehisa, M. (1991). Expert system for predicting protein localization sites in gram-negative bacteria. proteins. Struct. Funct. Genet. 11, 95–110.
Nakai, K. & Kanehisa, M. (1992). A knowledge base for predicting protein localization sites in eukaryotic cells. Genomics 14, 897–911.
Nothwehr, S. F., Roberts, C. J., & Stevens, T. H. (1993). Membrane protein retention in the yeast golgi apparatus: Dipeptidyl aminopeptidase a is retained by a cytoplasmic signal containing aromatic residues. J. Cell Biol. 121, 1197–1209.
Paul, C. & Rosenbusch, J. P. (1985). Folding patterns of porin and bacteriorhodopsin. EMBO J. 4, 1593–1597.
Pelham, H. R. B. (1990). The retention signal for soluble proteins of the endoplasmic reticulum. Trends Biochem. Sci. 15, 483–486.
Popowicz, A. M. & Dash, P. F. (1988). SIGSEQ: A computer program for predicting signal sequence cleavage sites. CABIOS 4, 405–406.
Pugsley, A. P. (1993). The complete general secretory pathway in gram-negative bacteria. Microbiol. Rev. 57, 50–108.

Rapoport, T. A. (1992). Transport of proteins across the endoplasmic reticulum membrane. Science 258, 931–936.
Rees, D. C., Komiya, H., Yeates, T. O., Allen, J. P., & Feher, G. (1989). The bacterial photosynthetic reaction center as a model for membrane proteins. Annu. Rev. Biochem. 58, 607–633.
Roise, D., Theiler, F., Horvath, S. J., Tomich, J. M., Richards, J. H., Allison, D. S., & Schatz, G. (1988). Amphiphilicity is essential for mitochondrial presequence function. EMBO J. 7, 649–653.
Sanders, S. L. & Schekman, R. (1992). Polypeptide translocation across the endoplasmic reticulum membrane. J. Biol. Chem. 267, 13791–13794.
Schatz, P. J. & Beckwith, J. (1990). Genetic analysis of protein export in *Escherichia coli*. Annu. Rev. Genet. 24, 215–248.
Schirmer, T. & Cowan, S. W. (1993). Prediction of membrane-spanning β-strands and its application to maltoporin. Prot. Sci. 2, 1361–1363.
Schneider, G., Rohlk, S., & Wrede, P. (1993). Analysis of cleavage-site patterns in protein precursor sequences with a perceptron-type neural network. Biochem. Biophys. Res. Commun. 194, 951–959.
Schwarz, E., Seytter, T., Guiard, B., & Neupert, W. (1993). Targeting of cytochrome-b2 into the mitochondrial intermembrane space - specific recognition of the sorting signal. EMBO J. 12, 2295–2302.
Shackleton, J. B. & Robinson, C. (1991). Transport of proteins into chloroplasts—The thylakoidal processing peptidase is a signal-type peptidase with stringent substrate requirements at the -3-Position and -1-Position. J. Biol. Chem. 266, 12152–12156.
Silver, P. A. (1991). How proteins enter the nucleus. Cell 64, 489–497.
Sipos, L. & von Heijne, G. (1993). Predicting the topology of eukaryotic membrane proteins. Eur. J. Biochem. 213, 1333–1340.
Swinkels, B. W., Gould, S. J., & Subramani, S. (1992). Targeting efficiencies of various permutations of the consensus c-terminal tripeptide peroxisomal targeting signal. FEBS Lett. 305, 133–136.
Townsley, F. M., Wilson, D. W., & Pelham, H. R. B. (1993). Mutational analysis of the human KDEL receptor: Distinct structural requirements for golgi retention, ligand binding and retrograde transport. EMBO J. 12, 2821–2829.
Treutlein, H. R., Lemmon, M. A., Engelman, D. M., & Brunger, A. T. (1992). The glycophorin A transmembrane domain dimer: sequence-specific propensity for a right-handed supercoil of helices. Biochemistry 31, 12726–12732.
van der Ley, P. & Tommassen, J. (1987). In: Phosphate Metabolism and Cellular Regulation in Microorganisms (Torriani-Gorini, A., Rothman, F. G., Silver, S., Wright, A., & Yagil, E., Eds.), pp. 159–163. American Society for Microbiology, Washington.
van't Hof, R., van Klompenburg, W., Pilon, M., Kozubek, A., de Kortekool, G., Demel, R. A., Weisbeek, P. J., & de Kruijff, B. (1993). The transit sequence mediates the specific interaction of the precursor of ferredoxin with chloroplast envelope membrane lipids. J. Biol. Chem. 268, 4037–4042.
von Heijne, G. (1986a). The distribution of positively charged residues in bacterial inner membrane proteins correlates with the transmembrane topology. EMBO J. 5, 3021–3027.
von Heijne, G. (1986b). Mitochondrial targeting sequences may form amphiphilic helices. EMBO J. 5, 1335–1342.
von Heijne, G. (1986c). A new method for predicting signal sequence cleavage sites. Nucleic Acids Res. 14, 4683–4690.
von Heijne, G. (1987). Sequence Analysis in Molecular Biology: Treasure Trove or Trivial Pursuit? Academic, San Diego, CA.
von Heijne, G. (1989). The structure of signal peptides from bacterial lipoproteins. Protein Eng. 2, 531–534.
von Heijne, G. (1990). The signal peptide. J. Membr. Biol. 115, 195–201.
von Heijne, G. (1992). Membrane protein structure prediction—Hydrophobicity analysis and the positive-inside rule. J. Mol. Biol. 225, 487–494.

von Heijne, G. (1994). Membrane proteins: from sequence to structure. Annu. Rev. Biophys. Biomol. Struct. 23, 167–192.
von Heijne, G. & Gavel, Y. (1988). Topogenic signals in integral membrane proteins. Eur. J. Biochem. 174, 671–678.
von Heijne, G. & Nishikawa, K. (1991). Chloroplast transit peptides—The perfect random coil? FEBS Lett. 278, 1–3.
von Heijne, G., Steppuhn, J., & Herrmann, R. G. (1989). Domain structure of mitochondrial and chloroplast targeting peptides. Eur. J. Biochem. 180, 535–545.
Weiss, M. S., Abele, U., Weckesser, J., Welte, W., Schiltz, E., & Schulz, G. E. (1991a). Molecular Architecture and electrostatic properties of a bacterial porint. Science 254, 1627–1630.
Weiss, M. S., Kreusch, A., Schiltz, E., Nestel, U., Welte, W., Weckesser, J., & Schulz, G. E. (1991b). The structure of porin from *Rhodobacter capsulata* at 1.8Å resolution. FEBS Lett. 280, 379–382.
Welte, W., Weiss, M. S., Nestel, U., Weckesser, J., Schiltz, E., & Schulz, G. E. (1991). Prediction of the general structure of OmpF and PhoE from the sequence and structure of porin from Rhodobacter-capsulatus—Orientation of porin in the membrane. Biochim. Biophys. Acta 1080, 271–274.
Wickner, W., Driessen, A. J. M., & Hartl, F. U. (1991). The enzymology of protein translocation across the *Escherichia coli* Plasma membrane. Annu. Rev. Biochem. 60, 101–124.
Yamaguchi, K., Yu, F., & Inouye, M. (1988). A single amino acid determinant of the membrane localization of lipoproteins in E. coli. Cell 53, 423–432.

COMPUTATIONAL APPROACH TO LIPID-PROTEIN INTERACTIONS IN MEMBRANES

Ole G. Mouritsen,[1] Maria M. Sperotto, Jens Risbo, Zhengping Zhang, and Martin J. Zuckermann

---

```
  I. Introduction ............................................. 16
     A. Biological Membranes and Integral Proteins ............ 16
     B. Lipid Bilayers as Model Membranes ..................... 18
     C. Phase Transitions and Membrane Organization ........... 18
     D. Principles of Lipid-Protein Interactions in Membranes . 19
     E. A Computational Approach to Lipid-Protein Interactions:
        Computer Simulations on Microscopic Interaction Models  21
 II. Microscopic Models of Lipid-Protein Interactions ......... 23
     A. Model of Pure Lipid Bilayers .......................... 23
     B. Model of Lipid Mixtures ............................... 25
     C. Model of Lipid-Protein Bilayers: The Mattress Model ... 25
     D. Model of Gramicidin-A Dimerization .................... 26
     E. Model of Polypeptide Aggregation and Channel Formation  28
```

[1]OGM is an Associate Fellow of *The Canadian Institute for Advanced Research*

Advances in Computational Biology
Volume 2, pages 15–64
Copyright © 1996 by JAI Press Inc.
All rights of reproduction in any form reserved.
ISBN: 1-55938-979-6

III. Computer-Simulation Techniques 30
    A. Monte Carlo Computer-Simulation Techniques 30
    B. Distribution Functions and Reweighting Techniques 30
    C. Finite-Size Scaling and Detection of Phase Transitions 32
IV. Phase Equilibria in One- and Two-Component Lipid Bilayers 34
    A. One-Component Lipid Bilayers 34
    B. Two-Component Lipid Bilayers 37
    C. Fluctuations, Lipid Domains and Dynamic Membrane Heterogeneity ... 38
V. Lipid Bilayers with Polypeptides and Integral Proteins 40
    A. Critical Mixing and Phase Diagram of Lipid-Polypeptide Mixtures 41
    B. Lipid-Acyl-Chain Order-Parameter Profiles Near Small Proteins 44
    C. Lateral Distribution of Proteins in Lipid Bilayers 45
    D. Compositional Lipid Profiles in Binary Lipid Mixtures
        Near a Protein Wall 51
    E. Structure and Thermodynamics of Binary Lipid Bilayers with
        Large Integral Proteins 53
    F. Relation Between Gramicidin-A Dimer Formation and Bilayer Properties .. 56
    G. Polypeptide Aggregation and Transmembrane Channel
        Formation in Lipid Bilayers 58
VI. Concluding Remarks and Future Perspectives 61
    Acknowledgments 62
    References 62

I. INTRODUCTION

A. Biological Membranes and Integral Proteins

Most biological activity takes place at organized templates and surfaces or near highly structured molecular assemblies. The biological membrane is such a template or interface which mediates a very large number of different biochemical and physiological processes on both the intra- and intercellular level. Considering the very small amount of 'free' water present in biological tissues, it is reasonable to postulate that most enzymatic reactions take place in close association with structured water and structured molecular interfaces such as membranes. The biological membrane is a highly stratified composite material consisting of several components: a fluid lipid bilayer as shown in Figure 1, a polymeric scaffolding network on the intracellular side (the cytoskeleton in eucaryotes) and a carbohydrate-based glycocalyx on the extracellular side. The lipid bilayer provides the membrane with a permeability barrier while the other structures maintain the cell shape and provide the proper interaction and communication with its environment.

It is standard dogma in cell molecular biology (Alberts et al., 1989) to regard the lipid bilayer as a passive structured solvent which in a fairly nonspecific way provides the necessary anchoring place for the active molecules of the cell, i.e., the proteins, enzymes, and receptors. A more modern viewpoint of the membrane (Gennis, 1989; Bloom et al., 1991; Mouritsen and Jørgensen, 1992) acknowledges

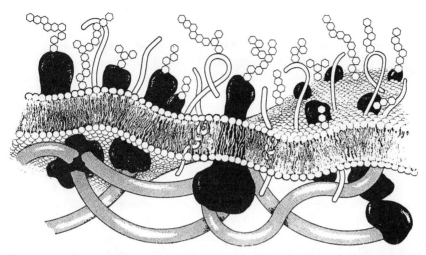

Figure 1. Schematic illustration of a eucaryotic cell membrane which shows the membrane as a composite of a fluid lipid bilayer sandwiched between the carbohydrate glycocalix on the outside and the cytoskeleton on the inside. Various integral and peripheral proteins and polypeptides are indicated as embedded in the bilayer or as superficially attached to the bilayer surface (illustration by Ove Broo Sørensen, The Technical University of Denmark).

that the lipid bilayer plays a much more active but also a somewhat more subtle role in the functioning of the membrane. Though the classical Singer-Nicolson model (Singer and Nicolson, 1972) of the membrane considered the bilayer membrane a mosaic many-component molecular mixture with a high degree of fluidity, it neglected to some extent the fact that the strong lateral correlations in the molecular mixture impart to the membrane a considerable degree of static and dynamic *heterogeneity*. In the context of protein and enzyme function associated with membranes, it is now realized that the physical properties in general, including the nature of the thermodynamic state, and the lateral organization and heterogeneity in particular, are of great importance (Mouritsen and Biltonen, 1993). There is a structure-function relationship hidden in the lipid-protein interactions and their manifestations in membranes in the following sense: the thermodynamic phase of the lipid bilayer and the conformational state of the lipids influence the molecular state of the proteins while at the same time the presence of the proteins affects the global and local organization of the lipid bilayer.

It is the viewpoint of this review that lipids are crucial in providing the proper substrate and structural solvent for the active molecules of the cell and that the lipids themselves play an active and functional role in biochemical processes associated with membranes.

B. Lipid Bilayers as Model Membranes

The lipid molecules which form the fluid bilayer of the biological membrane are amphiphilic molecules which spontaneously form lipid aggregates when dissolved in water (Cevc and Marsh, 1985). The morphology of these aggregates depends on the lipid species and the environmental conditions. Some lipids form lamellar bilayers which are the common lipid structural elements of biological membranes. However, nonlayer lipid phases, such as hexagonal and cubic phases (Tate et al., 1991), may also play a functional role under certain circumstances, such as fusion and vesiculation. In fact, the pure lipid extract of many biological membranes does not form lamellar bilayer phases in the absence of the membrane proteins which therefore must play a stabilizing role when the intact bilayer membrane is formed. We restrict ourselves in this review to situations involving the lamellar phase of lipid-water solutions only.

Lamellar lipid bilayers are unique models of the fluid lipid-bilayer component of cell membranes (Bloom et al., 1991). They can be formed by synthetic lipids or the lipid extract of real cell membranes. The composition of the bilayers can be varied in a systematic and well-controlled fashion and one can study them as a well-defined physical and physicochemical system. The bilayers can form into small or large single- or multilamellar vesicles (or liposomes) and they can be studied experimentally as either oriented or nonoriented assemblies.

Other membrane components can be incorporated into lipid bilayers, e.g., by reconstituting polypeptides or integral membrane proteins into the bilayers. In this fashion it is possible to prepare a series of model systems of biological membranes with an increasing level of complexity.

C. Phase Transitions and Membrane Organization

Lipid membranes are macroscopic or mesoscopic systems (large unilamellar vesicles typically contain 10^6 molecules) and should therefore be considered as many-body systems. This is a very important point often overlooked in the standard biochemical and biological literature on membranes. Due to the many-body nature of the system, it possesses properties which are highly nontrivial and which cannot be understood in terms of the properties of the individual molecules alone. The large assembly of molecules displays cooperative and correlated effects which are in no simple way dictated by the genetic code. This may be the reason why conventional molecular biologists often tend to neglect the role played by lipids in the molecular processes of life.

A clear manifestation of the many-particle nature of lipid bilayers is that they can undergo phase transitions. Indeed, the various morphological forms of the lipid aggregates are connected by phase transitions and the lamellar lipid bilayer state appears in several thermodynamic phases of different structure. The transitions between the different phases can be driven by temperature, electric field, composition, pH, ionic strength, as well as by interactions with other lipid bilayers. One of the more interesting lipid-bilayer phase transitions from the viewpoint of lipid-protein interactions, known as the main transition, takes the bilayer from a low-temperature solid (gel) phase to a high-temperature fluid (liquid-crystalline) phase (Mouritsen, 1991). This phase transition can also be driven by nonthermal effects. The gel phase is a two-dimensional crystalline phase in which the acyl chains have a high degree of conformational order. The crystalline lattice of lipid chains melts at the main phase transition and the bilayer becomes a two-dimensional liquid. The acyl chains have a considerable degree of conformational disorder in the fluid phase. During the main transition the hydrophobic thickness of the bilayer decreases dramatically, typically around 10–20%, depending on the lipids under consideration. This considerable change in thickness is of definite interest in connection with lipid-protein interactions because of the mechanical constraints imposed on the integral proteins by the lipid bilayer.

The main phase transition in one-component lipid bilayers closely resembles a first-order transition which is strongly influenced by density fluctuations (Mouritsen, 1991; Risbo et al., 1995). These density fluctuations are a direct consequence of the cooperative many-particle nature of the bilayer. They manifest themselves macroscopically in terms of thermal anomalies in response functions, such as the specific heat and the lateral compressibility. Microscopically, the fluctuations lead to dynamic formation of clusters or domains of highly correlated lipids, i.e., a substantial degree of dynamic heterogeneity. Such effects are crucial for the lateral organization of the lipid bilayer. Furthermore, when several different lipid species are present, the bilayer displays a very rich phase behavior with regions of static lateral phase separation. Due to the fluctuations in the bilayer, even the thermodynamic one-phase regions of the phase diagram of a multicomponent lipid bilayer mixture may sustain local heterogeneous structures due to density or compositional fluctuations (Jørgensen et al., 1993). The dynamic heterogeneity of lipid bilayers can also act as a vehicle for a variety of active and passive membrane functions (Mouritsen and Jørgensen, 1992; Mouritsen and Biltonen, 1993).

D. Principles of Lipid-Protein Interactions in Membranes

The pictorial description of the membrane in Figure 1 shows that there are basically two different classes of proteins associated with membranes: peripheral and integral proteins. In this review we focus on integral, or transmembrane, proteins and polypeptides and their interaction with membranes. The common structural element of these transmembrane molecules is one or several hydrophobic domains buried in the hydrophobic core of the lipid bilayer. The hydrophilic residues are in the aqueous medium, the polar-head group region of the lipid bilayer, or inside the protein, screened from the hydrophobic membrane core. Therefore membrane proteins themselves have to be amphiphilic like the lipid molecules which make up the fluid lipid bilayer. This imposes some basic mechanical constraints on the proteins which we regard as major determinants of lipid-protein interactions in membranes. These constraints are the reason for the difficulties encountered in obtaining high-resolution information on the three-dimensional structure of integral membrane proteins, because the information implies that the proteins denature when extracted from the membrane reluctantly forming two- or three-dimensional crystals (Kühlbrandt, 1988; 1992) suitable for diffraction analysis.

These considerations led us to propose a basic physical concept of lipid-protein interactions on which this review is focused, *hydrophobic matching* (Mouritsen and Bloom, 1984 and 1993; Mouritsen and Sperotto, 1993). This concept, illustrated schematically in Figure 2, emphasizes that part of lipid-protein interactions related to the mismatch between the hydrophobic thickness, d_L, of the lipid bilayer and the hydrophobic length, d_P, of the integral membrane protein or polypeptide. The hydrophobic mismatch is then defined as follows:

$$\text{Hydrophobic mismatch} = |d_P - d_L|. \qquad (1)$$

The concept embodied in Eq. (1) implies that a mismatch may arise from changes in the thickness of the lipid bilayer. These may be due to both global and local changes in the bilayer structure. The matching concept immediately suggests several interesting situations for study, such as the interaction of long and short polypeptides with a certain lipid bilayer or a specific protein in thick and thin bilayers. The thickness of the bilayer can be varied by changing either the molecular composition or a thermodynamic parameter such as temperature. It is clear from the picture in Figure 2 that lipid-protein interactions induced by hydrophobic mismatch will lead to an indirect protein-protein interaction via perturbations in the lipid matrix.

Without doubt there are other contributions to the interactions between lipids and proteins in membranes, such as electrostatic forces. We neglect these other forces in this review and only focus on the effects due to hydrophobic matching. This is both a strength and weakness of our approach.

The concept of hydrophobic matching has developed both as a means of rationalizing the manner in which lipid-protein interactions manifest themselves in spectroscopic studies (Bloom and Smith, 1985) and as a sufficiently simple theoretical concept leading to theoretical models from which one can obtain the type of information which makes contact with experimental observations. This paper deals with a particular approach to this kind of modeling lipid-protein interactions where computer simulations are crucial.

The physiological implications of the principle of hydrophobic matching for lipid-protein interactions have recently been reviewed (Mouritsen and Sperotto, 1993) and we shall not consider them further here.

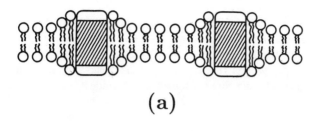

(a)

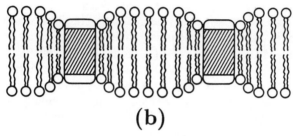

(b)

Figure 2. Schematic illustration of the concept of hydrophobic matching between the hydrophobic length of the transmembrane domain of a protein and the hydrophobic thickness of a lipid bilayer. (a) A 'thin' membrane. (b) A 'thick' membrane.

E. A Computational Approach to Lipid-Protein Interactions: Computer Simulations on Microscopic Interaction Models

The theory of lipid-protein interactions in membranes (Abney and Owicki, 1985; Mouritsen and Sperotto, 1993) has progressed in two major directions. One direction uses phenomenological models, such as solution theories and Landau-type expansions of the free energy. The other uses microscopic molecular and statistical mechanical interaction models. In contrast to phenomenological theoretical models, whose solution in terms of thermodynamic properties can be derived by straightforward minimization of the free energy, the microscopic statistical mechanical interaction models pose a computational problem because they are formulated in terms of the microscopic variables rather than macroscopic field variables. This computational problem can be solved by modern computer-simulation techniques (Mouritsen, 1984; 1990), e.g., Monte Carlo methods and molecular dynamics techniques. A statistical equilibrium ensemble of microstates of the model can be generated using these techniques. From this ensemble of states virtually any macroscopic thermodynamic quantity as well as any microscopic correlation function can be calculated. The accuracy of these numerical simulation methods is limited only by the computer resources available. The accuracy is dictated by the size of the system and the quality of the statistics making up the ensemble from which the thermal averages are derived.

The particular advantages of computer-simulation techniques to study the physical effects of lipid-protein interactions in membranes are manifold. First, the methods automatically account for the correct (nonideal) entropy of mixing and they can handle large assemblies of molecules exhibiting cooperative phenomena. Second, it is straightforward to perform a series of calculations along different paths in the phase diagram and then to calculate response functions from which the details of the phase diagram may be obtained. Third, the simulation provides typical equilibrium configurations of the model from which the lateral structure and organization of the membrane can be determined. Specifically, the membrane state can be assessed in terms of the lipid phase, protein-induced phase separation, and the state of aggregation of the proteins. Moreover, the structure of the lipid membrane at the protein-lipid interface can be studied at the same time.

In many respects a computer simulation resembles an experiment. It produces sets of raw data which must be analyzed and interpreted carefully. In connection with the study of phase equilibria, the two types of 'experiment' often suffer from the same drawback in that they seldom give access to the free energy itself but only derivatives of the free energy, e.g. densities and response functions. In both cases one is left with the problem of interpreting the variation of these derived quantities in terms of a plausible phase diagram. However, as we shall demonstrate in this review, it is possible by modern histogram techniques and finite-size scaling analysis to operate in some cases on the level of the free energy which makes very accurate determinations of phase equilibria possible (Zhang et al., 1993a). Computer simulation results often suffer from relaxation and nonequilibrium effects, in particular for mixtures where the equilibrium is limited by long-range lateral diffusion. In contrast to actual experiments, the computer simulation is always carried out under well-defined and controlled circumstances and in many cases it gives access to quantities difficult to arrive at experimentally. This is particularly important when complex situations, such as lipid-protein interactions, are being studied and where it may be difficult and often impossible under experimental conditions to vary the system parameters most optimally to reveal the nature of a given phenomenon.

This review should be considered topical in its description of a computational approach to lipid-protein interactions in membranes. It does not attempt to provide a full review of the field. A comprehensive list of references to theories and models of lipid-protein interactions can be found in recent reviews (Abney and Owicki, 1985; Mouritsen and Sperotto, 1993). The approach in this paper deals with a computational strategy involving Monte Carlo computer simulations only. Only by using these techniques in conjunction with statistical mechanical models is it feasible at present to treat sufficiently large particle assemblies to permit a study of cooperative phenomena in lipid-protein membranes. In particular, molecular dynamics techniques have recently been applied to aspects of lipid-protein interactions and detailed information has been obtained for membrane systems consisting of a small number of molecules (Edholm and Jähnig, 1988; Edholm and Johansson, 1987).

II. MICROSCOPIC MODELS OF LIPID-PROTEIN INTERACTIONS

The formulation of microscopic interaction models for lipid-protein interactions and their effects on membrane thermodynamics and phase equilibria requires a microscopic model which describes the pure lipid bilayer (Nagle, 1980; Mouritsen, 1991). One of the first models proposed to describe the main transition of the lipid bilayer is that of Marčelja (1974). A series of subsequent and related approaches, particularly suited for computer-simulation calculations, is based on the Pink model (Pink et al., 1980) for the main transition of the lipid bilayer. The Pink model, which will be described in more detail below, is a statistical mechanical lattice model, which takes detailed account of the interactions between the acyl chains of the lipid molecules.

A. Model of Pure Lipid Bilayers

The Pink model allows for a series of ten conformational states of the acyl chains of the lipid molecules. The ten-state model provides a reasonably accurate description of the phase behavior of pure lipid bilayers and the associated density fluctuations because it accounts for the most important conformational acyl-chain states of the acyl chains as well as their mutual interactions and statistics. In the Pink model the bilayer is considered as composed of two monolayer sheets which are independent of each other. Each monolayer is represented by a triangular lattice. The model is therefore a pseudo two-dimensional lattice model which neglects the translational modes of the lipid molecules and focuses on the conformational degrees of freedom of the acyl chains. Because we will be extending the formalism to include several molecular species, we label the lipid variables corresponding to a particular lipid species, A. Each acyl chain can take on one of ten conformational states m, each characterized by an internal energy E_m^A, a hydrocarbon chain length d_m^A (corresponding to half a bilayer, $2d_m^A \sim d_L$), and a degeneracy D_m^A, which accounts for the number of conformations with the same area A_m^A and the same energy E_m^A, where $m = 1, 2,..., 10$. The ten states can be derived from the all-*trans* state in terms of *trans*-gauche isomerism. The state $m = 1$ is the nondegenerate gel-like ground state, representing the all-*trans* conformation, and the state $m = 10$ is a highly degenerate, excited state characteristic of the melted or fluid phase. The eight intermediate states are gel-like states containing kink and jog excitations satisfying the requirement of low conformational energy and optimal packing. The conformational energies E_m^A are obtained from the energy required for a gauche rotation (0.45×10^{-13} erg) relative to the all-*trans* conformation. The values of D_m^A are determined by combinatorial considerations (Pink et al., 1980). The chain cross-sectional areas, A_m^A, are trivially related to the values of d_m^A because the volume of an acyl chain varies only slightly with temperature change. The saturated hydrocarbon chains are coupled by nearest-neighbor anisotropic forces which represent van der Waals and steric interactions. These interactions are formulated in terms of products of shape-dependent nematic factors. The lattice approximation automatically accounts for the excluded volume effects and to some extent for that part of the interaction with the aqueous medium which allows for bilayer integrity. An effective lateral pressure, Π, is included in the model to assure bilayer stability.

The Hamiltonian (which represents the total energy of a microconfiguration) for the pure lipid bilayer can then be written

$$\mathcal{H}^A = \sum_i \sum_{m=1}^{10} (E_m^A + \Pi A_m^A)\, \mathcal{L}_{mi}^A - \frac{J_A}{2} \sum_{\langle i,j \rangle} \sum_{m,n=1}^{10} I_m^A I_n^A \mathcal{L}_{mi}^A \mathcal{L}_{nj}^A , \qquad (2)$$

where J_A is the strength of the van der Waals interaction between neighboring chains, and $I_m^A I_n^A$ is an interaction matrix which involves both distance and shape dependence (Pink et al., 1980). $\langle i,j \rangle$ denote nearest-neighbor indices on the triangular lattice. The Hamiltonian is expressed in terms of site occupation variables $\mathcal{L}_{mi}^A : \mathcal{L}_{mi}^A = 1$ if the chain on site i is in state m, otherwise $\mathcal{L}_{mi}^A = 0$. The model parameters J_A and Π are chosen so as to reproduce the transition temperature and transition enthalpy for a pure dipalmitoyl phosphatidylcholine (DPPC) bilayer (Mouritsen, 1990). The values for other phospholipids were then determined by simple scaling (Jørgensen et al., 1993; Risbo et al., 1995).

Extensions of the ten-state Pink model in Eq. (2) have recently been made by Zhang et al. (1992a; 1992b) to study effects of acyl-chain mismatch among lipid species of the same type and effects due to interactions between the two monolayers of the bilayer. The lipid mismatch model of the main phase transition (Zhang et al., 1992a) in lipid bilayer membranes was introduced to clarify the nature of this transition which has been controversial, experimentally (Biltonen, 1990) and theoretically (Mouritsen, 1991). We shall briefly introduce this model here because it forms the basis for simulation studies of critical mixing of polypeptides in lipid bilayers (Zhang et al., 1993b). The Hamiltonian of the mismatch model for lipid species A can be written as follows:

$$\mathcal{H}_{\text{mis}}^A = \mathcal{H}^A + \frac{\Gamma_A}{2} \sum_{\langle i,j \rangle} \sum_{m,n=1}^{10} |d_m^A - d_n^A|\, \mathcal{L}_{mi}^A \mathcal{L}_{nj}^A . \qquad (3)$$

The lipid mismatch parameter, Γ_A, is related to the hydrophobic effect. Its value was determined to be 0.005×10^{-13} erg/Å (Zhang et al., 1993b).

B. Model of Lipid Mixtures

The ten-state Pink model for a pure lipid bilayer has been extended to binary lipid mixtures (Sperotto and Mouritsen, 1993; Jørgensen et al., 1993; Risbo et al., 1995) by explicitly incorporating a mismatch term which accounts for the incompatibility of acyl chains of different hydrophobic lengths of the two species. The Hamiltonian for a binary mixture of the two lipid species A and B is written

$$\mathcal{H} = \mathcal{H}^A + \mathcal{H}^B + \mathcal{H}^{AB}, \tag{4}$$

where the two first terms describe the interaction between like species and the last term the interaction between different species. The composition of the mixture is given by $x_B = \Sigma_m \langle \mathcal{L}_{im}^B \rangle = 1 - x_A$. The interaction between different lipid species is described by the Hamiltonian

$$\mathcal{H}^{AB} = \frac{-J_{AB}}{2} \sum_{\langle i,j \rangle} \sum_{m,n=1}^{10} (I_m^A I_n^B \mathcal{L}_{im}^A \mathcal{L}_{jn}^B + I_m^B I_n^A \mathcal{L}_{im}^B \mathcal{L}_{jn}^A)$$

$$+ \frac{\Gamma_{AB}}{2} \sum_{\langle i,j \rangle} \sum_{m,n=1}^{10} (|d_{im}^A - d_{jn}^B| \mathcal{L}_{im}^A \mathcal{L}_{jn}^B + |d_{im}^B - d_{jn}^A| \mathcal{L}_{im}^B \mathcal{L}_{jn}^A). \tag{5}$$

The first term in $\mathcal{H}^{AB}$ describes the direct van der Waals hydrophobic contact interaction between different acyl chains. The corresponding interaction constant is taken to be the geometric average $J_{AB} = \sqrt{J_A J_B}$. Γ_{AB} in the second term of Eq. (5) represents the mismatch interaction and has been shown to be 'universal' in the sense that its value does not depend on the lipid species in question (Risbo et al., 1995). The value of the mismatch parameter used in the simulation approach to the statistical mechanics of the model in Eq. (5) was found to be $\Gamma_{AB} = 0.038$ erg/Å (Jørgensen et al., 1993).

C. Model of Lipid-Protein Bilayers: The Mattress Model

Several extensions of the ten-state Pink model have been proposed to account for the lipid-protein interactions in a specific fashion which depends on the conformational states of the lipid chains (for references, see Mouritsen and Sperotto, 1993). Here we consider a particular model in which the lipid-protein interactions have been incorporated into the microscopic Pink model by identifying part of the interaction parameters in terms of hydrophobic matching between the hydrophobic length of the lipid chain and hydrophobic protein length. This is implemented in the spirit of the phenomenological mattress model of lipid-protein interactions (Mouritsen and Bloom, 1984 and 1993; Fattal and Ben-Shaul, 1993). The lipid-protein interactions were included in the model by assuming that the hydrophobic membrane-spanning part of the protein or polypeptide molecule is a stiff, rodlike, and hydrophobically smooth object with no appreciable internal flexibility. In this way the protein is characterized geometrically only by a cross sectional area, A_P (or circumference ρ_P), and a hydrophobic length, d_P. The protein can occupy one or more sites of the lipid lattice depending on its actual hydrophobic circumference.

The Hamiltonian of the microscopic version of the mattress model for a lipid bilayer of species A interacting with a protein P is now written (Sperotto and Mouritsen, 1991a; 1991b)

$$\mathcal{H}^{AP} = \Pi A_P \sum_i L_{Pi}$$

$$+ \frac{\Gamma_{AP}}{4} \left(\frac{\rho_P}{z}\right) \sum_{<i,j>} \sum_{m=1}^{10} (|2d_{im}^A - d_P| \mathcal{L}_{im}^A L_{Pj} + |2d_{jm}^A - d_P| \mathcal{L}_{jm}^A L_{Pi})$$

$$- \frac{J_{AP}}{4} \left(\frac{\rho_P}{z}\right) \sum_{<i,j>} \sum_{m=1}^{10} (\min(2d_{im}^A, d_P) \mathcal{L}_{im}^A L_{Pj} + \min(2d_{jm}^A, d_P) \mathcal{L}_{jm}^A L_{Pi}). \quad (6)$$

The parameter J_{AP} is related to the direct lipid-protein van der Waals-like interaction which is associated with the interfacial hydrophobic contact of the two molecules, while the parameter Γ_{AP} is related to the hydrophobic effect. $L_{Pi} = 0,1$ is the protein occupation variable. $\mathcal{L}_{im}^A$ and L_{Pi} satisfy a completeness relation at each lattice site, $\Sigma_m \mathcal{L}_{im}^A + L_{Pi} = 1$. The model in Eq. (6) can readily be extended to binary lipid mixtures using Eqs. (4) and (5).

The appropriate values of the lipid-protein interaction parameters, Γ_{AP} and J_{AP}, will be discussed below.

D. Model of Gramicidin-A Dimerization

From an experimental point of view, the cation permeability of lipid bilayers increases in the presence of the linear channel forming pentadecapeptide, gramicidin-A (Wagner et al., 1972; Andersen et al., 1992). In this subsection we present a model for forming such gramicidin channels. The model requires us to consider the bilayer as explicitly composed of two monolayers each of which are described by the model for pure bilayers described in Section II.A. This is because gramicidin channels are dimers and every dimer is formed from two monomers with one from each monolayer of the bilayer. It is known that the two monomers of gramicidin-A forming a dimeric channel are linked by six hydrogen bonds (Hladky and Haydon, 1972; Bamberg and Laüger, 1974). When two molecules of gramicidin A, one from each monolayer, form a linear dimer, an ion-specific channel through the bilayer is created. The channel loses its ionic conductivity when it dissociates into monomeric units. Studies of the statistics of channel opening by conductance measurements show that the dimers (channels) and monomers are in thermal equilibrium (Hladky and Haydon, 1972; Bamberg and Laüger, 1973; Zingsheim and Neher, 1974)

$$G + G \rightleftharpoons G_2 \tag{7}$$

where G and G_2 represent the gramicidin monomers and dimers, respectively.

Most experimental results for the mean lifetime of a single channel support the assumption that the mismatch between the gramicidin dimer (a channel) and the lipid bilayer mainly accounts for the dissociation of the gramicidin dimer (Elliott et al., 1983). For example, it was found that the mean lifetime of gramicidin channels in monoacylglycerol-squalene bilayers increases as the bilayer thickness decreases from 28.5 to 21.7 Å while the hydrophobic length of the channel is assumed to be 21.7 Å. This is consistent with the fact that a decrease in mismatch makes the dimeric state more stable.

There have been many theoretical studies on the kinetics of the channel formation and its dependence on membrane structure. A theoretical model was proposed for the relation between the mean lifetime of gramicidin channels and the thickness of lipid bilayers by Elliott et al. (1983) and was modified by Huang (1986) in terms of an elastic bilayer deformation. The basic idea of the model is as follows: When a dimeric gramicidin channel is formed in a membrane of thickness greater than the length of the channel, the membrane deforms locally to accommodate the channel. The restoring force of the deformed membrane will then reduce the stability of the dimer. The dissociation constant k_D can, in this case, be estimated from (Huang, 1986)

$$k_D = \nu e^{-g^*/k_B T}, \tag{8}$$

where g^* is the free energy of activation, which is the energy required to break the hydrogen bonds linking two monomers, and ν is a frequency factor almost independent of temperature.

Here we analyze dimer formation on a phenomenological microscopic level. The assumption that the hydrocarbon chains of lipids adapt to the thickness of the hydrophobic core of a protein in the membrane is the basis of the mattress model of Section II.C., which allows us to formulate a microscopic model for gramicidin channels in the lipid bilayers using the bilayer model discussed in Section II.A. suitably modified to account for the coupling to lipid molecules of the gramicidin monomers or dimers incorporated in the bilayers. By analogy with the interactions between the acyl chains, the interactions between lipid molecules of species A and gramicidin monomers G are then written as

$$\mathcal{H}^{AG} = \sum_{\alpha=1}^{2} \left[\sum_{i} \Pi A_G \mathcal{L}_{Gi}^\alpha - \frac{J_{AG}}{2}\left(\frac{d_G}{d_1}\right) \sum_{\langle i,j \rangle} \sum_{m=1}^{10} I_m^A \mathcal{L}_{im}^\alpha \mathcal{L}_{jG}^\alpha \right] - E_H \sum_i \mathcal{L}_{iG}^1 \mathcal{L}_{iG}^2 +$$

$$+ \frac{\Gamma_{AG}}{2} \sum_{\langle i,j \rangle} \sum_{mm'=1}^{10} |d_m^{A,1} + d_{m'}^{A,2} - 2d_G| \mathcal{L}_{im}^1 \mathcal{L}_{im'}^2 \mathcal{L}_{jG}^1 \mathcal{L}_{jG}^2. \tag{9}$$

$\mathcal{L}_{iG}^{\alpha}$ is the occupation variable for gramicidin monomers which is unity when the i-th site of the α-th monolayer is occupied by a gramicidin monomer and zero otherwise. In this approximation, the hydrophobic part of the gramicidin monomer is assumed to be a stiff rodlike object with no internal flexibility. It can therefore be characterized by a cross-sectional area, A_G, and the length of the hydrophobic core, d_G. J_{AG} is the direct lipid-protein interaction constant, which depends on the properties of gramicidin hydrophobic surface. The parameter, Γ_{AG}, is related to the hydrophobic effect describing the hydrophobic area exposed to water by the longer species. E_H is the strength of the hydrogen bonding between the two monomers of a dimer.

E. Model of Polypeptide Aggregation and Channel Formation

In Section D we proposed a model for a specific case of ionic channel formation in a lipid bilayer environment: the monomer-dimer process for the formation of the gramicidin-A channel. There are many possible mechanisms for channel formation depending on the nature of the components of the system under consideration. In some cases the original molecular structure of the molecule is sufficient to give a complete channel and no self-assembly in the bilayer is therefore required. Another possibility is the self-assembly of channel-forming protein 'monomers' (already dissolved in the bilayer) via lateral diffusion and nonspecific protein-protein association. The latter includes steric interactions between the channel-forming monomers themselves and between channel-forming monomers and the surrounding lipid matrix.

For ease of calculation, we assume that the channel-forming monomers which are subject to lateral diffusion in the bilayer are all identical, i.e. we consider only one type of protein or polypeptide. We also assume that, even though the monomers can be incorporated in the bilayer, they possess some hydrophilic residues in the hydrophobic sequence which prefer to avoid contact with the hydrophobic lipid chains. This requires that the channel-forming monomers interact with one another and with the surrounding lipid matrix via selective anisotropic interactions. A channel would therefore self-organize via the association of several channel-forming monomers into a closed structure with the buried hydrophilic residues residing in the channel interior. The collective interaction from these residues should then be sufficient to allow a conducting water channel to form.

The details of our model for describing channel formation in a lipid bilayer of species A are as follows:

1. The bilayer is represented as a triangular lattice and each channel-forming protein or polypeptide component (referred to here as a protein) as a hexagon which occupies seven lattice sites. The proteins are allowed to diffuse translationally and to rotate on the lattice which can be regarded as a uniform

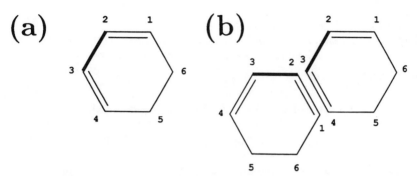

Figure 3. Anisotropic nature of a channel-forming protein component. (a) Model of a single protein showing labeling of corners and indicating sides (1,2) and (3,4) which can share sites with neighboring proteins. Side (2, 3) represents the interior sequence which contains hydrophilic residues. (b) A protein cluster for $n = 2$ as defined in the text.

lipid background because the structure of the lipid molecules are not treated in detail in this model.

2. Each site on the lattice can be occupied by the corners of three proteins at most but the overlap of the centers of two proteins is forbidden. If a protein shares the same site or the same lattice bond with a neighboring protein, neither protein is allowed to rotate.
3. Only contact interactions, either between two neighboring proteins or between a protein and the surrounding lipids, are taken into consideration.
4. The interactions between proteins are assumed to be highly anisotropic as follows. Let us label the corners (lattice sites) on each protein hexagon from 1 to 6 as shown in Figure 3(a). The corners of two neighboring hexagons occupying the same site interact with a mutual repulsion, J_{P1}. If either of the sites are 5 or 6, then $J_{P1} = \infty$. The important interaction for channel formation is the highly anisotropic interaction between the sides of two proteins which are allowed to share the same bond. This is only advantageous when one side is (1, 2) of the first protein and the second side is (3, 4) of the second protein with corner 1 on the same site as corner 4, as shown in Figure 3(b). The interaction in this case is attractive and given by $-J_{P2}$.
5. The outer sites (corners) of proteins interact anisotropically with the uniform lipid background except when more than one such site occupies the same site on the lattice. This protein-lipid interaction is a repulsive interaction with coupling constant J_{AP}. For sites 5 and 6, $J_{AP} = 0$. Therefore a channel or a 'hole' surrounded by proteins self-assembles naturally in the sense that no protein-lipid interaction exists inside the 'hole.'

In this model the lipid bilayer is treated as a uniform medium of no internal molecular structure. This is reasonable because ionic channels are usually observed in the fluid phase of lipid bilayers and the channel-forming monomers are usually much larger than lipid molecules.

III. COMPUTER-SIMULATION TECHNIQUES

A. Monte Carlo Computer-Simulation Techniques

A full description of the implementation of standard Monte Carlo simulations will not be given, but we refer the reader to general references on this subject (Mouritsen, 1984 and 1990). We shall, however, concentrate on describing several new techniques which are extremely powerful for characterizing phase equilibria in lipid bilayers and lipid-protein systems when combined with Monte Carlo simulations.

The fundamental problem in numerical investigations of cooperative phenomena is associated with an unambiguous assessment of the nature of phase transitions as well as a determination of the precise location of the phase boundaries in a phase diagram. Numerical simulations share this problem with laboratory experiments. The root of the problem is that most approaches do not function on the level of the free energy but rather in terms of derivatives of the free energy, such as densities and order parameters (first derivatives) or response functions (second derivatives). In the case of one-component systems, these derivatives are seldom known with sufficient accuracy to discern between, for example, a fluctuation-dominated first-order phase transition and a continuous transition (critical point). In the case of phase coexistence in, e.g., a two-component system, there is no direct way of relating features in the response functions to the precise location of equilibrium phase boundaries (Risbo et al., 1995). However, by novel techniques (Zhang et al., 1993a), this problem can be numerically circumvented and access obtained to that part of the free energy necessary to locate the phase equilibria. These techniques involve calculating distribution functions (histograms) of thermodynamic functions, e.g., order parameters (composition) or internal energy, thermodynamic reweighting of the distribution functions to locate the phase transition or the phase equilibria, and then a subsequent analysis of the size dependence of the reweighted distribution functions by finite-size scaling theory.

B. Distribution Functions and Reweighting Techniques

To use the finite-size scaling method effectively for studying phase equilibria in mixtures requires that a multidimensional distribution function, $\mathcal{P}(\{y_i\}_i, T, L)$, can be calculated very close to coexistence for a set of thermodynamic variables, $\{y_i\}_i$. L is the linear extension of the two-dimensional membrane system which therefore consists of $N = L^2$ particles. For one-component cases, e.g., pure lipid bilayers, this set of thermodynamic variables, $\{y_i\}_i$, contains the internal energy and the order parameter, which usually is taken to be the bilayer area. For a two-component system, the set also includes the relative chemical potential, μ, for the two species. The distribution function for any given variable can be obtained from the complete distribution function as follows:

$$\mathcal{P}(y_j, T, L) = \sum_{i(\neq j)} \mathcal{P}(\{y_i\}_i, T, L). \tag{10}$$

Without losing generality, we now focus on the joint distribution function (or histogram), $\mathcal{P}(E, x, T, L)$, for the internal energy, E, and the composition, x. This distribution function, for a given system size L, is now calculated very accurately close to coexistence (see below for detecting coexistence). For a given set of thermodynamic parameters (T, μ), the distribution function is defined by

$$\mathcal{P}_\mu(E, x, T, L) = \frac{n(E, x, L)\exp[-(E+\mu x)/k_B T]}{\sum_{E,x} n(E, x, L)\exp[-(E+\mu x)/k_B T]}, \tag{11}$$

where $n(E, x, L)$ is the density of states which is independent of temperature.

If $\mathcal{P}(E, x, T, L)$ is calculated with sufficient accuracy for a specific set (T, μ), the distribution function can be determined for a nearby set of parameters (T', μ') employing the reweighting method of Ferrenberg and Swendsen (1988). The method is based on standard thermodynamics, and it leads to the reweighted distribution function

$$\mathcal{P}_{\mu'}(E, x, T', L) = \frac{\mathcal{P}_\mu(E, x, T, L)\exp\left[-\left(\frac{1}{k_B T'} - \frac{1}{k_B T}\right)E - \left(\frac{\mu'}{k_B T'} - \frac{\mu}{k_B T}\right)x\right]}{\sum_{E,x} \mathcal{P}_\mu(E, x, T, L)\exp\left[-\left(\frac{1}{k_B T'} - \frac{1}{k_B T}\right)E - \left(\frac{\mu'}{k_B T'} - \frac{\mu}{k_B T}\right)x\right]}. \tag{12}$$

The two one-dimensional distribution functions can readily be derived from the combined two-dimensional distribution function, $\mathcal{P}(E, x, T, L)$, as follows:

$$\mathcal{P}(E, T, L) = \sum_x \mathcal{P}(E, x, T, L) \tag{13}$$

and

$$\mathcal{P}(x, T, L) = \sum_E \mathcal{P}(E, x, T, L). \tag{14}$$

These distribution functions allow us to calculate average quantities, such as internal energy, order parameter, and response functions at any set of thermodynamic parameters provided that the originally sampled distribution function is accurate enough to contain sufficient statistical information about the relevant part of phase space. Data sets very dense in, e.g., temperature, can therefore be obtained which is a prerequisite for locating the transition point and for performing a detailed finite-size scaling analysis at the transition point.

C. Finite-Size Scaling and Detection of Phase Transitions

Based on the distribution functions described above, phase equilibria can be examined by the powerful method of Lee and Kosterlitz (1991). This method, which is based on the finite-size scaling analysis of distributions derived from computer simulations, constitutes an unambiguous technique for numerically detecting first-order transitions and phase coexistence (Zhang et al., 1993a). The method involves calculating a certain part of the free energy using the distribution functions in Eqs. (12)–(14). Free energylike functions, e.g., $\mathcal{F}(x, T, L)$, can be defined as follows from these distribution functions:

$$\mathcal{F}(x, T, L) \sim -\ln \mathcal{P}(x, T, L). \tag{15}$$

The quantity $\mathcal{F}$ differs from the bulk free energy by a temperature- and L-dependent additive quantity. However, the shape of $\mathcal{F}(x, T, L)$ at fixed T and L is identical to that of the bulk free energy and, furthermore, $\Delta\mathcal{F}(\mu, T, L) = \mathcal{F}(x, T, L) - \mathcal{F}(x', T, L)$ is a correct measure of free energy differences. At a first-order transition, $\mathcal{F}(x, T, L)$ has pronounced double minima corresponding to two coexisting phases at $x = x_1$ and $x = x_2$ separated by a barrier, $\Delta\mathcal{F}(\mu, T, L)$, with a maximum at x_{max} corresponding to an interface between the two phases. The height of the barrier measures the interfacial free energy between the two coexisting phases and is given by

$$\Delta\mathcal{F}(\mu, T, L) = \mathcal{F}(x_{max}, T, L) - \mathcal{F}(x_1, T, L) = \gamma(\mu, T)L^{d-1} + O(L^{d-2}), \tag{16}$$

where d is the spatial dimension of the system ($d = 2$ in the present case). $\gamma(\mu, T)$ is the interfacial free-energy density or interfacial tension. Therefore $\Delta\mathcal{F}(\mu, T, L)$ increases monotonically with L at a first-order transition (or at phase coexistence) corresponding to a finite interfacial tension. The detection of such an increase is an unambiguous sign of a first-order transition and two-phase coexistence. In contrast, $\Delta\mathcal{F}(\mu, T, L)$ approaches a constant at a critical point, corresponding to vanishing interfacial tension in the thermodynamic limit. In the absence of a transition, $\Delta\mathcal{F}(\mu, T, L)$ tends to zero.

The iterative computational method for detecting coexistence proceeds as follows: A trial value of the chemical potential at coexistence, μ_m, for a given temperature is guessed or estimated from a short simulation on a very small lattice. The accurate value of μ_m for that system size is then determined by a long simulation (typically 10^6 Monte Carlo steps per lattice site) at this trial value of μ, and an accurate value of μ_m is determined by the reweighting technique; cf. Eq. (12). If the trial value turns out to be too far from coexistence to allow a sufficiently accurate sampling of the two phases, a new trial value is estimated from the long simulation. The system size is then increased and the value of μ_m for the previous system size is used as trial value for the long simulation on the larger system. This rather lengthy iterative procedure assures a very accurate determination of the chemical potential at coexistence and hence leads to an accurate determination of the compositional phase diagram via the finite-size scaling analysis.

The Ferrenberg–Swendsen reweighting technique becomes more troublesome the larger the size of the system studied because the distribution functions are narrower for larger systems. The method works best for small systems subject to large fluctuations allowing more information about a larger part of phase space to be sampled. A particular problem is associated with the present application of the reweighting to mixtures which involves an extra thermodynamic variable, the chemical potential. Reweighting of data obtained for one value of the chemical potential to another value gives a larger uncertainty for the larger system simply because it is the extensive composition which is involved in the Hamiltonian and hence in the reweighting procedure; cf. Eq. (11). This implies that a knowledge of the value of the chemical potential is required at coexistence with an uncertainty which decreases approximately as L^{-2}. Another more physical bottleneck is the time, τ, associated with the crossing of the free-energy barrier between the two phases. This time increases exponentially with the linear dimension of the two-dimensional model system, $\tau \sim \exp(\gamma L)$. In order to facilitate the reweighting, the actual simulation time has to be much larger than τ. For a given model, this sets a well-defined upper bound for the system sizes which can be studied by these simulation techniques.

First-order transitions in one-component systems are characterized by discontinuities in the first derivatives of the free energy. In the thermodynamic limit, this results in a δ-function singularity for the specific heat, $C(T)$, and a singularity in the susceptibility, $\chi(T)$, at the transition. In a finite system, however, the transition region is broadened, the peaks in $C(T)$ and $\chi(T)$ are finite, and their height increases with increasing linear lattice size, L. Furthermore, the locations of the maxima vary in a size-dependent manner. The maxima grow as L^d in d spatial dimensions and a δ-function is obtained in the thermodynamic limit because the width decreases as L^{-d}. The maximum values of the response functions for a finite d-dimensional lattice scale are

$$C_L^{\max} = a_C + b_C L^d \tag{17}$$

and

$$\chi_L^{\max} = a_\chi + b_\chi L^d, \tag{18}$$

where the a and b parameters are size independent but model-dependent quantities. This finite-size scaling behavior at a first-order phase transition is qualitatively different from that of a continuous transition where, for example, the specific heat maximum diverges as $C_L^{max} \sim L^{\alpha/\nu}$. Here α and ν are the critical exponents characterizing the singularities of the specific heat and the correlation length, respectively. The phase transition temperature, $T_m(L)$, in the finite system can be estimated in several different ways: e.g., from the position of the maximum in the specific heat, $C(L)$, from the position of the maximum in the susceptibility, $\chi(L)$, or from the criterion that the two minima in the free energy are equally deep. In the thermodynamic limit, all three measures of transition temperature should approach the same value, T_m.

IV. PHASE EQUILIBRIA IN ONE- AND TWO-COMPONENT LIPID BILAYERS

A. One-Component Lipid Bilayers

We shall briefly describe the results of model calculations of the mismatch model for one-component lipid bilayers in order to illustrate the strength of the present computational approach to phase transitions in lipid bilayer membranes. The mismatch model is defined by the Hamiltonian in Eq. (3). It has been found that the mismatch model in the case of DPPC has a critical point for a specific value of the mismatch parameter, Γ_A (Zhang et al., 1992a). Above this critical value, the main transition is of first order and below this value there is no phase transition. The free energy functional, $\mathcal{F}(A, T, L)$, derived from the distribution function for the membrane area, A, was calculated for several values of the mismatch parameter to examine the phase behavior of the system and to locate the critical point. It was found that the model did not exhibit a phase transition for values of Γ_A below 4×10^{-16}erg/Å. At this value of Γ_A, the system is at, or extremely close to, the critical point. Above this value of Γ_A, the transition is of first order. The data on which these conclusions are based are presented in Figure 4 which shows that the free energy as a function of area per lipid chain exhibits two minima separated by an energy barrier. The height of this barrier, $\Delta \mathcal{F}(L)$, changes with system size very differently for different values of Γ_A. This can be seen in Figure 5, where $\Delta \mathcal{F}(L)$ is shown as a function of system size for the three values of Γ_A corresponding to Figures 4(a–c). Figure 5 shows that $\Delta \mathcal{F}(L)$ decreases with increasing L for $\Gamma_A = 3 \times 10^{-16}$erg/Å, implying the absence of a transition, and increases with increasing L for $\Gamma_A = 5 \times 10^{-16}$erg/Å, implying the occurrence of a first-order phase transition. For $\Gamma_A = 4 \times 10^{-16}$erg/Å, the height of the barrier does not depend on system size to within calculational error, indicating that the transition for this parameter value is almost at a critical point.

The temperature and size dependence of the specific heat are shown in Figure 6(a) for $\Gamma_A = 5 \times 10^{-16}$erg/Å corresponding to a first-order transition. As the system size increases, the peak height of $C(T)$ increases while the width of the peaks decreases. The peak height scales as expected for a first-order transition, Eq. (17),

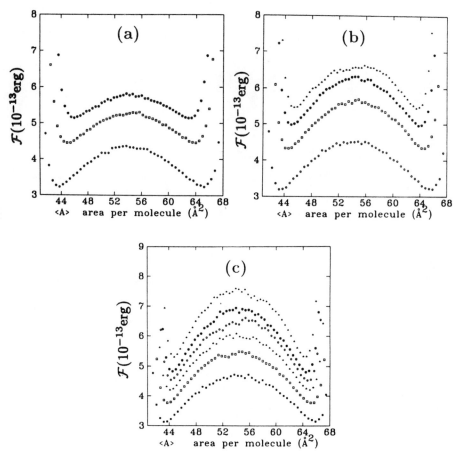

Figure 4. Free energy functional, $\mathcal{F}(T_m(L))$, for the mismatch model, Eq. (3), of the main transition in DPPC lipid bilayers as a function of average molecular area, A, shown for three different values of the mismatch parameter, Γ_A, and for different linear lattice dimensions, L. (a) $\Gamma_A = 3 \times 10^{-16}$ erg/Å and $L = 8, 16, 24$ (from bottom to top). (b) $\Gamma_A = 4 \times 10^{-16}$ erg/Å and $L = 8, 16, 24, 32$ (from bottom to top). (c) $\Gamma_A = 5 \times 10^{-16}$ erg/Å and $L = 8, 12, 16, 20, 24, 32$ (from bottom to top).

as seen in the inset of Figure 6(a). The full finite-size scaling behavior of $C(T)$ is shown in Figure 6(b). The expected scaling behavior for a first-order phase transition is clearly confirmed by this figure. The equilibrium phase transition temperature can be estimated from these data.

The results from this numerical study of the nature of the main transition in the mismatch model for pure lipid bilayers were used to interpret a series of different experimental measurements (Zhang et al., 1992a). Experimentally, it is virtually

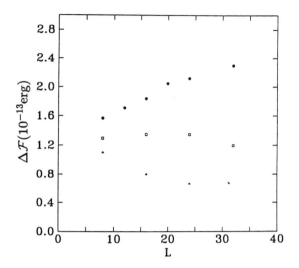

Figure 5. Mismatch model of the main phase transition in DPPC bilayers; cf. Eq. (3). Barrier height, $\Delta\mathcal{F}(T_m(L))$, in the free energy functional at the transition for mismatch values (from bottom to top) $\Gamma_A = 3$ (no transition), 4 (critical point), and 5 (first-order transition) as a function of linear lattice size, L. Γ_A is in units of $\times 10^{-16}$ erg/Å.

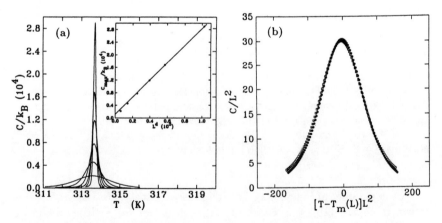

Figure 6. Mismatch model of the main phase transition in DPPC lipid bilayers; cf. Eq. (3). (a) Temperature dependence of the specific heat, $C(T)$, for $L = 8, 12, 16, 20, 24, 32$ and $\Gamma_A = 5 \times 10^{-16}$ erg/Å. (b) Corresponding scaling function for the specific heat. The specific heat is in units of k_B. The insert in (a) shows the finite-size scaling of the maximum of the specific heat; cf. Eq. (17).

Computation of Lipid-Protein Interactions

impossible to distinguish between a weak first-order transition and a pseudotransition dominated by very strong fluctuations. The numerical work presented here supports the hypothesis that lipid bilayers exhibit pseudocritical behavior and are close to a critical point (Biltonen, 1990; Mouritsen, 1991).

B. Two-Component Lipid Bilayers

The possible absence of a phase transition in one-component lipid bilayers poses some intriguing problems as to the existence of phase equilibria in lipid bilayers incorporated with other species, such as other lipids, cholesterol, or proteins. It has recently been shown (Risbo et al., 1995) by a computational approach of the type described in this review that a first-order transition, which is absent in a one-component lipid bilayer, may be induced by mixing in small amounts of another species so as to approach the critical points and the related closed coexistence regions. It is important to come to terms with such phase equilibria before the more complicated problem of lipid-protein interactions are considered in detail.

Figure 7 shows selected results from calculations on the lipid bilayer model, Eq. (2), extended to account for interactions between saturated phospholipid acyl chains with different hydrophobic length; cf. Eqs. (4) and (5), specifically dimyristoyl

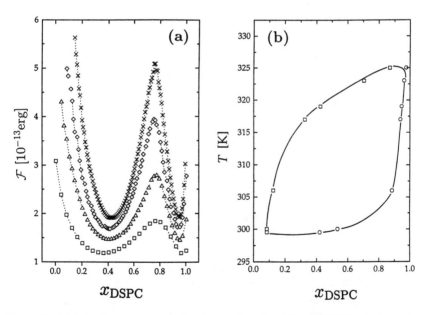

Figure 7. Monte Carlo computer-simulation data for a DMPC-DSPC mixture described by the model in Eqs. (4) and (5). (a) Size dependence of the free energy functional, $\mathcal{F}(x, L)$, evaluated at coexistence at temperature $T = 319$ K. The linear sizes of the systems correspond to $L = 5$ (□), 7 (△), 9 (◇) and 11 (×). (b) Phase diagram.

phosphatidyl choline-distearoyl phosphatidylcholine (DMPC-DSPC) mixtures (Risbo et al., 1995). The size dependence of the compositional distribution function is displayed in Figure 7(a). The family of distribution functions is shown at the appropriate size-dependent chemical-potential values corresponding to phase coexistence. It is seen that, as the system size is increased, there is a dramatic increase in the free energy barrier between the two phases and that the composition of the mixture at the given thermodynamic conditions is given by $x^g_{DSPC} = 0.95$ and $x^f_{DSPC} = 0.42$. This is unambiguous evidence of phase coexistence in the thermodynamic limit and that the interfacial tension tends towards a nonzero value; cf. Eq. (16). From data of this type, the entire phase diagram can be determined very accurately, as shown in Figure 7(b). This phase diagram shows that, even though bilayers of the two pure components, DMPC and DSPC, do not have a phase transition in a strict thermodynamic sense, a transition is induced by mixing the two components. This leads the way to two critical mixing points in the phase diagram. Between these two critical points, there is a dramatic two-phase coexistence region. The phase diagram found by these techniques is rather close to experimental results. For all practical purposes in the context of lipid-protein interactions, which we will consider next, this peculiar nature of the phase diagram will have negligible influence on the phase equilibria in lipid-protein systems. In contrast, the strong fluctuations near the pure lipid limit will strongly influence the manifestations of the lipid-protein interactions.

C. Fluctuations, Lipid Domains and Dynamic Membrane Heterogeneity

Because computer simulation techniques applied to statistical mechanical models fully allow for the correlations in the thermal and compositional fluctuations, they can provide a direct picture of the microscopic phenomena which underlie macroscopic events, e.g., in terms of lateral membrane organization (Jørgensen et al., 1993; Huang and Feigenson, 1993). Figure 8 gives examples of lateral organization obtained from computer simulations near the main transition in a pure DPPC lipid bilayer and for a binary DMPC-DSPC lipid mixture in the fluid phase. The two frames illustrate the effect of density (area) fluctuations and compositional fluctuations, respectively.

The density fluctuations in Figure 8(a) manifest themselves in the formation of domains or clusters of correlated lipids of a structure and a density different from that of the bulk equilibrium lipid matrix. These domains are dynamic and highly fluctuating entities which are consequences of the cooperative fluctuations of the bilayer. The range over which the fluctuations are operative is described by a coherence length which is a measure of the average domain size. The average domain size depends on temperature and attains a maximum at the gel-to-fluid phase transition. For example for DPPC bilayers, the actual lipid-domain size may become as large as several hundred lipid molecules. Hence, the dynamic membrane heterogeneity leads to membrane organization on a mesoscopic length scale. The

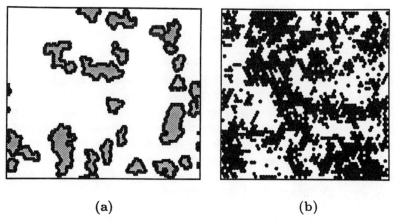

(a)   (b)

Figure 8. Dynamic membrane heterogeneity: Planar organization of lipid-bilayer model systems as obtained from computer-simulation calculations. The one-component DPPC bilayer in (a) displays density (area) fluctuations and the binary DMPC-DSPC lipid mixture in (b) displays compositional fluctuations. In (a) dark and light areas denote the gel and the fluid phase, respectively, and in (b) dark and light areas denote fluid-phase DMPC and fluid-phase DSPC, respectively.

average domain size depends on the lipid species in question and increases as the acyl-chain length decreases (Ipsen et al., 1990).

The compositional fluctuations in Figure 8(b) are striking characteristics of lipid mixtures in the fluid phase (Jørgensen et al., 1993), here revealed by a direct computer-simulation calculation. Hence, thermodynamic one-phase regions of mixtures may imply a considerable lateral structure of the membrane and a dynamically heterogeneous organization. The coherence length of the compositional fluctuations increases as the temperature approaches the phase boundary. Moreover, the local structure in the fluid phase is more pronounced the more nonideal the lipid mixture becomes.

One macroscopic consequence of the dynamic membrane heterogeneity is that response functions, such as the specific heat or the compositional susceptibility, exhibit anomalies as a function of temperature. In particular, the specific heat has a sharp peak at the main transition. The peak intensity decreases for decreasing chain length whereas the intensity in the wings of the transition increases. Hence, away from the main transition, shorter acyl-chain lengths lead to a higher degree of membrane heterogeneity. Another consequence of the dynamic heterogeneity is an enhancement of the passive membrane permeability (Ipsen et al., 1990).

The effect of cholesterol incorporated into pure lipid bilayers in large amounts, $\geq 20\%$, is to lower the passive permeability. However, it has been shown both experimentally and theoretically that cholesterol at low concentrations increases the passive ion permeability in the neighborhood of the phase transition (Corvera et al., 1992). The computer simulations have revealed that this increased permeability is related to enhanced density fluctuations and a higher degree of dynamic heterogeneity.

Addition of foreign molecules to lipid bilayers can in some cases have a dramatic influence on both the phase equilibria and the bilayer dynamic heterogeneity (Mouritsen and Jørgensen, 1992). Cholesterol and drugs, as well as polypeptides and proteins, couple strongly to the heterogeneity which may either be suppressed or strongly enhanced. In particular the interaction of proteins and polypeptides with lipids takes advantage of the lateral density and compositional fluctuations, and the coherence length of the perturbation on the lipid matrix due to proteins is directly related to the correlation length characterizing the dynamic membrane heterogeneity, as we shall demonstrate in Sections V.B.–V.E. below.

V. LIPID BILAYERS WITH POLYPEPTIDES AND INTEGRAL PROTEINS

The major effect of incorporating integral proteins and transmembrane polypeptides into membranes is a dramatic change in the phase equilibria. Usually, proteins are predominantly soluble in the fluid lipid-bilayer phase and a protein-induced phase separation occurs in the gel phase (Mouritsen and Sperotto, 1993). The presence of the proteins leads to structural changes in the adjacent lipid molecules as well as a modification of the local composition. Discussions of the physical effects of protein-lipid interactions on lipid-membrane structure have to a large extent been dominated by the experimentalists' preoccupation with spectroscopic order parameters. The spectroscopic order parameters refer to acyl-chain conformational order (or hydrophobic membrane thickness) and the influence of the proteins is sometimes signalled by the occurrence of additional spectral features. Such features will clearly depend on the intrinsic time scale of the spectroscopic technique used and on the diffusional characteristics of the spectroscopic probe. Therefore, it is difficult to infer the nature of the local structure and heterogeneity in membrane-protein systems from spectroscopic experiments alone.

One of the mechanisms proposed to relate protein-induced, lipid-bilayer phase equilibria to the fundamental physical properties of lipid-protein interfacial contact is based on the hydrophobic matching concept of Eq. (1) (Mouritsen and Bloom, 1993) between the lipid-bilayer and the hydrophobic region of the protein. This concept has enjoyed considerable success in predicting phase diagrams for lipid bilayers reconstituted with specific proteins. Here we place special emphasis on this mechanism and we shall describe some of the results which have been obtained from computer-simulation calculations on the molecular interaction model known as the mattress model (Mouritsen and Bloom, 1984) which is based on this mechanism. According to the concept of hydrophobic matching, a major contribution to protein-lipid interactions is controlled by a hydrophobic matching condition which requires that the hydrophobic lipid-bilayer thickness matches the length of the hydrophobic domain of the integral membrane protein. From the results described above on the occurrence of membrane heterogeneity in phospholipid bilayers of different kinds, cf. Figure 8, it can be anticipated that integral membrane proteins, which couple to the membrane lipid acyl-chain order (area density) and/or the membrane composition via a hydrophobic matching condition, are going to influence the degree of heterogeneity. Conversely, a certain degree of membrane heterogeneity will couple to the conformational state of the individual proteins and to the aggregational state of an ensemble of proteins.

A. Critical Mixing and Phase Diagram of Lipid-Polypeptide Mixtures

In this section we show how numerical simulations can be used to model the gel-fluid coexistence loop and the lower critical mixing point found in lipid-polypeptide mixtures. The model was a combination of the microscopic interaction model of Eqs. (3) and (6), the mattress model described above for the mismatch between the lipid bilayer and the polypeptide and a model for the mismatch between the hydrophobic thickness of the lipid bilayer in its gel and fluid phases (Zhang et al., 1993b). The simulation results agreed well with the experimental data of Morrow et al. (1985) and Morrow and Davis (1988). The geometric parameters of the polypeptide were chosen to be $A_P = 68.0 \text{Å}^2$ and $d_P = d_{10}^A = 11.25 \text{Å}$, where d_{10}^A is the acyl-chain length of DPPC in the fluid state. The value of the direct lipid-polypeptide interaction parameter was fixed at $J_{AP} = 0.25 \times 10^{-13}$ erg. In the present version of the model, the pure DPPC bilayer has a fluctuation-dominated first-order transition. Extremely long simulations were performed at the chemical potential corresponding to the transition at the chosen temperature.

Figure 9 shows the free energy functional, $\mathcal{F}(\mu_m, T, L)$, defined in Eq. (15), for four different temperatures, $T = 313.0$ K, 310.0 K, 304.0 K, and 303.5 K, all of which are below the pure DPPC lipid-bilayer main phase transition temperature, $T_m = 313.7$ K, of the present model. c denotes the concentration of the peptides which only occupy single sites of the lipid lattice within the model. The free energy is presented for a value of the chemical potential, $\mu = \mu_m$, at which the two lipid phases coexist, and hence the two minima of the free energy are equally deep. From the size dependence of the data in Figure 9, the following conclusions can be drawn. For the two higher temperatures, T = 313.0 K and 310.0 K, the free energy barrier, $\Delta\mathcal{F}(L)$ in Eq. (16), separating the two minima, increases with system size indicating that the two phases are separated by a first-order phase transition, i.e., that the two phases coexist in the thermodynamic limit. In contrast, at $T = 304.0$ K in Figure 9(c), the barrier height does not depend on system size within the numerical accuracy indicating that the system is close to a continuous transition, in this case a lower critical mixing point. For even lower temperatures, e.g., $T = 303.5$ K in Figure 9(d), the barrier decreases with system size indicating that the difference between the two minima, and hence the two phases, vanishes in the thermodynamic

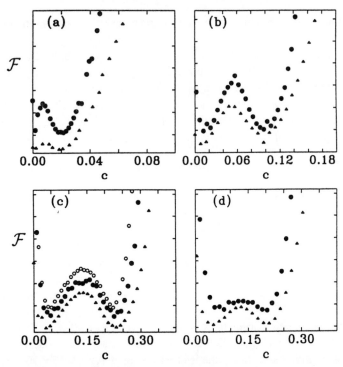

Figure 9. Mismatch model, Eq. (6), of the main phase transition of a DPPC lipid bilayer with intercalated amphiphilic polypeptides. Results from computer-simulation calculations are shown for the free energy functional, $\mathcal{F}(\mu_m(N), T, L)$, as a function of bilayer composition, c, for different lattice sizes, $L \times L$, $L = 16$ (△), 24 (●), 32 (○), for three different temperatures, (a) $T = 313.0$ K, (b) $T = 310.0$ K, (c) $T = 304.0$ K, and (d) $T = 303.5$ K. $\mu_m(L)$ refers to the chemical potential at phase coexistence.

limit. At these temperatures the mixture is therefore in a one-phase region at all compositions.

The phase diagram spanned by temperature and chemical potential was obtained from data of the type presented in Figure 9 and is shown in Figure 10(a). The solid line in Figure 10(a) indicates the line of coexistence between the gel and fluid lipid phases. The concentration variable, x_p, in this diagram has been recealed from c to account for the polypeptide Lys$_2$-Gly-Leu$_n$-Lys$_2$-Ala-amide which has been studied experimentally by NMR techniques (Morrow et al., 1985). The experimental data, also reproduced in Figure 10 shows that the lipid-polypeptide interactions of the present microscopic model produce the characteristic 'teardrop'-shaped, closed-coexistence loop of a binary mixture with a lower critical mixing point. The agreement between the experimental data and the theoretical predictions is quite satisfactory considering the approximation which underlies the scaling of concentration. This

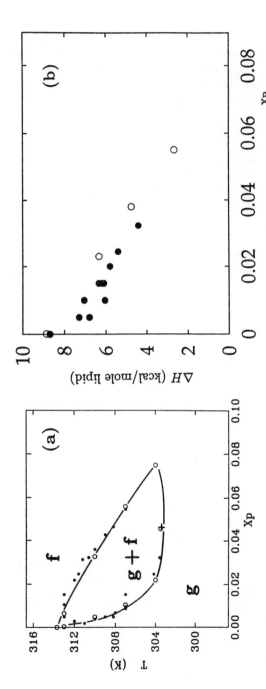

Figure 10. Mismatch model, Eq. (6), of the main phase transition of a DPPC lipid bilayer with intercalated amphiphilic polypeptides. Results from computer-simulation calculations. (a) Rescaled theoretical $T - x_P$ phase diagram (○). The rescaling is performed to facilitate a comparison with a specific membrane system: DPPC bilayers mixed with α-helical amphiphilic transmembrane polypeptides of the type Lys$_2$-Gly-Leu$_r$-Lys$_2$-Ala-amide. The experimental data (●) for mixtures of DPPC with Lys$_2$-Gly-Leu$_{16}$-Lys$_2$-Ala-amide (Morrow et al., 1985) are shown for comparison. The critical mixing point is indicated by +. The solid line connecting the theoretical points is drawn as a guide to the eye. (b) Corresponding transition enthalpy, ΔH, per lipid molecule as a function of peptide concentration, x_P, as obtained from simulations (○) and from experiments (●).

suggests that the microscopic interaction model has captured the essentials of the lipid-polypeptide interactions in the present membrane system. It specifically suggests that the direct lipid-polypeptide interactions are responsible for the location of the critical mixing point. Figure 10(b) shows a comparison between the computer simulation results and the experimental data for the enthalpy of melting, ΔH, as a function of peptide concentration.

B. Lipid-Acyl-Chain Order-Parameter Profiles Near Small Proteins

The question of the shape and the coherence length of the acyl-chain order parameter near integral membrane proteins was originally addressed within the conventional Landau–de Gennes theory (Abney and Owicki, 1985). Mean field theory of this type suppresses the lateral fluctuations in the transition region, and it will, therefore, lead to estimates of the correlation length that are too low. It is only possible to obtain reliable values for the correlation length near the transition by using computer simulation on microscopic models. Using a Monte Carlo simulation, Scott (1986) studied a continuum model of a lipid membrane with a small number of hydrocarbon chains and a single polypeptide (or small protein). For different models of the protein, it was found that the disturbance of the neighboring lipids was only marginal. A similar result was obtained by Edholm and Johansson (1987) who performed a molecular dynamics study of a model membrane with 64 lipid molecules incorporated in a single polypeptide chain. Both of these extremely detailed simulation studies were performed on quite small systems at temperatures well above the pure bilayer phase transition. In order to determine the lipid profile closer to the transition, much larger systems have to be studied which, by computational limitations, implies that one has to restrict the model to a lattice.

The microscopic version of the mattress model, Eq. (6), has been studied by Monte Carlo simulations to determine the coherence length, $\xi_P(T, x_P)$, for the spatial fluctuations of the lipid order parameter profiles around integral membrane proteins

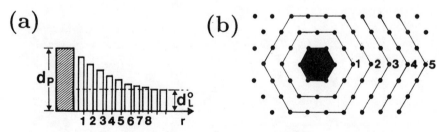

Figure 11. (a) Schematic drawing of the lipid acyl-chain length profile near an integral membrane protein. r is the distance parameter in units of lattice spacings. d_P is the protein hydrophobic length and d_L^0 is the unperturbed lipid bilayer hydrophobic thickness. (b) Lattice model of lipid layers around an isolated protein.

for a distribution of proteins in a large lipid-bilayer array in the transition region (Sperotto and Mouritsen, 1991a). The model is studied at low protein concentrations where the overlap between the lipid profiles from neighboring proteins is negligible. The profile is calculated around each protein embedded in the lattice model in a geometry shown in Figures 11(a) and (b). The protein is assumed to be immobile, to have a hexagonal shape, and to comprise n_p sites of the lipid lattice.

The coherence length is found to have a dramatic temperature dependence with a sharp peak at the transition, as illustrated in Figure 12 in the case of DPPC bilayers. For this extremely dilute system with nonmobile proteins, it is observed that there is a sharp transition and that the transition point of the mixture is very close to that of the pure bilayer. The systematics revealed in Figures 12(a)–(c) show the importance of the degree of hydrophobic matching for the coherence length. First of all, the protein-induced disturbance of the lipid bilayer is seen to extend beyond the first few molecular layers in a wide temperature range. Secondly, the coherence length becomes very large close to the transition. Comparison of the different cases in Figures 13(a)–(c) shows that $\xi_p(T)$ is strongly suppressed at the transition for $d_p = 30$Å relative to $\xi_p(T)$ for $d_p = 24$ Å and 36 Å. For the short protein, the decay length is very large in the transition region, up to about seven molecular layers at T_m. Far away from T_m, the decay length in all cases is very short, less than one molecular distance, in accordance with the vanishing lipid-lipid density fluctuations.

In addition to the temperature dependence and protein-length dependence of $\xi_p(T)$, there is a strong dependence on the protein size (circumference ρ_p, i.e., n_p within the present model). This is demonstrated in Figure 12(d) by data which refer to a single-site protein ($n_p = 1$), a seven-site protein ($n_p = 7$), and a protein wall ($n_p = \infty$) for $d_p = 32$ Å. It is seen that the correlation length increases as the protein size is increased. The reason for this is that the curvature of the protein decreases as n_p increases, and the influence of a given part of the protein extends to more lipid sites. In this way, a given lipid chain interacts with a greater portion of the protein hydrophobic surface as the protein becomes larger. This clearly implies that nonsmooth variations on the surface of the protein will have a larger effect on the lipid environment the more comparable these variations are to the lipid-length scale.

The long coherence length found in these calculations provides a mechanism for indirect lipid-mediated, protein-protein, long-range attraction and hence may play an important role in regulating protein segregation as discussed in Section V.C. below.

C. Lateral Distribution of Proteins in Lipid Bilayers

Several factors control the lateral distribution of proteins in the lipid membrane plane. Among these, the following are of major importance: a) the protein concentration x_p; the higher x_p is, the greater is the probability that a protein is next to another protein and hence forms an aggregate, b) the temperature; the higher the

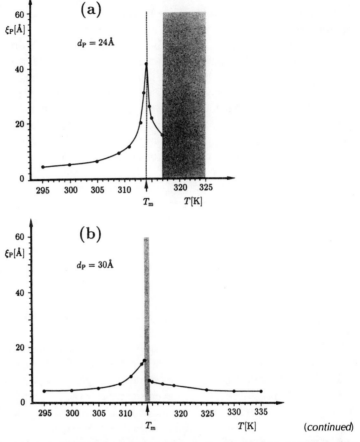

Figure 12. Temperature dependence of the coherence length, ξ_P, for a model protein of cross-sectional area $n_P = 7$ (relative to the cross-sectional area of a lipid-acyl chain) and hydrophobic lengths of $d_P > 24$Å (a), 30Å (b), and 36Å (c) embedded in DPPC lipid bilayers. (d) Temperature dependence of ξ_P for proteins with $n_P = 1$ (dotted line), $n_P = 7$ (dashed line), and $n_P = \infty$ (solid line).

temperature is, the higher is the effect of the entropy which will tend to randomize the protein distribution, c) the lipid-protein interactions and the lipid-mediated, protein-protein interactions, and d) the direct protein-protein interactions which may be long range due to extramembrane moieties.

The microscopic version of the mattress model, Eq. (6), has been used systematically to study the lateral distribution in model membranes as controlled by factors b) and c). By a computer simulation calculation (Sperotto and Mouritsen, 1991b), the relative importance for protein aggregation of different contributions to the lipid-protein interactions has been studied, specifically the attractive lipid-protein

Computation of Lipid-Protein Interactions

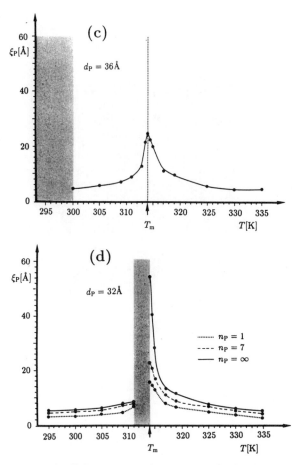

Figure 12. (Continued) All three cases refer to protein hydrophobic lengths of $d_P = 32$Å. The shaded areas indicate temperature regions where $d_L^0 \approx d_P$ very close to the protein and the decay cannot be resolved within the numerical accuracy to yield a reliable value of ξ_P. The different data sets have been obtained from computer-simulation calculations on the microscopic mattress model of protein-lipid interactions; cf. Eq. (6).

hydrophobic contact interaction and the repulsive interaction due to a possible mismatch between lipid bilayer and protein hydrophobic thicknesses. The results suggest that the formation of protein aggregates in the membrane plane is predominantly controlled by the strength of the direct van der Waals-like lipid-protein interaction. Whereas the mismatch is of prime importance for determining the phase equilibria, it is found that a mismatch may not be the only reason for protein aggregation within each of the individual phase. Depending on the strength of the van der Waals-like interaction associated with the direct lipid and protein hydrophobic contact, the proteins may still remain dispersed in the fluid phospholipid

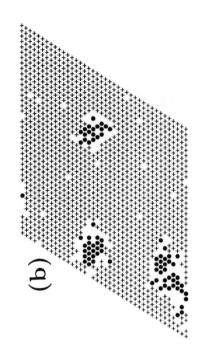
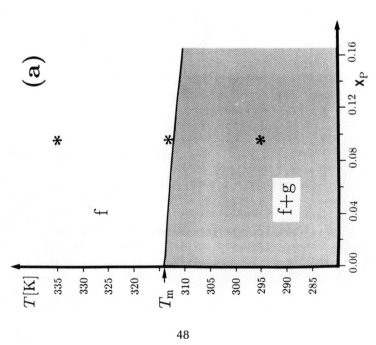

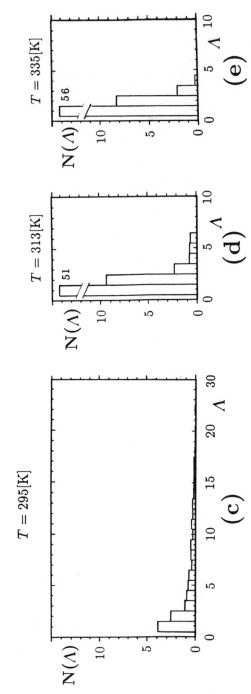

Figure 13. (a) Phase diagram in temperature, T, versus protein concentration, x_P, for a mixture of DPPC lipids and small proteins with a hydrophobic length of 24 Å and a very weak hydrophobic protein-lipid interaction. $T_m = 314$ K is the transition temperature of the pure lipid bilayer. The labels **f** and **g** refer to the fluid and gel lipid phases and the shaded region **f + g** indicates the fluid-gel coexistence region. The points indicated by * denote the points in the phase diagram investigated by computer simulations on the microscopic version of the mattress model. (b) Snapshot of a typical microconfiguration of the lattice at $T = 295$ K. The proteins are indicated by dots, and gel and fluid lipid regions are denoted by grey and white areas, respectively. (c), (d), and (e) Protein cluster size distribution, $N(\Lambda)$, as a function of temperature for a lipid bilayer matrix with 40×40 acyl chains, 80 of which have substituted with a small proteins of a hydrophobic length of 24 Å. The data have been obtained from computer-simulation calculations on the microscopic mattress model of protein-lipid interactions; cf. Eq. (6).

49

bilayer, even if the mismatch between the lipid and protein bilayer thicknesses is as high as 12 Å.

The type of data on which the above conclusions are based is exemplified in Figure 13 in the case of a DPPC bilayer. The figure shows a collection of results for a small, rather short protein, $d_P = 24$ Å, in a DPPC bilayer membrane at low concentration. The phase diagram in Figure 13(a) shows that massive phase separation occurs below T_m. In this figure the solidus line is so close to the temperature axis that it cannot be discerned. In the phase-separated region, the proteins are dissolved almost exclusively in the fluidlike regions of the bilayer, because the protein length is closely matched to the fluid bilayer thickness and solution of proteins in the gel phase would, therefore, be very costly. However, because the attractive interaction between the lipids and the proteins in this case is assumed to be very low, the solubility of the protein is also low in the fluid phase, and, therefore, one might expect a tendency for protein aggregation within the coexistence region at low temperatures where the entropy is low. These expectations are confirmed by the results from the simulations. At $T = 295$ K, for the chosen protein concentration $x_P = 0.095$, the system is in the phase separation region. The aggregate-size distribution function, $N(\Lambda)$, in Figure 13(c), shows that the number of isolated proteins is low and large protein aggregates are formed. $N(\Lambda)$ is a measure of the number of protein aggregates consisting of Λ proteins. The appearance of the protein aggregates almost exclusively in the fluid region of the lattice is demonstrated by the snapshot of Figure 13(b). As the temperature is raised toward $T = 313$ K, the system leaves the phase separation region. The proteins are no longer only dissolved in a limited region, and the number of isolated proteins increases strongly, as can be seen from Figure 13(d). A number of small protein aggregates remain at the temperature just above the pure lipid transition temperature. At $T = 335$ K, the entropy disordering effect allows only a small number of protein dimers and trimers to remain in the system, as shown in Figure 13(e).

The simulation results indicate that, when the value of the lipid-protein interaction parameter, J_{AP}, is sufficiently small, protein aggregates form in the fluid region of the phase diagram just above the phase boundary due to dynamic aggregation induced by the lipid density fluctuations. This is a highly nontrivial effect caused by the dramatic density fluctuations accompanying the main transition, cf. Section IV.C. and Figure 8. The lipid domains are characterized by a coherence length which is related to the distance over which lipid-mediated protein-protein attractive forces are operative. Hence by this mechanism, lipid fluctuations can induce dynamic protein aggregation which should be most pronounced close to the phase boundaries. As the strength of the direct lipid-protein interaction is increased, the tendency for forming protein aggregates is diminished.

It thus appears from the simulations on the microscopic mattress model that the hydrophobic mismatch interaction has only a marginal effect on the protein aggregation (except for the lipid fluctuation-induced dynamic aggregation near the phase boundaries). Moreover, the attractive lipid-protein interactions tend to keep the proteins apart by 'spacing' them with lipids. Such 'spacer' lipids are bound in the sense that they are dynamically trapped. Hence, the direct lipid-protein van der Waals-like interactions tend to reduce the degree of protein aggregation. By focusing on a model without direct protein-protein interactions we have therefore managed to isolate the effects of purely lipid-facilitated protein aggregation. It is important to know the nature of these effects before more realistic and complicated models involving long-range, direct, protein-protein interactions are invoked.

D. Compositional Lipid Profiles in Binary Lipid Mixtures Near a Protein Wall

We now consider the next level of complexity where a large protein is embedded in a lipid bilayer composed of two different lipid species, A and B. The lipid-lipid interactions are described by Eqs. (4) and (5), and the interaction between the lipids and the proteins are given by the mattress model in Eq. (6). This combined model was studied by computer simulation to determine to what extent bare physical effects may be responsible for lipid selectivity and lipid specificity of membrane proteins (Sperotto and Mouritsen, 1993). The basic idea behind the calculation is that, via the hydrophobic matching condition mentioned above, the lipid chains of varying length will feel the perturbation of the protein surface to different extents and the lipid species which most easily adapts to the matching condition will be selected, on a statistical basis, and have an increased probability of being close to the protein-lipid interface. This is an example of interface enrichment.

The fact that such a selectivity can result from the hydrophobic matching condition is demonstrated by the data in Figure 14 in the case of a DMPC-DSPC mixture. The data in this figure refer to a very large protein ($n_p \sim \infty$) of length $d_p = 20$ Å and an equimolar lipid mixture at a temperature, $T = 340$ K, well above the coexistence region of the mixture, i.e., in the fluid lipid phase; cf. the phase diagram in Figure 7(b). The value of d_p is chosen to be close to the hydrophobic thickness of fluid DMPC bilayers. Figure 14 shows the lipid-concentration profiles of DMPC and DSPC as a function of distance, r, from the protein. The protein is seen to select the lipid species (in this case DMPC) which most easily wets the hydrophobic surface of the protein. Conversely, the protein-lipid interface is depleted in the other species. Hence, the protein-lipid interactions lead to compositional heterogeneities. In other words, the protein-lipid interactions couple to the compositional fluctuations of the binary lipid mixture in the fluid phase; cf. Figure 8(b).

A very striking observation made from the model simulations (Sperotto and Mouritsen, 1993) of protein-induced compositional heterogeneity is related to a nonequilibrium transient effect found in the different concentration profiles of the two lipid species as these profiles establish themselves in the course of time. This effect, which may have some important consequences for steady-state membrane organization, refers to a situation where a thermally equilibrated binary lipid mixture is prepared in the fluid phase and then suddenly is made subject to the

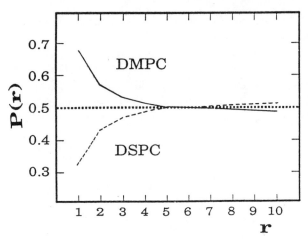

Figure 14. Protein-lipid interface enrichment and physical lipid specificity in a binary lipid mixture. Lipid concentration profiles, $P(r)$, for the two lipid species are shown as a function of distance, r, from a very large integral membrane protein. The data refer to computer simulation on an equimolar binary mixture of DMPC and DSPC at 340 K which is well above the coexistence region. The protein hydrophobic length is taken to be close to the hydrophobic thickness of a fluid DMPC lipid bilayer.

boundary condition imposed by the presence of the proteins. In response to the proteins, the mixture has to reorganize itself laterally and decompose locally. This reorganization proceeds via long-range diffusional processes. The interdiffusion of the species is, however, limited by the conservation law imposed by the global composition of the mixture. As a consequence, the composition profile away from the protein surface displays a pronounced oscillatory behavior dictated by the diffusional processes and the mass conservation law. After introduction of the proteins to the initially equilibrated mixture, the protein surfaces are, on a time scale corresponding to short-range diffusion, enriched in the appropriate species whose hydrophobic acyl-chain length is compatible with the protein thickness. However, on this time scale, the mixture does not have time to reorganize fully and to compensate for the excess mass of the enriched species. Therefore, a depletion layer of the same species is formed next to the enrichment layer near the protein. Because the other species have to follow suit by the opposite series of local depletion and enrichment layers, a full oscillatory behavior develops as shown in Figure 15. As time elapses, the nodes of the oscillations move toward larger values of r, eventually dampen out, and the equilibrium concentration profiles in Figure 14 are recovered.

The results presented above refer to the case of immobile model proteins, such as proteins bound to specific positions of the membrane, for example, via the cytoskeleton, or to proteins which diffuse very slowly relative to the lipids. However, in the case of mobile proteins, it can be anticipated from the general

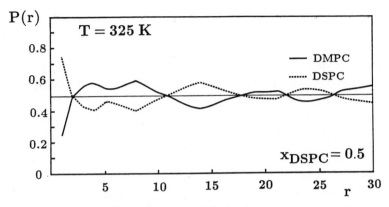

Figure 15. Transient oscillatory behavior of the lipid concentration profiles, $P(r)$, for an equimolar mixture of DMPC and DSPC, as a function of the distance, r, from a very large integral membrane protein of hydrophobic length $d_P = 26\text{Å}$. The data are obtained from computer simulations on a system with 60×60 lipid chains and refer to a temperature, $T = 325$ K.

nature of the results for static proteins that the structured concentration profiles in Figure 15 will facilitate a medium-range, lipid-mediated, indirect, protein-protein attraction which will influence the state of protein aggregation. This observation may have biological relevance for those proteins whose biological activity depends on their aggregational state.

It is interesting to note that for a nonequilibrium system, say a protein-lipid membrane driven by external sources of energy which couple to protein conformational changes, the oscillatory profile in Figure 15 may be dynamically maintained. The mobile proteins in the driven system may organize themselves laterally to fit into the part of the profile which is enriched in the lipid species with the higher affinity for the protein. This picture may straightforwardly be generalized to systems of different proteins with different lipid selectivity. The possibility exists that there may be parts of the phase diagram in which the enrichment equilibrium profiles in Figure 14 develop into a complete wetting phenomenon implying that the enriched layer grows macroscopically large. Wetting phenomena would have a pronounced effect on the heterogeneous membrane structure.

E. Structure and Thermodynamics of Binary Lipid Bilayers with Large Integral Proteins

The simulations described in Section V.D. above were performed in the extreme limit of a very large protein and under the assumption that the protein is immobile. For smaller laterally mobile proteins, the overlap in compositional profiles from adjacent proteins results in an effective lipid-mediated, protein-protein interaction which can give rise to an attractive force and hence protein aggregation. The mattress model for lipid-protein interactions, Eq. (6), has recently been used (Sperotto and Mouritsen, 1994) in conjunction with the model for binary mixtures of phospholipids with different acyl-chain lengths, Eqs. (4) and (5), to study the thermodynamic properties as well as the local compositional structure in the case of dilauroyl phosphatidylcholine-distearoyl phosphatidylcholine (DLPC-DSPC) mixtures incorporated in a protein of a size corresponding to that of bacteriorhodopsin. The model simulations have been inspired by recent experimental work on the effect of bacteriorhodopsin on lipid bilayers of different thickness (Piknova et al., 1993).

In this model, the large proteins are mobile and each is assumed to be of hexagonal form occupying $n_P = 19$ sites of the lipid lattice. The interaction parameters of the model are taken to be $\Gamma_{AB} = 0.038 \times 10^{-13}$ erg $Å^{-1}$ (Jørgensen et al., 1993), which reproduces the nonideal phase behavior of DLPC-DSPC mixtures, and $d_P = 30Å$ together with $A_P = 19 \times 34 = 646Å^2$ in order to mimic bacteriorhodopsin. The lipid-protein interaction constants are estimated to be $\Gamma_{AP} = 0.027 \times 10^{-13}$ erg $Å^{-1}$ and $J_{AP} = 0.04 \times 10^{-13}$ erg $Å^{-1}$ by comparison with the experimental phase behavior of the DMPC-bacteriorhodopsin mixture (Piknova et al., 1993; Sternberg et al., 1989).

Results from Monte Carlo simulations of this model of bacteriorhodopsin in equimolar DLPC-DSPC mixtures have been obtained for a protein/lipid molar ratio of P/L = 0.025 (Sperotto and Mouritsen, 1994). The simulations were carried out on a lattice with 40×40 sites. Figure 16(a) shows the results for a specific heat scan with and without bacteriorhodopsin incorporated. Due to the extreme degree of nonideality for the DLPC-DSPC mixture, which exhibits peritectic phase behavior (Mabrey and Sturtevant, 1976; Ipsen and Mouritsen, 1988), the liquidus and solidus (three-phase line) of the phase diagram clearly manifest themselves in two well-separated thermal anomalies in the specific heat. In the presence of bacteriorhodopsin, the gel-fluid coexistence region is getting more narrow and the specific heat anomalies decrease in intensity.

The lateral distributions of the proteins in two typical cases are also shown in Figure 16. The two cases correspond to the gel-gel coexistence region (b) and the gel-fluid coexistence region (c). The higher solubility, caused by a better hydrophobic matching of bacteriorhodopsin in the gel phase to the lipid species with the shorter chain length, DLPC, is clearly seen in Figure 16(b). In the gel-fluid coexistence region, where bacteriorhodopsin is again expelled from the predominantly DSPC gel phase due to poor matching, the protein in the fluid region prefers to be wetted by the well-matched fluid DSPC lipids. The fluid DLPC lipids, which are too short, are repelled by the protein.

The local lipid compositional structure around bacteriorhodopsin in the fluid phase is analyzed quantitatively in Figure 17. This figure clearly shows that there is

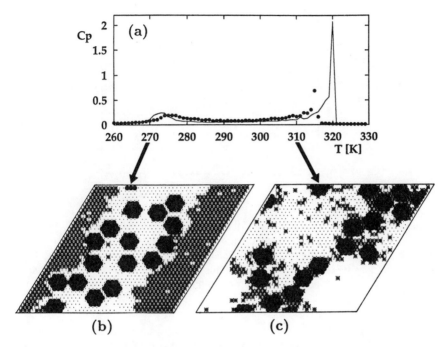

Figure 16. (a) Computer simulation data for the specific heat of a DLPC-DSPC mixture without (——) and with (●) a dispersion of bacteriorhodopsin in a molar protein-lipid ratio of 0.025. The data are obtained from simulations on the microscopic mattress model of lipid-protein interactions, Eq. (6), used together with the binary lipid mismatch model in Eqs. (4) and (5). (b) Typical microconfigurations of the mixture at $T = 270$ K. The proteins are shown as solid hexagons. · denotes gel DLPC lipids and * denotes gel DSPC lipids. Empty spaces denote lipid molecules in the fluid state. (c) Typical microconfiguration at $T = 310$ K. The proteins are shown as solid hexagons. · denotes fluid DLPC lipids, * denotes fluid DSPC lipids, and empty spaces denote lipid molecules in gel states.

a statistical dynamic lipid annulus for this mobile protein. The well-matched fluid DSPC lipids are enriched in concentration near the protein, whereas there is a concomitant depletion of the poorly matched fluid DLPC lipids. The figure also shows the protein-protein correlation function which has a highly nontrivial distance dependence due to the delicate demixing phenomenon in the lipid mixture near the protein surface. Hence, bacteriorhodopsin within this model has a distinct statistical preference for DSPC lipids in the fluid phase. This type of physically controlled lipid selectivity of integral membrane proteins has yet to be investigated experimentally.

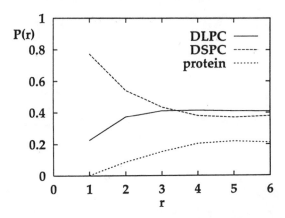

Figure 17. Computer simulation data for the lipid composition profiles, $P(r)$, for DLPC and DSPC as a function of the distance, r, from the lipid-protein interface. The results refer to $T = 330$ K for an equimolar DLPC-DSPC mixture. The system is in the fluid phase. The data are obtained from simulations on the mattress model of lipid-protein interactions, Eq. (6), used together with the binary lipid mismatch model in Eqs. (4) and (5). The protein-protein correlation function is also shown.

F. Relation Between Gramicidin-A Dimer Formation and Bilayer Properties

We now apply the model described by Eq. (9) to gramicidin-A dimer formation in a lipid bilayer. In this model it is assumed for computational convenience that the monomers are single-site objects on the lattice, substituting for a single acyl chain. It is however possible to relate the results of this model to those of the experimental situation, where the monomers occupy about seven adjacent lattice sites, by a simple scaling, though this is clearly an approximation. From a scaling argument, we therefore choose $E_H \approx 2 \times 10^{-13}$ erg corresponding to a single hydrogen bond instead of six or seven hydrogen bonds in the actual gramicidin-A dimer. Although the approximation of single-site peptides overestimates the entropy of mixing, it works quite well in the lipid-protein model described in Section V.A. which was used to calculate the phase diagram of polypeptide-DPPC bilayers. Here we only consider the case of $d_G = d_{10} = 11.25$Å because experimental results (Elliott et al., 1983; and Huang, 1986) suggest that the hydrophobic thickness of gramicidin channels closely matches the thickness of the membrane in the fluid phase. Consequently the value of Γ_{AG} is chosen to be the same as Γ_{AP} in Section V.A..

The equilibrium dimer probability, P_d, is defined as the percentage of dimers in thermodynamic equilibrium and is therefore a measure of the tendency of the gramicidin monomers to form ionic channels in a lipid bilayer. The dimer probability is affected by several factors, such as the temperature, the mismatch condition between the gramicidin dimer and the lipid chains, and the direct lipid-protein interaction. An increase in temperature may cause the monomer-monomer bonding in a channel to break, thereby lowering the dimer probability. Temperature also indirectly affects the probability via changes in the hydrophobic thickness of the lipid bilayer. Furthermore, the mismatch between the dimers and lipids tends to break a dimer into two monomers. However, the effect of mismatch is dependent on the lateral distribution of the dimers, which is controlled by the strength of the direct van der Waals interactions between lipids and proteins. For the case where this attractive interaction is very weak, the gramicidin monomers and dimers aggregate in the bilayer plane so that the contact between lipids and dimers decreases maximally and the effect of mismatch on the channel is therefore suppressed. For the case where the lipid-protein interaction is sufficiently strong, i.e., close to the strength of van der Waals interaction between lipid molecules, the gramicidin dimers or monomers disperse considerably so that the environment of each channel will be similar to that of an isolated channel.

The dimer probabilities, $P_d(T)$, as calculated by Monte Carlo simulations are shown in Figure 18 as functions of temperature, T, for the cases with and without the direct van der Waals interaction between lipids and gramicidin. The gramicidin-lipid mixture was simulated on two 40×40 triangular lattices with 3% of the sites occupied by gramicidin monomers. It is found that the equilibrium channel probabilities in the fluid phase of the bilayer are almost the same for the two cases because the lipid-gramicidin interaction in the fluid phase is very weak. However, they are quite different in the gel phase of the bilayer. Indeed, the direct lipid-protein interaction, together with the effect of entropy, overcomes the mismatch between the hydrophobic thickness of gramicidin dimers and that of the lipid bilayer and disperses the dimers randomly in the bilayer. In this situation, Figure 18 shows that the mismatch is very effective so that the channel probability in the gel phase decreases. In the opposite case, when the direct lipid-protein interaction is very weak, the dimers aggregate in the bilayer so that the mismatch does not affect the channels inside the gramicidin clusters, and the entropy effect on the channels decreases. Therefore the channel probability in this case is even higher than in the fluid phase where the hydrophobic thickness of dimers is closely matched to that of the lipid bilayer. Overall, the dimer probability has a large value (about 80%) over a large temperature region in the gel phase because the hydrogen bonding in dimers is considerably stronger than the van der Waals interactions in DPPC bilayers. The equilibrium constant of dimerization, K_d, can be related to the channel probability P_d by $K_d \approx AP_d(1 - P_d)^{-2}$ (Cohen et al., 1971), where A is the mean area of lipid bilayer. The analysis of the dimerization constant is still in the preliminary stages but is planned to be extended to bilayers of different hydrophobic thickness.

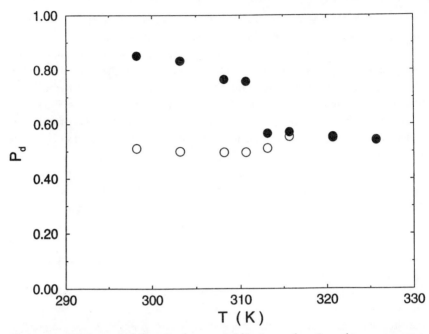

Figure 18. Dimer probability, P_d, for gramicidin-A as a function of temperature. ○ refers to the data with the van der Waals lipid-gramicidin interaction. ● refers to the data without such an interaction.

G. Polypeptide Aggregation and Transmembrane Channel Formation in Lipid Bilayers

In Section II.E. we presented a model for the self-assembly of channel-forming protein 'monomers' via lateral diffusion and anisotropic intercomponent interactions. A computer-simulation study of this model performed by the Monte Carlo method with Kawasaki dynamics was used to move proteins laterally, and Glauber dynamics was used for protein rotation. No lipid diffusion is included because the lipids are assumed to form a uniform background. The simulation data was obtained for a lattice with $L^2 = 60 \times 60$ sites. However, by comparing with results obtained from larger lattices, it was found that the results do not depend on lattice size.

We define $p(n)$ as the equilibrium probability of finding a cluster of n protein monomers. Such a protein cluster is defined as a set of proteins in which each protein shares one or two bonds with neighboring proteins. $n = 6$ corresponds to a closed transmembrane channel. The protein monomer mole fraction, x_P, is given by $x_P = 7N_P/L^2$ where N_P is the number of proteins in the bilayer. Figure 19(a) shows $p(n)$ versus n for several values of x_P between 4.7% and 25.3%. In this figure,

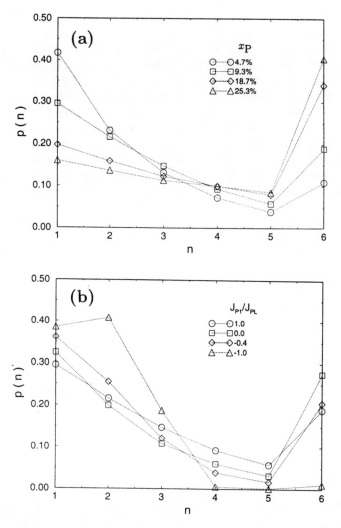

Figure 19. (a) The equilibrium probability, $p(n)$, of finding a cluster of n proteins, as defined in the text, for different concentrations of protein, x_P = 4.7%, 9.3%, 18.7%, and 25.3%. (b) The equilibrium probability of finding a cluster with n proteins for different values of J_{P1}. In both (a) and (b) the temperature, the concentration of proteins, and the parameters of other interactions are set to $k_B T = J_{AP}$, n_p = 9.3%, $J_{P2} = -4J_{AP}$, and $J_{AP} = 1$, respectively. The dotted lines connecting the data points are drawn as guides to the eye.

$k_BT/J_{AP} = 1.0$, $J_{P1} = J_{AP}$ and $J_{P2} = -4J_{AP}$. Figure 19(a) shows that the tendency to form channels increases with concentration while the probability of finding protein monomers decreases correspondingly. However the values of $p(n)$ for intermediate clusters of size $n = 2$ to $n = 5$ hardly change with increasing protein concentration. This indicates a strong tendency to form channels for this set of parameters. This situation changes dramatically when the reduced temperature, k_BT/J_{AP}, is decreased to 0.5. In this case, monomers are not favored and $p(n)$ has almost the same value for all other values of n. Furthermore $p(n)$ hardly changes with increasing x_P. In this case the channel-forming tendency is absent, and the results indicate that a transition may occur with increasing temperature. Figure 19(b) shows the effect of decreasing the direct protein interaction, J_{P1}. Figure 19(b) demonstrates that $p(n)$ is almost unchanged when J_{P1} remains repulsive. However, the channel-forming tendency is removed completely when this interaction becomes attractive.

Figure 20 shows the distribution of lifetimes, $p(\tau)$, of self-assembled channels for $n = 6$. The values of the parameters are the same as for Figure 19 and $x_P = 9.3\%$. The data for $p(\tau)$ shows that the channels are long-lived in the sense that the value

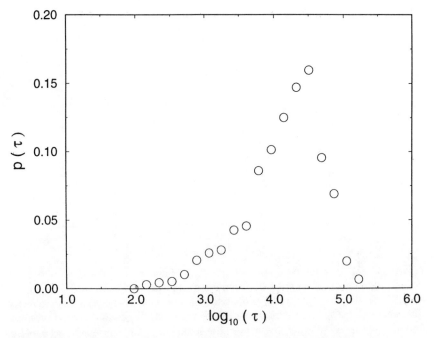

Figure 20. A semilog plot for the equilibrium distribution of lifetimes, $p(\tau)$, of self-assembled channels (protein clusters with $n = 6$). The lifetime is measured in Monte Carlo steps per protein molecule. The temperature, the protein concentration, and the interaction parameters are set to $k_BT = J_{AP}$, $n_p = 9.3\%$, $J_{P1} = J_{AP}$, $J_{P2} = -4J_{AP}$, and $J_{AP} = 1$, respectively.

of $p(\tau)$ peaks at 30,000 Monte Carlo steps, but this value clearly depends on the interaction strength. The data of Figures 19 and 20 are encouraging but must be regarded as preliminary.

VI. CONCLUDING REMARKS AND FUTURE PERSPECTIVES

In this topical review of a computational approach to lipid-protein interactions in membranes, we have focused on the use of statistical mechanical models to describe the interactions and the cooperative phenomena in a simple and well-defined setting. We have shown how powerful modern computer-simulation methods can provide a solution to the thermodynamic properties of these models, and we have demonstrated in a number of specific cases that highly nontrivial phenomena may arise due to the many-particle character of the membrane assembly.

In many respects our approach is admittedly very simple, and it neglects a number of aspects of lipid-protein interactions which in several cases are of crucial importance for the biochemical implications of lipid-protein interactions. This is the weakness of the approach. However, by focusing on simple models and general physical principles we have gained an advantage which is the strength of our computational approach. The modeling is sufficiently simple to reveal the effects of central features of lipid-protein interactions without the disturbing influence of 'noisy' details. Hence, our modeling may lack some important details but it provides a transparent framework for interpreting experimental data and for guiding a more well-focused experimental study of lipid-protein interactions. This latter feature is probably the most important contribution of today's computational approach to study the mutual influence of lipids and proteins in biological membranes.

Some of the future perspectives in the part of computational molecular biology related to biological membranes may be seen as follows. On the one hand we will see that access to faster computers will allow modeling larger molecular assemblies in more detail. This implies that processes of self-assembly, as well as interactions between amphiphilic molecules and water and ions, can be investigated. This should facilitate a deeper study of interactions between different membranes and lead to insight into how this interaction is controlled by the physical properties of the lipid bilayer and the interactions between proteins and lipids. On the other hand, deeper insight into the details of the molecular interactions within membranes, and in particular the interactions in the hydrophobic-hydrophilic interface of the membrane, will allow a study of conformational transitions in receptors and enzymes interacting with membranes, a study which ultimately will be important not only for lipid-protein interactions but also for the way these interactions can be modulated by interaction with foreign molecular agents, such as small drug particles or large virus particles.

ACKNOWLEDGMENTS

This work was supported by the Danish Natural Science Research Council under grant J.no. 11-0065-1, by the Carlsberg Foundation (MMS), by le FCAR du Quebec under a centre and team grant, and by the NSERC of Canada.

REFERENCES

Abney, J. R. & Owicki, J. S. (1985). Theories of protein lipid and protein-protein interactions. In: Progress in protein lipid interactions (Watts, A. & de Pont, S. S., Eds.), pp. 1–60, Elsevier, Amsterdam.

Alberts, A., Bray, D., Lewis, J., Raff, M., Roberts, K., & Watson, J. D. (1989). Molecular Biology of the Cell, 2nd ed., Garlang, New York.

Andersen, O. S., Sawyer, D. B., & Koeppe II, R. E. (1992). In: Biomembrane Structure and Function—The State of the Art (Gaber, B. P. & Easwaren, K. R. K., Eds.), Adenine, pp. 227–244.

Bamberg, E. & Laüger, P. (1973). Channel formation kinetics of gramicidin-A in lipid bilayer membranes. J. Membr. Biol. 11, 177–194.

Bamberg, E. & Laüger, P. (1974). Temperature-dependent properties of gramicidin-A channels. Biochim. Biophys. Acta 367, 127–133.

Biltonen, R. L. (1990). A statistical-thermodynamic view of cooperative structural changes in phospholipid bilayer membranes: their potential role in biological function. J. Chem. Thermodyn. 22, 1–19.

Bloom, M., Evans, E., & Mouritsen, O. G. (1991). Physical properties of the fluid-bilayer component of cell membranes: a perspective. Quart. Rev. Biophys. 24, 293–397.

Bloom, M. & Smith, I. C. P. (1985). Manifestations of lipid–protein interactions in deuterium NMR. In: Progress in Protein–Lipid Interactions, Vol. 1 (Watts, A. & De Pont, J. J. H. H. M., Eds.) pp. 61–88, Elsevier, Amsterdam.

Cevc, G. & March, D. (1985). Phospholipid Bilayers: Physical Principles and Models, Wiley, New York.

Cevc, G. & March, D. (1987). Phospholipid Bilayers. Physical Principles and Models, Wiley-Interscience, New York.

Cohen, G. H., Atkinson, P. H., & Summers, D. F. (1971). Interactions of vesicular stomatitis virus structural proteins with HeLa plasma membranes. Nat. New Biol. 231, 121–123.

Corvera, E., Mouritsen, O. G., Singer, M. A., & Zuckermann, M. J. (1992). The permeability and the effects of acyl-chain length for phospholipid bilayers containing cholesterol: theory and experiment. Biochim. Biophys. Acta 1107, 261–270.

Edholm, O. & Jähnig, F. (1988). The structure of a membrane-spanning polypeptide studied by molecular dynamics. Biophys. Chem. 30, 279–292.

Edholm, O. & Johansson, J. (1987). Lipid bilayer polypeptide interactions studied by molecular dynamics simulations. Eur. Biophys. J. 14, 203–209.

Elliott, J. R., Needham, D., Dilger, J. P., & Haydon, D. A. (1983). The effects of bilayer thickness and tension on gramicidin single-channel lifetime. Biochim. Biophys. Acta 735, 95–103.

Ferrenberg, A. M. & Swendsen, R. H. (1988). New Monte Carlo technique of studying phase transitions. Phys. Rev. Lett. 61, 2635–2638.

Fattal, D. R. & Ben-Shaul, A. (1993). A molecular model for the lipid-protein interaction in membranes: the role of hydrophobic mismatch. Biophys. J. 65, 1795–1809.

Gennis, R. B. (1989). Biomembranes. Molecular Structure and Function. Springer, London.

Hladky, S. B. & Haydon, D. A. (1972). Ion transfer across lipid membranes in the presence of gramicidin-A: I. studies of the unit conductance channel. Biochim. Biophys. Acta 274, 294–312.

Huang, H. W. (1986). Deformation free energy of bilayer membrane and its effect on gramicidin channel lifetime. Biophys. J. 50, 1061–1070.

Huang, J. & Feigenson, G. W. (1993). Monte Carlo simulation of lipid mixtures: finding phase separation. Biophys. J. 65, 1788–1794.

Ipsen, J. H. & Mouritsen, O. G. (1988). Modelling the phase equilibria in two-component membranes of phospholipids with different acyl-chain lengths. Biochim. Biophys. Acta 944, 121–134.

Ipsen, J. H., Jørgensen, K., & Mouritsen, O. G. (1990). Density fluctuations in saturated phospholipid bilayer increase as the acyl–chain length decreases. Biophys. J. 58, 1099–1107.

Jørgensen, K., Sperotto, M. M., Mouritsen, O. G., Zhang, Z., Ipsen, J. H., & Mouritsen, O. G. (1993). Phase equilibria and local structure in binary lipid bilayers. Biochim. Biophys. Acta 1152, 135–145.

Kühlbrandt, W. (1988). The three dimensional crystallization of membrane proteins. Quart. Rev. Biophys. 21, 429–477.

Kühlbrandt, W. (1992). The two-dimensional crystallization of membrane proteins. Quart. Rev. Biophys. 25, 1–49.

Lee, J. & Kosterlitz, J. M. (1991). Finite-size scaling and Monte Carlo simulations of first-order phase transitions. Phys. Rev. B 43, 3265–3277.

Mabrey, S. & Sturtevant, J. M. (1976). Investigation of phase transitions of lipids and lipid mixtures by high sensitivity differential scanning calorimetry. Proc. Natl. Acad. Sci. USA 73, 3862–3866.

Marčelja, S. (1974). Chain ordering in liquid crystals. II. Structure of bilayer membranes. Biochim. Biophys. Acta 367, 165–176.

Morrow, M. R., Hushilt, J. C., & Davis, J. H. (1985). Simultaneous modelling of phase and calorimetric behavior in an amphiphilic/phospholipid model membrane. Biochemistry 24, 5396–5406.

Morrow, M. R. & Davis, J. H. (1988). Differential scanning calorimetry and ^{2}HrNMR studies of the phase behavior of gramicidin-phosphatidylcholine mixtures. Biochemistry 27, 2024–2032.

Mouritsen, O. G. (1984). Computer Studies of Phase Transitions and Critical Phenomena, Springer, Heidelberg.

Mouritsen, O. G. (1990). Computer simulation of cooperative phenomena in lipid membranes. In: Molecular Description of Biological Membrane Components by Computer Aided Conformational Analysis, Vol. 1 (Brasseur, R., Ed.), CRC, Boca Raton, FL, pp. 3–83.

Mouritsen, O. G. (1991). Theoretical models of phospholipid phase transitions. Chem. Phys. Lipids 57, 178–194.

Mouritsen, O. G. & Bloom, M. (1984). Mattress model of lipid-protein interactions in membranes. Biophys. J. 46, 141–153.

Mouritsen, O. G. & Jørgensen, K. (1992). Dynamic lipid-bilayer heterogeneity: a mesoscopic vehicle for membrane function? BioEssays 14, 129–136.

Mouritsen, O. G. & Biltonen, R. L. (1993). Protein-lipid interactions and membrane heterogeneity. In: Protein-Lipid Interactions (Watts, A., Ed.), pp. 1–39, Elsevier Sci., Amsterdam.

Mouritsen, O. G. & Bloom, M. (1993). Models of lipid-protein interactions in membranes. Annu. Rev. Biophys. Biomol. Struct. 147–171.

Mouritsen, O. G. & Sperotto, M. M. (1993). Thermodynamics of lipid-protein interactions in lipid membranes: the hydrophobic matching condition. In: Thermodynamics of Surface Cell Receptors (Jackson, M., Ed.), pp. 127–181, CRC, Boca Raton, FL.

Nagle, J. F. (1980). Theory of the main lipid bilayer phase transition. Ann. Rev. Phys. Chem. 31, 157–195.

Piknova, B., Perochon, E., & Tocanne, J.-F. (1993). Hydrophobic mismatch and long-range protein-lipid interactions in bacteriorhodopsin/phosphatidylcholine vesicles. Eur. J. Biochem. 218, 358–396.

Pink, D. A., Green, T. J., & Chapman, D. (1980). Raman scattering in bilayers of saturated phosphatidylcholines. Biochemistry 19, 349–356.

Risbo, J., Sperotto, M. M., & Mouritsen, O. G. (1995). Theory of phase equilibria and critical mixing points in binary lipid bilayers. J. Chem. Phys. 103, 3643–3656.

Scott, H. L. (1986). Monte Carlo calculations of order parameter profiles in models of lipid-protein interactions in bilayers. J. Chem. Phys. 67, 6122–6126.

Singer, S. J. & Nicolson, G. L. (1972). The fluid mosaic model of the structure of cell membranes. Science 173, 720–731.

Sperotto, M. M. & Mouritsen, O. G. (1991a). Monte Carlo simulation studies of lipid order-parameter profiles near integral membrane proteins. Biophys. J. 59, 261–270.

Sperotto, M. M. & Mouritsen, O. G. (1991b). Mean-field and Monte Carlo studies of the lateral distribution of proteins in membranes. Eur. Biophys. J. 19, 157–168.

Sperotto, M. M. & Mouritsen, O. G. (1993). Lipid enrichment and selectivity of integral membrane proteins in two-component lipid bilayers. Biophys. J. 22, 323–328.

Sperotto, M. M. & Mouritsen, O. G. (1996). Lipid selectivity of integral membrane proteins in two-component lipid bilayers. Bacteriorhodopsin in DLPC-DSPC mixtures (in preparation).

Sternberg, B., Gale, P., & Watts, A. (1989). The effect of temperature and protein content on the dispersive properties of bacteriorhodopsin from *H. halobium* in reconstituted DMPC complexes free of endogenous purple membrane lipids: a freeze-fracture electron microscopy study. Biochim. Biophys. Acta 980, 117–126.

Tate, M. W., Eikenberry, E. F., Turner, D. C., Shyamsunder, E., & Gruner, S. M. (1991). Nonbilayer phases of membrane lipids. Chem. Phys. Lipids 57, 147–164.

Wagner, R. R., Prevee, L., Brown, F., Summers, D. F., Sokol, F., & MacLeod, R. (1972). Classification of rhabdovirus proteins: a proposal. J. Virol. 10, 1228–1230.

Zhang, Z., Laradji, M., Guo, H., Mouritsen, O. G., & Zuckermann, M. J. (1992a). Phase behavior of pure lipid bilayers with mismatch interaction. Phys. Rev. A 46, 7560–7567.

Zhang, Z., Zuckermann, M. J., & Mouritsen, O. G. (1992b). Effect of intermonolayer coupling on the phase behavior of lipid bilayers. Phys. Rev. A 46, 6707–6713.

Zhang, Z., Mouritsen, O. G., & Zuckermann, M. J. (1993a). Detecting phase equilibria in models of thermotropic and lyotropic liquid crystals. Mod. Phys. Lett. B 7, 217–232.

Zhang, Z., Sperotto, M. M., Zuckermann, M. J., & Mouritsen, O. G. (1993b). A microscopic model for lipid/protein bilayers with critical mixing. Biochim. Biophys. Acta 1147, 154–160.

Zingsheim, H. P. & Neher, E. (1974). The equivalence of fluctuation analysis and chemical relaxation measurements: A kinetic study of ion pore formation in thin lipid membranes. Biophys. Chem. 2, 197–207.

PROTEIN MODEL BUILDING USING STRUCTURAL SIMILARITY

Raúl E. Cachau

Abstract 66
I. Introduction 66
II. The Feedback-Restrained Procedure (FRMD) 68
    A. The Archetype Hypothesis 68
    B. Methodology 69
    C. Application: RGD-Containing Peptides 75
III. Positive Biased Averaging 83
    A. Application to Degenerated Intermolecular Interactions: Taxol 83
    B. Application to Degenerated Molecular Conformations: Defensins 90
IV. Modeling Large Molecular Aggregates 93
    A. Methods 93
    B. The Microdomain Description: Modeling of Glutathione S-Transferases . . 93
    C. The Validation Problem: Modeling of Heat Shock
        Proteins-70 from HLA Motifs 108
V. Remarks 114
    Acknowledgments 115
    References 155

ABSTRACT

An approach is described for modeling the three-dimensional structure of a protein from the tertiary structure of a homologous protein determined by X-ray or NMR analysis. In this approach we will consider a known structure as an *archetype* for the unknown structure of another protein. The structure of a protein of homologous sequence can then be modeled based on the similarity with its archetype. Two problems arise in this modeling approach: 1) the a priori univocal assignment of the correspondence between model and tethering structure equivalent atoms and 2) the most efficient search algorithm to describe the changes from the initial structure to the homologous one. A novel technique is described which simultaneously solves the problem of assignment and search by relying on a self-consistent procedure of restraint assignment during a biased molecular dynamics (MD) trajectory. The technique is exemplified by its application to several unrelated systems ranging from very small peptides to domain movements in large proteins.

I. INTRODUCTION

A review of the recent literature on protein modeling studies (Doolittle and Blundell, 1993; DeGrado and Matthews, 1993; Wodak and Rooman, 1993; Benner, 1992) shows an explosive increase in activity in this area of research. A combination of factors is responsible for this renewed activity. Four main components can be recognized: 1) the increased number of protein sequences deposited in public databanks, most of which are of unknown structure, 2) the large number of protein structures available at the Brookhaven Protein Databank and the average high resolution of the newer structures, 3) the large number of recent mutagenesis studies, many of which are combined with structural and calorimetric studies, and 4) the increasing availability of computer power and the maturation of molecular modeling techniques.

As a consequence, the computational modeling of protein structure has evolved as an interdisciplinary exercise. The combination of sequence alignment methods, three-dimensional structural information analysis, and approaches from computational chemistry and statistical mechanics are commonplace.

On the other hand, the recent studies of protein structure and stability have modified some of the traditional dogmas in this area. The degenerate nature of the sequence-structure relationship has been established by numerous experimental mutagenesis studies (Sondek and Shortle, 1990), which have demonstrated that the overall fold of a protein is much more tolerant to sequence modifications than previously suspected. The analysis of known, high-resolution, three-dimensional protein structures reveals similarities for proteins with very different sequences and functions (Farber and Petsko, 1990; Kabsch et al., 1990; Argos, 1991). Theoretical analysis common to polymer studies indicates that the most relevant properties of the protein native state (i.e., compactness (Hue Sun and Dill, 1991), uniqueness

(Shaknovich and Gutin, 1990), and characteristic folding motifs (Finkelstein and Ptitsyn, 1987)) result from general physical properties of the polypeptide chain more than from specific sequence features. The particular *details* found in protein folds fine-tune these properties so as to fulfill specific functional needs.

An immediate consequence of this new body of information is the realization that naïve statistical approaches will not suffice to properly model the relevant details of a complex protein. The initial trend towards protein folding/protein structure simulations consisted of procedures which reduced the most relevant three-dimensional information to a simplified form of sequence-specific amino acid properties. Thus, the three-dimensional information is drastically simplified by a one- or two-dimensional mapping statistical procedure. The sequence of results mentioned early on confuses the 1D/2D statistical approaches by describing a) similar sequences for different function, b) similar sequences for different folding, and c) different sequences for similar folding/functions.

The current trend in modeling studies consists of three-dimensional mappings of three-dimensional information in several different forms. We can mention, for instance, the recent attempts at deriving and testing *effective* force fields. These are the so-called knowledge-based potentials (KBP) (Blundell, 1987), frequently used for evaluating protein conformations. The KBPs are drastic simplifications of the molecular mechanics force field of a biomolecule. The parametrization of energy functions in terms of *reduced* descriptions of the protein conformations has been common in protein computer simulations and structure predictions, even though the general lack of accuracy of protein crystallography as a source of data for the fitting of *detailed* potential energy parameters has been discussed in the literature (Weber, 1992). On the other hand, the field of KBPs recently received new impetus with the increasing number of protein sequence and structure data available, which is especially relevant to these techniques, given the intrinsic *statistical* nature of the KBPs' force fields. KBP potential parameters might be adjusted either by optimization (Maiorov and Crippen, 1992; Crippen, 1991; Goldstein et al., 1992a) or by associative memory procedures (Goldstein et al., 1992b; Friedrichs and Wolynes, 1989; Friedrichs et al., 1991) including, for instance, the use of neural network protocols (Hopfield, 1982).

The use of an effective force field, however, carries the intrinsic difficulty of minimization search procedures in a multidimensional space. Large efforts have been devoted to overcoming this *sampling* bottleneck. The solution to this problem is in general viewed as requiring the development of *smart* search strategies consisting of a set of simple rules. The implicit assumption in these rule-based games is, of course, the validity of the energy potential.

Because the force fields are valid approximations within a range of conformations, there is also a limit to the range of use of the rules, and some form of *boundary* restriction, delimiting the valid region of the potential energy function, must be imposed on the search protocol. For instance, a protocol that results in a sequence of breaking and building events (e.g., search procedures which overcome local minima by splicing the protein in parts) might form arrangements which, even though favorable in energy, might not be realistic. This is true of the Genetic Algorithm search protocols (Dandekar and Argos, 1992), which have proven to be a very fast alternative to more traditional approaches, however, with a tendency to find deeper minima that do not correspond to real conformations of the protein (Shulze–Kremer, 1992). Thus, a smart algorithm must combine the overall features of a potential energy function with the known structure of a homologous protein to model a system of unknown structure. In what follows, we will describe a general procedure to transfer three-dimensional information from a known structure to an unknown structure using a combination of molecular mechanics and a rule-based, self-consistent algorithm.

II. THE FEEDBACK-RESTRAINED PROCEDURE (FRMD)

A. The Archetype Hypothesis

The explosive increase in the number of high-resolution protein structures (about 800 coordinate sets are currently available in the latest version of the Brookhaven Protein Databank) and the even more significant increase in the number of known protein sequences, (whose number currently exceeds 60,000), have been an important driving force in the recent development of this area. A rapid inspection of these numbers indicates a rate of more than 75 known sequences per known three-dimensional structure. If the identical structures are removed from this list, and the very close mutations removed from the list of sequences, the ratio improves only slightly to 65:1. A more revealing analysis consists of clustering the structures by folding homology and sequences in families after removing redundancies. The analysis of the results reveals that large groups of sequences, for example, similar proteins from different plants or animals, cluster in tight families (Chothia, 1992) with over 50% amino acid composition conservation. On the other hand, after three-dimensional alignment the tight folding homologies reveal sequence similarities which frequently retain a few percentages of identity (Blundell and Johnson, 1993). This is the case, for instance, of the large family of jellyroll topologies (Chelvanayagam et al., 1992). Moreover, the development of advanced DNA sequencing technologies (Lipshutz and Fodor, 1994) make the advent of genome sequencing projects possible. To appreciate the amount of information encoded in these genomes, we can compare the length of the typical nematode genome with the amount of information accessible today in the protein sequence data banks (Bork et al., 1994); a nematode genome contains about five times more information than the information contained in *all* of the known sequences of proteins today. It is expected that a complete nematode genome sequence will be known by the turn of the century. Around this time the human genome sequence will also be known. The human genome sequence is about six times larger than that of a nematode. The complete genome sequence of yeast will be known in a few years. This sequence alone will contain as much information as all of the currently known protein structures. In short, unless an unpredictable technological change takes place, the ratio of the number of known protein sequences to the number of sequences for which the three-dimensional structure is known will drastically increase in the coming decades.

Given these statistics, an obvious alternative to the traditional homology modeling approach is the structure-recognizes-sequence protocol, proposed earlier as the inverse protein folding problem (Sander, 1991). In short, the approach involves finding sequences compatible with a given structure. In this sense, the fewer three-dimensional structures (when compared with the lengthy list of known sequences) are considered archetypes, and their value is reduced when averaged with other structures. However, the use of single structures for homology modeling exercises, results in the impossibility of validating the model by statistical means (i.e., by comparing the resulting model with several parent structures). Because a single structure can not be used for statistical fitting exercises, we must rely on an a priori adjusted molecular mechanics force field to guide elaborated search procedures (e.g., the annealing technique of conformation optimization; Brünger, 1992) in matching the known structure with the building model as well as possible. However, as with the example of the GA algorithms mentioned before, we must carefully limit the boundaries of the search. A self-consistent procedure devised for this purpose in our laboratory, employs a template-forcing technique where the problem molecule is loosely tethered to the known structure (used as a rigid framework) and the three-dimensional information is transferred to the unknown structure during the course of a biased molecular dynamics (MD) trajectory. This technique can be better explained if we first dissect it by describing its implementation in studying small molecules and later apply it in studying larger molecular aggregates.

B. Methodology

For simplicity, we will first consider the case of determining the conformation of a small peptide, using NMR restrained molecular dynamics.

In a typical NOE (Nuclear Overhauser Effect) distance-restrained molecular dynamics simulation, NOE information is applied as a set of harmonic restraints between pairs of hydrogen atoms for which the NOE distances serve as equilibrium values (Scheek et al., 1989; Brünger, 1990a). This procedure has several drawbacks owing to the use of arbitrary values for the restraint force constants. If the force constant is set too high, intermediate conformations become severely distorted and realistic minima are difficult to find. If the force constant value is set too low, the molecular mechanics potentials tend to dominate the conformation search and prevent the model from exploring regions of configuration space consistent with the NOE distances (Withka et al., 1992). In either case, realistic conformations may not be properly sampled. To overcome these limitations, a feedback-scaling procedure that applies variable scale factors to the individual nuclear Overhauser effect (NOE) distance restraints *during* the molecular dynamics trajectory can be used. This procedure, termed Feedback-Restrained Molecular Dynamics (FRMD), facilitates sampling the conformations that are energetically relaxed and which *simultaneously* agree with the experimental NOE distances throughout the molecular dynamics trajectory.

In FRMD calculations a time-dependent (τ), NMR-derived, interproton distance restraint is incorporated as an energy restraint, E_{NMR}, using the following relationship:

$$E_{NMR}(\tau) = 1/2 \sum S_{ij}(\tau) K_{ij} \Delta R_{ij}^2 \qquad (1)$$

where $S_{ij}(\tau)$ are the scale factors for NOE interproton distances computed over all i,j protons, K_{ij} is a force constant (typical value for K_{ij} = 15.0 Kcal/mol-Å^2 for all i,j), and $\Delta R_{ij}(\tau)$ is the difference between the interproton distance $R_{ij}(\tau)$ for the current trajectory structure and the experimentally derived NOE interproton distance R_{ij}^0. $S_{ij}(\tau)$ are calculated at discrete intervals, n, using a feedback function of the form,

$$S_{ij}(\tau_n) = S_{ij}^0 (|\Delta R_{ij}(\tau_n)| + 1)^{-4} \qquad (2)$$

where τ_n is the update time, and S_{ij}^0 is the maximum scale value for the pair (i,j). τ_n is calculated as a function of the time interval Δt, where $\tau_n(\tau) = \tau_{n-1} + \Delta \tau(\tau)$ for all τ so that $[\tau_n - \Delta \tau(\tau)/2 < \tau < \tau_n + \Delta \tau(\tau)/2]$.

We may describe the expected behavior of a MD trajectory using Eqs. (1) and (2) by comparing it with some of the most traditional techniques. If the initial geometry is far from the geometry best matching the restraints, some of the unsatisfied restraints (Eq. (1) with $S_{ij} = 1 \forall i,j$) may deform the molecule. Figure 1 describes this problem. Because regions of the molecule are limited in their mobility due to their interaction with the surrounding atoms, the naïve application of Eq. (1) ($Sij = 1 \ \forall$ pair i,j) would result in an inefficient search path. One of the frequent solutions to this problem is the use of *gaps* in the potential energy form (Flores and Moss, 1990):

$$E_{NMR} \begin{cases} 1/2 \ \sum S_{ij}(\tau) K_{ij} \Delta R_{ij}^2 & \forall R_{ij} > Rcut1 \\ 0.0 & \forall Rcut1 < R_{ij} < Rcut2 \\ 1/2 \ \sum S_{ij}(\tau) K_{ij} \Delta R_{ij}^2 & \forall R_{ij} < Rcut2 \end{cases} \qquad (3)$$

The use of gaps (schematically described in Figure 2) ensures some extra flexibility of the restraint potential energy term at the cost of a larger unrestricted region of the conformation space that can be explored. Moreover, the extreme deformation due to the severely stretched restraints is not removed unless extremely large gaps are used. In this context, if we imagine the MD trajectory as a peculiar path followed by a minimization algorithm, we might conclude that a *smart* search

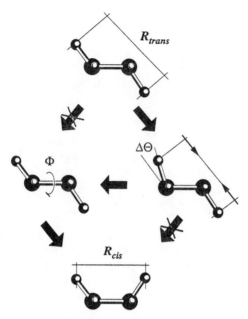

Figure 1. Scheme of molecular distortion during a restrained MD trajectory. The molecule, initially in a *cis*-conformation, is strained by an NMR-like restraint, which agrees with a *trans*-conformation. During a MD trajectory, sets of isolated atoms which could move by *soft* degrees of freedom are frequently caged by the surrounding atoms of the molecule. Under these circumstances, the MD trajectory, which is impeded from evolving through the dihedral angle Φ, will instead distort the molecule (e.g., deforming the bond angles which are frequently less affected by the molecular caging). The molecule, in consequence, is deformed, exploring conformations unlikely to occur in an unrestrained MD trajectory.

will restrict the path from the initial to the final conformation just enough to *guide* the search in the proper direction but not enough to induce it to explore forbidden regions of the conformation space. In this sense, Eq. (3) is not satisfactory, because large gaps allow too large a portion of the configuration space to be explored and small gaps force the molecule through conformations with very high energy, thus exploring uninteresting portions of its configurational space.

This problem is partially solved by using Eq. 2, which guides the search algorithm through a subregion of the configuration space *compatible* with as much of the experimental information available as possible, but not resulting in deformations of the molecule by imposing overstretched restraints. As the MD progresses, it is expected that the FRMD search will carry the molecule through some of the conformations bordering the global solution. When this happens, some other

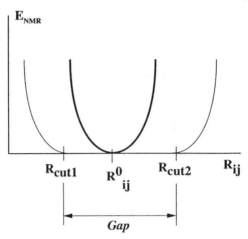

Figure 2. Schematic representation of the *gap* insertion in the NMR-like restraint function. Two semigaps are frequently defined (R_{cut1} and R_{cut2}) which describe the differing flexibilities of the region to compression or expansion. Different force constants for short and large distances are also frequently used. The values for these parameters must be calibrated for families of compounds of known structure.

restraints will automatically be turned on, and the molecule will be retained with minimal distortion around this conformational neighborhood.

However, another problem becomes evident by simple inspection of Eqs. (2) and (3). It is unlikely that a MD path will bring the molecule to the correct conformation unless the energy barriers that must be overcome in the path from the initial to the final coordinates are very small. Eq. (3) will force the molecule to initial high-energy conformations, which the very stretched restraints will help hold together.

A simple solution to this problem is the so-called simulated annealing (SA) protocol (Brünger, 1992, 1991a, 1991b). SA consists of heating the molecule during the MD trajectory, later slowly cooling it to room temperature, and finally minimizing it. The expected behavior of the molecule during this procedure is as follows: the high temperature dynamics must help bring the molecule over the medium to high energy barriers, thus overcoming the difficulties described before with Eq. (3); the ensuing slow cooling step must give the structure enough time to accommodate to the most favorable conformation and eventually reach the global minimum. Because the difficulty in trespassing a barrier is proportional to the MD temperature, a long enough trajectory must assure the detection of the *wider* minimum in the conformation space. Following this same reasoning, as the temperature decreases, the molecular conformation will continue to evolve toward the most preferred conformations, (now in a subregion of the configurational space) and will, finally, be trapped in the global minimum, assuming that this minimum lies within the boundaries of the widest *global* minima, which, as we mentioned earlier, will be detected in the first steps of this procedure. For SA to succeed, an

Protein Model Building

extremely low temperature gradient must be imposed during the cooling step. Considering that most cooling experiments last less than a nanosecond, this condition is seldom satisfied. To understand this behavior better we can imagine a process during which a protein is heated to 900 K and then cooled to room temperature (300 K) in 400ps. This represents typical temperatures and times for

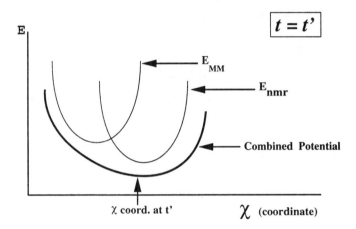

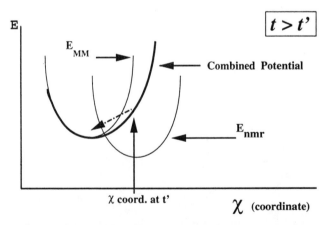

Figure 3. Time evolution of the generic coordinate χ affected by the combination of molecular mechanics and a restraint function. At $t = t'$ the force field is a combination of both energy terms. As Eq. (4) acts on the restraint term, the χ coordinate approaches the minima defined by the molecular mechanics force field. The periodic nature of Eq. (4) ensures that the coordinate χ will explore the surroundings of the molecular mechanics minima, in case of disagreement between the molecular mechanics and the restraint parts of the force field.

MD annealing procedures (Brünger, 1991a, 1991b). It is well known that, for temperatures above 400 K, the dihedral angles described in a MD program must behave, essentially, as free rotors (Hill, 1968). For simplicity, let us assume a linear temperature ramp. Therefore, the MD must reach equilibration on the dihedral degrees of freedom in *less* than 70ps, because it takes 333ps to reach 400 K in this simulation. Thus, the most common distortion in MD annealing procedures is that, at the end of the trajectory, the dihedral degrees of freedom will be frozen in a *high temperature* conformation.

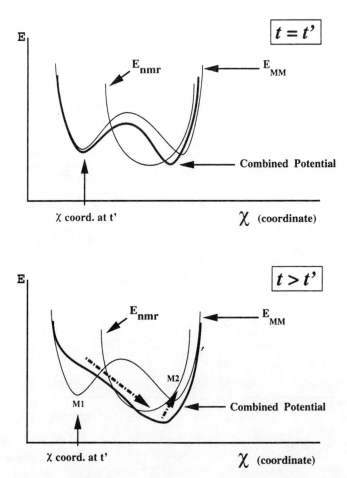

Figure 4. In this example the MD trajectory carries the coordinate χ between two minima M1 and M2 dominated by the molecular mechanics force field. In this example Eqs. (1) to (4) periodically *deform* the potential energy hypersurface helping the system escape from a local minimum (see Figure 3).

In short, slow cooling protocols are expected to reach conformations near global minima, overcoming some of the difficulties posed by Eq. (3). However, the ideal simulation conditions required by a SA technique are seldom attainable. Thus, in many cases, the SA technique will give only a first approximation to the final conformation. Note that part of the difficulty in using SA protocols is due to the use of high temperatures which in itself works to *flatten* the dihedral conformation space of the molecule, greatly extending the total conformation area of search and making location of the global minimum difficult. Hence, a *low temperature* protocol is required to overcome the local minimum problem.

Going back to equation (2), this again presents the local minimum problem; as the trajectory progresses, new restraints will arise. Thus, the trajectory will eventually be trapped in a local minimum. It must be noted though that, differing from the SA treatment, this conformation might not result in a high energy conformation, because unsatisfied restraints are not turned on. A simple solution to escape the local minimum in this formulation consists in affecting S_{ij} in Eq. (2) by a periodic *quenching* function. Thus $S_{ij}(\tau_n)$ will now read

$$S_{ij}(\tau_n) = (S_{ij}^0 (|\Delta R_{ij}(\tau_n)| + 1)^{-4})(|\sin(\alpha\tau + \delta)| + \kappa_{ij}) \quad (4)$$

where α and δ determine the periodicity and phase of the sine function. α and δ can be functions of τ in a more generic treatment. κ_{ij} is a lower bound for $S_{ij}(\tau_n)$. κ_{ij} can be used to bias the trajectory using it as a *memory weight* factor. Note that, if $\kappa_{ij} = 0$, the sine function will *quench* the restraint to $S_{ij}(\tau_n) = 0$ every π/α. Thus, the structure can escape the local minimum and explore the surrounding configuration space without high temperatures. However, a careful analysis reveals that Eq. (4) actually *pumps* energy into the system. In fact, Eq. (4) helps the molecule explore the surrounding environment by freeing it and by locally heating the atoms involved in the restrained interactions in a very particular manner. Figures 3 and 4 describe how the local heating of the molecule, by using Eq. (4), helps the MD trajectory *jump* over barriers. To keep the molecule cool, a continuous thermalization must be used.

C. Application: RGD-Containing Peptides

The procedure above described will be first illustrated by the study of small flexible peptides containing the sequence RGD. The sequence Arg-Gly-Asp (RGD) is one of the recognition determinants in the integrin family (Ruhoslahti, 1988). Short peptides containing RGD as the central sequence retain the ability to bind to receptors, almost independently of the nature of the terminal residues. These peptides have been shown to inhibit the interactions of cell adhesion proteins with their receptors and to have dramatic effects on developmental processes involving cellular recognition (Ruhoslahti, 1988). In proteins of known three-dimensional structure, the RGD sequence is found either in β-turns (α-lytic protease, Fujinaga et al., 1985; thermolysin, Monzigo and Matthews, 1982) or in an extended conformation (γ-crystallin, Summers et al., 1984). These structures reveal a high degree of flexibility for the RGD segment. This flexibility is confirmed by peptide NMR experiments (Cachau et al., 1989). The combination of Monte Carlo and NMR experiments on GRGDS peptides suggest the presence of a type II β-turn as the dominant conformation in solution with a large contribution from an extended conformation (Cachau et al., 1989). The type II β-turn conformation has also recently been found in the experimental structure of Fibronectin (Leahy et al., 1992). The conformation of the RGD sequence bound to receptors of the integrin

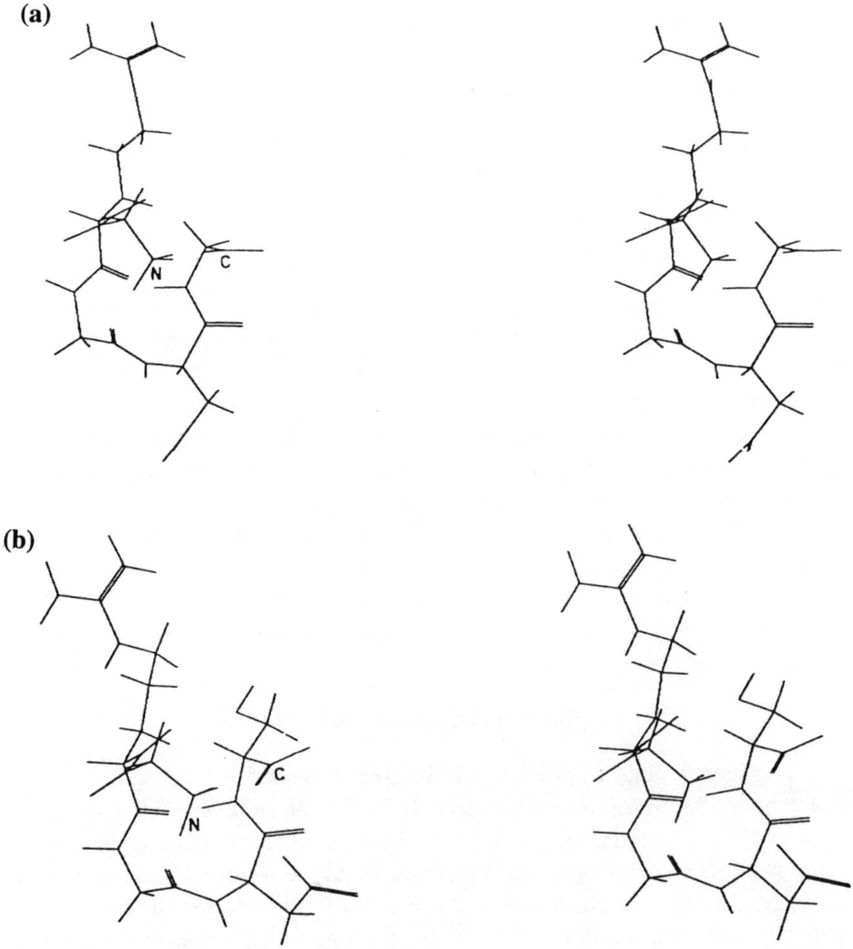

(continued)

Figure 5. RGD dominant conformations. a) to c) are the stereo diagrams of the dominant conformations, corresponding, in order to conformations T, T' and A. Figures d), e), and f) describe secondary minima.

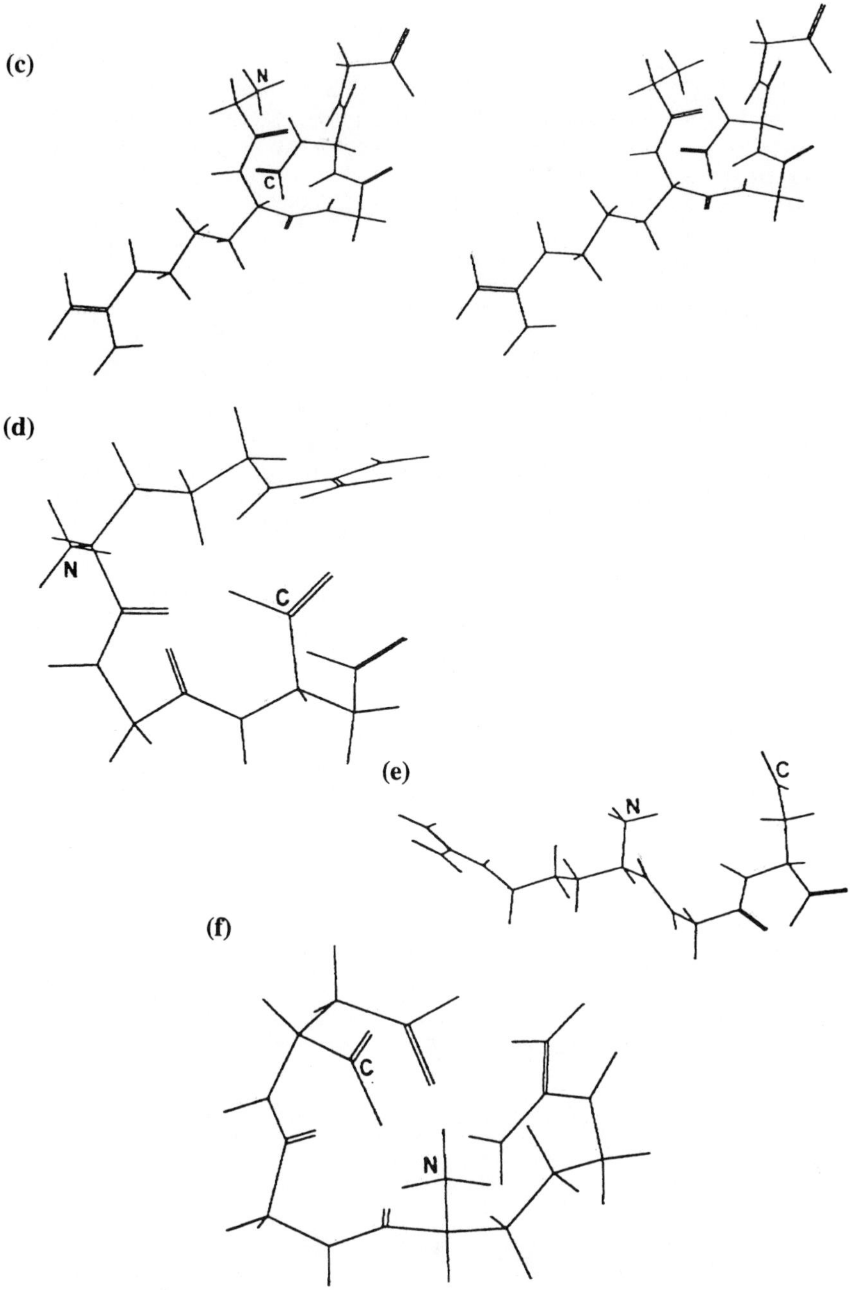

(c)

(d)

(e)

(f)

Figure 5. (Continued)

Table 1. GRGDS Dominant Conformations (Cachau et al., 1989)

Atoms	A		G		T		T'	
	Range	<R_{NMR}>	Range	<R_{NMR}>	Range	<R_{NMR}>	Range	<R_{NMR}>
R2βH-S5NH	5.5–6.0	5.0	9.2–9.2	8.0	3.4–4.4	3.2	3.4–4.7	3.2
R2γH-S5NH	6.2–6.2	5.8	7.0–8.7	6.2	5.7–6.2	5.0	3.7–3.7	3.4
G3αH-D4NH	2.8–3.6	2.7	2.2–3.3	2.1	2.3–3.3	2.2	2.7–3.3	2.6
D4NH-G3αH	2.8–3.6	2.7	2.2–3.3	2.1	2.3–3.3	2.2	2.7–3.3	2.6
D4NH-R2αH	4.3–4.3	4.1	6.0–6.0	5.3	6.1–6.1	5.6	6.5–6.5	5.9
D4βH-G3αH	5.8–6.5	5.4	4.7–6.5	4.4	4.7–6.4	4.3	5.3–6.0	4.7

family, however, remains unclear. Direct measurements on appropriate receptor-bound ligands will be necessary to elucidate this problem further.

The GRGDS conformation space has been thoroughly explored and the dominant conformations characterized (Cachau et al., 1989). Figure 5 presents the most important conformations accessible to this molecule at room temperature. Tables 1 and 2 describe these conformations in terms of the most relevant NOE signals used for their characterization. Note that none of the preferred conformations completely satisfies the experimentally determined NOE intensities (Table 2). The Monte Carlo technique used to generate the conformations described in Figure 5 and Table 1 does not contain any bias from the experimental NOE (Cachau et al., 1989).

FRMD and SA Simulation of GRGDS

Two sets of MD calculations are carried out on GRGDS, an SA optimization and a FRMD. All calculations are carried out using Xplor 3.1 (Brünger, 1992) and the standard set of parameters for the united atom representation.

a) Simulated annealing. The heating time is set to 100ps at 1000 K. The cooling time is set to 400ps with a linear temperature ramp to 300 K. The

Table 2. NMR-NOE Signal Intensity Reconstructed from Monte Carlo Experiments (Cachau et al., 1989)

Atoms	A	G	T	T'	NOE
R2βH-S5NH	–	–	w	m/w	w
R2γH-S5NH	–	–	–	m/w	s
G3αH-D4NH	m	s	m	m	m
D4NH-G3αH	m	s	s/m	m	m
D4NH-R2αH	w	–	–	–	w
D4βH-G3αH	–	–	w	w	m

Table 3. NMR ($1/R^6$) Averaged Distances for the Dominant Conformations of GRGDS during a FRMD Trajectory and for T' during the Final Segment of a SA MD Calculation

Atoms	A	G	T	T'	T'(SA)
R2βH-S5NH	5.3	8.2	3.1	3.3	3.0
R2γH-S5NH	5.9	6.8	5.6	3.5	3.5
G3αH-D4NH	3.0	2.2	2.3	2.6	2.9
D4NH-G3αH	3.0	2.3	2.3	2.6	2.9
D4NH-R2αH	4.0	5.7	5.8	6.0	4.7
D4bH-G3aH	5.2	4.4	4.4	5.0	4.2

conformation is finally minimized using the standard Powell minimization procedure. The NMR restraints are maintained during all stages of the simulation. A gap of 0.5 Å is allowed equally spaced around the expected equilibrium distance [Eq. (3)] for each of the NOE-restrained distances. The determined conformation closely matches T', however, showing some noticeable degree of distortion due to the D4NH-R2αH restraint, which is not completely satisfied by this conformation. The average NOE signals, reconstructed from the statistics along the MD trajectory for the different conformations, are presented in Table 3.

b) FRMD simulation. This simulation is carried out using the same molecular mechanics parameters used for the SA experiment, except that the NMR restraints are applied following Eqs. (1), (2) and (4). The simulation trajectories consist of a 10ps initialization period (T ramp from 10 K to 300 K), followed by an equilibration step (30 ps, 300 K) and a production run of 400 ps. No restraints are applied during the initialization and equilibration steps.

One of the main drawbacks of the SA technique is the impossibility of identifying multiple conformations from the same simulation run. This can be accomplished during a FRMD run by inspecting the overall agreement between the structure of the molecule at a given time (τ) with respect to the set of restraints. The simplest way of doing this is by inspecting the total agreement function,

$$F1(\tau_n) = \sum_{ij} S_{ij}(\tau_n) \qquad (5)$$

or its average over a period α [Eq. (4)], $<F1>$. Another useful parameter is the product of the molecular total energy, E_T, from the molecular mechanics calculation, (which can be averaged over a period α: $<E_T>$) and $<F1>$:

$$F2 = <F1><E_T>. \qquad (6)$$

The *relaxed* conformations that best match the experimental restraints can then be scooped out of the trajectory by using <$F1$> and $F2$ as the leading parameters. These selected conformations can then be clustered and analyzed.

Results

The FRMD trajectory described above resulted in three main conformations satisfying most of the experimental restraints, which in turn correspond to the T, T' and A conformation. The G conformation is also detected but as too small a contribution to the FRMD trajectory to be considered statistically relevant. The SA trajectory correctly found the T' dominant conformation. The average NOE signals, reconstructed from the statistics along the MD trajectory for the different conformations, are presented in Table 3. The advantage of FRMD over the more traditional SA in locating multiple conformations is evident in this example. The average interproton distances presented in Table 3 for the SA calculation are estimated during the last 30ps of the slow cooling trajectory. Because only T' is characterized during SA, the final geometries do not report enough information for comparison purposes. The average geometry is a more significant parameter set for comparison purposes because it contains, in part, information on the shape of the conformation space being explored by both procedures during the trajectory. As seen in Table 3, the average distance values (in the sense of NMR statistics) for the SA trajectory are biased toward too short distances during the final segment of the SA trajectory. This behavior can be expected because all restraints are fully applied during this simulation.

Another interesting aspect that differentiates the SA and FRMD trajectories is shown in the series of Ramachandran plots described in Figure 6. These plots present a collection of the Phi-Psi angles for the four central peptide bonds of the GRGDS peptide, collected over a period of 50ps of trajectory. The Ramachandran plot for the SA trajectory, Figure 6(a), reveals a very slight structure, resulting from the high temperature used during the simulation which, as mentioned before, *flattens* the dihedral conformation space of the molecule during the trajectory. Thus, a high temperature simulation results in an inefficient search of the configuration space because the molecule scans uninteresting areas during most of the simulation time. The Ramachandran plot for the restrained MD trajectory, Figure 6(b), shows how the imposition of the restraints and low temperature MD result in a trajectory where the molecule is trapped in a single conformation. The Ramachandran plot for the FRMD trajectory, Figure 6(c), shows an intermediate picture. The plot is biased when compared with the free trajectory, Figure 6(d). However, the characteristic of the Ramachandran plot of the free molecule, Figure 6(d), are still clearly recognizable.

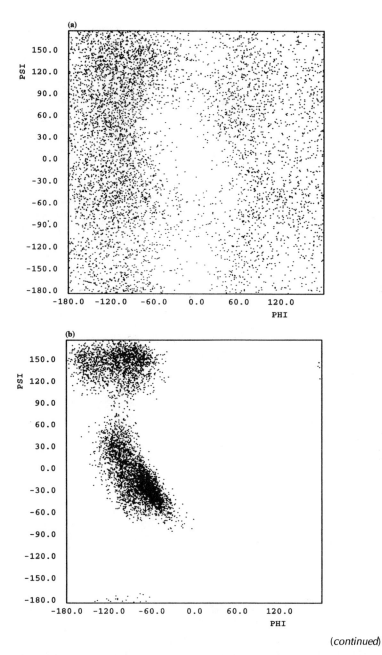

(continued)

Figure 6. Ramachandran diagram for the MD trajectory of a GRGDS peptide using different simulation conditions. a) High temperature MD (1000 K). At this temperature most dihedrals act as near free rotors. Note that the Ramachandran regions are poorly defined. b) Low temperature (300 K) restrained MD.

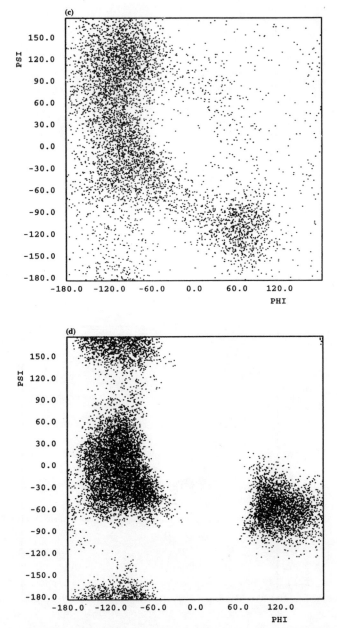

Figure 6. (Continued) The restraints hold the structure near a local minimum for the total span of the trajectory. c) FRMD. During this trajectory the molecule explores favorable regions of the Ramachandran diagram (compare with Figure d). However, the trajectory is biased towards a subset of conformations. Compare with figures (b) and (d). d) Low temperature (300 K) free MD.

III. POSITIVE BIASED AVERAGING

As shown in the previous section, because multiple conformations are common in small, flexible molecules, a mechanism that could help us distinguish these minima during a MD trajectory is desirable. On the other hand, a MD calculation could, in principle, be useful for automatically selecting false NOE signals, by analyzing which restraints are more frequently satisfied during the trajectory. The main drawback of this approach is that we must complete a simulation to find which restraint can be discarded, which in turn carries us to a self-consistent series of simulations to ensure the proper selection (i.e., we should run trajectories to determine the valid set of restraints for a subsequent experiment and so on).

Simpler techniques to accomplish this goal, that can be easily implemented during a MD trajectory, are based on positive bias averaging of restraint signals. In what follows we will present one such technique which consists, in short, of normalizing the total restraint signal by a modified calculation of $<F1>$. This normalization coefficient now *excludes* the S_{ij} corresponding to the signal to be normalized, thus ensuring a positive bias toward the best subset of signals. As a direct consequence of this averaging procedure, NMR-NOE signals resulting in restraints inconsistent with the subregion of conformations compatible with most of the restraints being used will be downscaled. In this sense, the procedure self-evaluates the degree of agreement of the complete set of restraints. Clusters of signals can then be easily detected and spurious NMR restraints isolated. Given the practical significance of this technique, the best illustration consists of its application to some simple systems.

A. Application to Degenerated Intermolecular Interactions: Taxol

Taxol is a complex diterpene (Figure 7), derived from the Pacific yew *Taxus brevifolia*, that exhibits antitumor activity (Wani et al., 1971). This drug is currently being used in the treatment of several types of cancer, including ovarian and breast cancer (McGuire et al., 1989; Holmes et al., 1991). Taxol's antitumor properties are exerted via its unique ability to bind to and stabilize microtubules and, thereby, interfere with microtubule depolymerization (Horwitz, 1992; Schiff et al., 1979; Schiff and Horwitz, 1980). The limited availability of taxol from natural sources has prompted a search for semisynthetic derivatives that would retain the antitumor activity of taxol (Kingston et al., 1990; Kingston, 1991). Optimization of the therapeutic potential of taxol using medicinal chemistry strategies has been difficult due to the structural complexity of the molecule, and the bioactive conformation of taxol is unknown. Knowledge of the three-dimensional structure of taxol may lead to a better understanding of the structural requirements that relate to its biological activity and could provide pharmacophore-based targets for the development of new semisynthetic or fully synthetic strategies for taxol analogs. Comparison of crystallographic data for taxotere (Gueritte-Voegelein et al., 1990), a

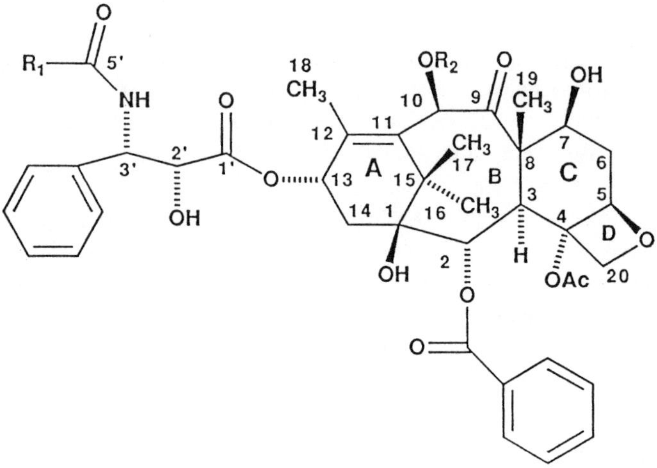

Figure 7. Chemical structure of taxol. R1 is an acetyl group in taxol and a *tert*-butyl in taxotere.

taxol analogue (Figure 7), and NMR-NOE data for taxol in $CDCl_3$ (Hilton et al., 1992; Chmurny et al., 1992; Falzone et al., 1992) indicates that the baccatin nucleus (Figure 7, rings A–D) for these compounds is conformationally rigid. However, there is evidence from the NMR studies that the A-ring side chain is flexible, and a unique conformation for taxol in solution is not identified in these studies.

FRMD Protocol for Taxol

An FRMD experiment using the scaling procedure described above [Eqs. (1), (2) and (4)] with previously determined NOE-based distance restraints is carried out as follows (Cachau et al., 1994a): An initial taxol model is built based on the Taxotere crystal structure (Gueritte-Voegelein et al., 1990) using the program INSIGHT. The model is minimized using the DISCOVER molecular mechanics force field (Hagler et al., 1979). A total of 47 interproton distances for taxol in $CDCl_3$ are derived from two-dimensional NMR-NOE experiments (Hilton et al., 1992; Chmurny et al., 1992). FRMD is performed in three stages for a total of two nanoseconds. Stage I comprised heating the structure from 50 K to 2000 K for 300 ps. During this stage, the NOE distance restraints are deactivated to maximize conformation exploration of the native internal force field representation of the molecule. In stage II, the NOE restraints are applied during a 1400 ps simulation at 2000 K. In stage III the structure is cooled from 2000 K to 50 K over 300 ps using NOE distance restraints. Note that in this particular example a very long high-temperature trajectory is used to obtain an exhaustive sampling of the NOE signal agreement with the conformation space accessible to this molecule.

For degenerate distances, (e.g., interproton distances related to methyl hydrogens) all equivalent protons are treated as independent atoms rather than as average atoms or pseudoatoms. A positive bias function is then used to select which of the equivalent i,j proton pairs might dominate the NOE signal:

$$S_{ij}(\tau_n) = S_{ij}(\tau_n) / \left[\sum S_{i'j'}(\tau_n) \right] \qquad (7)$$

where the summation is over the equivalent set of proton pairs (i',j') excluding the case where $i,j = i',j'$. The complete set of $S_{ij}(\tau_n)$ obtained is normalized so that $0.0 < S_{ij}(\tau_n) < 1.0$. After Eq. (7) is applied, a similar procedure is used to scale the complete set of restraints, thus favoring the dominant restraints.

Two Possible Side Chain Conformations for Taxol

After the dominant structures are determined, DISCOVER energy minimizations are performed on them, using steepest descents and conjugate gradients until the norm of the gradient is less than 1.0×10^{-3}. The consistent valence force field with automatic torsion, valence, bond, out-of-plane, and cross-term parameters are employed for both molecular mechanics and molecular dynamics studies (Hagler et al., 1979).

The FRMD experiment produced two frequently recurring conformations (designated CI and CII) that agreed with a majority of the NOE distance data (Figure 8). Correlation coefficients for the agreement of calculated and observed distances are 0.57 and 0.78 for the CI and CII conformers, respectively (Figure 9). The largest structural differences between CI and CII are in the A-ring side chain conformation (Figure 8). The taxol side chain exhibited limited rotation about its ester linkage, whereas the distal segments of the side chain appeared to be more flexible judging from examination of the conformations sampled during the molecular dynamics trajectory. In CI, the 2'-OH forms a hydrogen bond with the carbonyl oxygen of the 3'- amide. In CII, the 2'-OH is hydrogen bonded to the 4-OAc moiety of the oxetane ring. The CI and CII NOE-minimized structures are fully optimized without NOE distance restraints at a semiempirical molecular orbital level of calculation using MNDO/PM3 (Stewart, 1990) to confirm the stability of the conformations found (Cachau et al., 1994a).

Validation of the FRMD Procedure

The main objective of this simulation is to study the effect of bias averaging in the FRMD treatment of MD NMR-restrained studies. The validity of applying Eq. (7) to the degenerated H-H interactions can be described by calculating the mean force potential for the interaction between equivalent hydrogens. The stress ratio (measured as the ratio of the average module of the mean force acting on each atom in a free MD trajectory vs. a FRMD trajectory) for restrained and unrestrained atoms shows a mere 7.2% increase in the stress suffered by the restrained atoms through

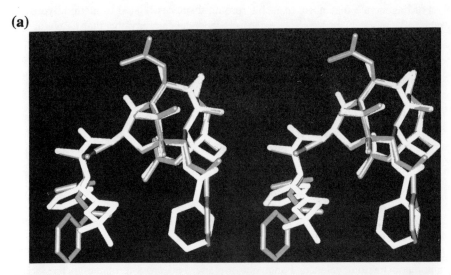

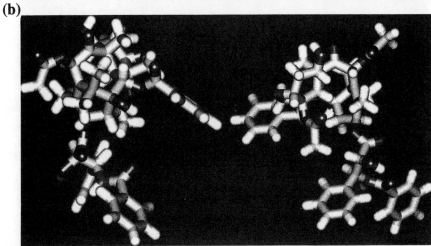

Figure 8. Taxol dominant conformations. a) CI is compared with the crystal structure of taxotere. Note the similar disposition of the extended side chains. b) Conformations CI (left) and CII (right) are compared. Note the different packing of the phenyl groups in both conformations. Conformation CII is rotated 160° to reveal an interesting packing of the groups which determine the taxol activity in a single face (Cachau et al., 1994a).

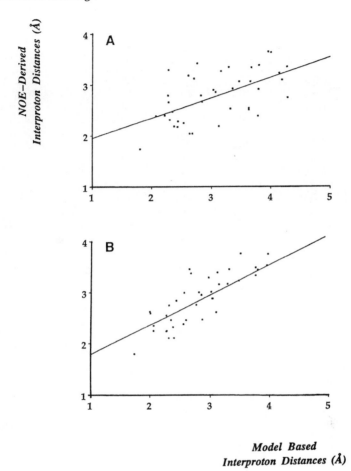

Figure 9. Scatter plots of interproton distances computed from the NMR-NOE data (NMR) and from the CI (A) and CII (B) structures. Although the agreement is high, it is not perfect. Thus the protocol of refinement must be tolerant to absorb the tension imposed by the use of distance restraints.

this procedure. Independent experiments show that this value can be as high as 300% when using a SA protocol (Cachau, unpublished).

An even more revealing insight can be obtained from the analysis of the individual scaling factors.

The FRMD protocol initially used a total of 47 unique proton-proton distance restraints: 30 distances are assigned to the baccatin core, and 17 involved side chain-side chain and side chain-core distances (Table 4). The FRMD procedure identified 7 of the input distance restraints (6 core and 1 side chain) as incompatible

Table 4. NOE Distance Restraint Used during the Taxol FRMD Simulation. Average Distances Corresponding to Structure CI and CII and Taxotere and Scale Factors S_{ij} for the Given Pair of Atoms. In Bold the Rejected Restraints.

		Distances (Å)				
		NOE	Taxotere	CI	CII	S_{ij}
H-2	**H-20a**	**3.21**	**4.00**	**3.90**	**4.39**	**0.00**
H-2	H-20b	3.07	2.83	3.65	3.32	0.69
H-2	Me-16	2.28	1.94	2.43	2.06	0.93
H-2	Me-19	2.40	2.27	2.21	2.89	0.87
H-3	H-5	3.13	3.60	2.70	3.52	0.85
H-3	H-7	2.04	2.19	2.67	2.32	0.91
H-3	H-10	2.53	2.89	3.31	3.13	0.17
H-3	Me-16	3.63	3.45	4.00	4.02	0.83
H-3	Me-18	2.79	2.87	2.82	3.06	0.79
H-3	Me-19	3.34	3.69	3.67	3.84	0.82
H-5	H-6a	2.18	1.95	2.36	2.28	0.95
H-5	H-6b	2.86	2.57	3.06	2.87	0.99
H-5	**Me-19**	**3.22**	**3.64**	**4.22**	**4.05**	**0.03**
H6a	H7	2.25	2.84	2.52	2.39	0.97
H6a	H-6b	1.74	1.74	1.80	1.74	0.98
H-6b	Me-19	2.39	2.40	2.06	2.35	0.93
H-7	Me-18	2.90	2.56	3.03	2.57	0.95
H-7	Me-19	3.38	3.63	3.80	3.81	0.78
H-7	H-10	2.04	2.39	2.62	2.40	0.96
H-10	Me-18	2.17	2.19	2.42	2.27	0.89
H-13	Me-17	2.18	1.64	3.12	2.06	0.93
H-13	Me-18	3.19	3.11	2.60	3.00	0.90
H-14a	H-3	2.39	2.22	3.78	2.61	0.87
H-14b	**H-3**	**2.81**	**3.59**	**4.40**	**3.90**	**0.00**
Me-16	Me-17	2.48	2.38	2.34	2.27	0.93
Me-16	Me-19	3.29	2.65	2.27	3.15	0.85
H-20a	**H-6a**	**3.18**	**4.79**	**3.67**	**5.00**	**0.00**
H-20a	Me-19	2.56	3.36	2.05	3.67	0.01
H-20b	**H-6a**	**3.26**	**4.68**	**4.81**	**4.67**	**0.00**
H-20b	Me-19	2.51	2.06	3.63	2.01	0.19
4-Ac	H-2'	2.75	3.98	4.28	2.44	0.63
4-Ac	H-3'	2.67	2.64	2.83	2.78	0.82
4-Ac	o-Ph1	3.42	3.13	2.75	4.00	0.11
4-Ac	o-Ph2	3.35	2.70	4.27	2.66	0.20
H-2'	H-3'	2.31	2.33	2.29	2.56	0.67
H-2'	Me-18	3.35	2.54	3.29	3.35	0.89
H-2'	o-Ph2	2.92	2.75	3.35	3.04	0.92
H-3'	**m-Ph2**	**3.22**	**4.60**	**4.63**	**4.64**	**0.00**
H-3'	o-Ph2	2.66	2.42	2.27	2.32	0.81
3'-NH	H-2'	3.23	3.68	4.14	3.81	0.17
3'-NH	H-3'	2.79	2.90	2.26	3.07	0.62
3'-NH	o-PH2	2.91	2.08	3.79	2.83	0.75
3'-NH	o-Ph3	2.54	na	3.63	2.00	0.16
10-Ac	H-13	3.10	na	4.19	4.91	0.10
10-Ac	Me-16	3.64	na	3.96	3.54	0.77
H-20a	o-Ph1	3.06	3.37	3.46	3.17	0.83
H-20b	o-Ph1	3.27	4.04	3.16	2.69	0.28

with any realistic structure. The algorithm assigned S_{ij} values for these constraints of <0.01, so that the corresponding distance restraints did not contribute to the conformation search. It is noteworthy that these same distance restraint assignments had been assigned as false transverse NOE cross-peaks (Hilton et al., 1992; Chmurny et al., 1992). This independent assessment of the true NOE-derived distances illustrates an important advantage of the feedback scaling procedure. S_{ij} values for the remaining 39 distance restraints at the end of the trajectory ranged from 0.10–0.99 Table 4.

The ability of FRMD to determine several conformations partially matching the experimental information is one of its main advantages. In the particular example of taxol, this results in relevant information that can be used in the structure-based modeling of new drugs. In fact, both conformations CI and CII determined in this simulation are required to explain the known SAR information.

SAR studies have shown that the R configuration of the 2′-OH group is critical when the amide substituent is present in the side chain (Swindell et al., 1991). In both conformers, as well as in taxotere, the 18-Me group of the core is within 3.3 Å of the 2′-hydrogen. Visual inspection reveals that a 2′-OH in the S configuration would destabilize the side-chain conformation in CII due to the steric interaction with 18-Me (Figure 8(a)). In addition, the hydrogen bond between the 2′-OH and 3′-carbonyl oxygen observed in the CI conformation would be disrupted using the S configuration for the 2′-OH group.

The opposite face of the side chain [Figure 8(b)] provides direct access to the 2′-OH and 3′-NH moieties of the side chain and to the 10-OAc and 18-Me groups of the baccatin core. However, SAR studies, which indicate that deacetylation at the 10-position has little or no effect on bioactivity (Gueritte-Voegelein et al., 1991), make this site less attractive for microtubule binding. Thus, efforts to make nonbaccatin analogues of taxol should focus on structures that would mimic the hydrophobic surface as observed in taxol CII and taxotere. In particular, it may be useful to design alternative structural frameworks that would position the appropriate pharmacophores in a fashion similar to taxol CII. Our conclusions are consistent with previous suggestions for a hydrophobic taxol binding cleft on microtubules (Swindell et al., 1991), but differ on the importance of the baccatin core as an organizing influence on the A-ring side chain. An analogy for the role of the baccatin core might be the role of a protein polypeptide backbone fold for placing key amino acid side chains together to form a substrate or ligand binding site. A direct role for CI in binding cannot be ruled out and should be evaluated in the context of future, structure-based SAR studies. While our analysis points to CII as the bioactive conformer, the ultimate verification of any hypotheses for the bioactive structure of taxol must await the structural determination of taxol in its receptor-bound conformation.

Remarks

This study demonstrates that the FRMD procedure can be used to identify multiple conformation minima that agree with a portion of the experimental NOE distance restraints. In the present example of taxol, two interconverting conformations, each contributing to the total NOE signal, dominated the molecular dynamics trajectory. Thus, this technique addresses the traditional problem in NMR structural determination of identifying multiple, rapidly interconverting conformations that may contribute to the total signal (Cachau et al., 1994a; Cachau et al., 1989; Withka et al., 1992). The analysis of the FRMD trajectory selected a low energy conformer, CII, as the preferred structure for taxol in solution. Recently reported solution structures of taxol based on NMR data and molecular mechanics energy minimization (Baker, 1992) appear to be in general agreement with the CII conformation, and, in a new report, a conformation described as a *hydrophobic collapse* of the side chains (which resembles CII) is also found.

The taxol FRMD example show an important advantage of FRMD over static scaling methods which is its objective ability to filter out false NOE cross-peaks that may lead to erroneous structural interpretations.

B. Application to Degenerated Molecular Conformations: Defensins

Another interesting family of problems, where FRMD can be applied, is in the resolution of problems of incomplete knowledge of the covalent molecular structure. In these cases several different bonding schemes (or topologies as they are usually called) are feasible, none of which can be a priori preferred.

One such a case is the determination of disulfur bond patterns in small peptides. In what follows we will analyze one example of this family of problems, namely the determination of probable disulfur bonding patterns in defensins.

Defensins are a family of cationic peptides isolated from mammalian granulocytes and believed to permeabilize membranes (Fujii et al., 1993). They are 29–34 amino acid residues long, found in rats (Eisenhauer et al., 1989), rabbits (Selsted et al., 1985a), guinea pigs (Selsted and Harwig, 1987), and humans (Ganz et al., 1985; Selsted et al., 1985b; Wilde et al., 1989). Interactions between other peptides and membranes have been studied extensively (Subbarao et al., 1987), but the defensins are unusual because their lipid bilayer-perturbing properties are not derived from amphiphilic, α-helical secondary structures, as with most membrane-associating peptides. Instead, it seems likely that the β-structure, stabilized by three disulfide bonds, is preserved during interaction with the membrane (Fujii et al., 1993). The three-dimensional structure of several defensins, including a purified form from human neutrophil cells (Hill et al., 1991), and two rabbit forms (Pardi et al., 1992), have been characterized. All of these forms are closely related (Fujii et al., 1993), and the three-dimensional structures show the critical role of disulfur bonds in stabilizing the structure (Pardi et al., 1992).

Cyclopsychotride A (from *Psycotria viridis*) and Circulin A and B (from *Chassalia parvifolia*) are members of a recently discovered family of defensins which present a distinctive sequence (Gustavson et al., 1994). The defensins of the latter family do not present some of the most characteristics sequence patterns observed in the previously characterized families. For instance, the characteristic three pairs of adjoining amino acids (-Cys-Cys-, -Asn-Arg- and -Cys-Arg-; Fujii et al., 1993) are missing. In spite of this, the sequence of Circulin B (Figure 10) has all of the main features described previously for defensins (Pardi et al., 1992), namely, a large number of Cys residues, properly alternated to facilitate their engagement in disulfur bridges, and a large number of hydrophilic residues intercalated with large hydrophobic ones. However, given the sequence differences between these peptides and those whose three-dimensional structures are known, a proper alignment is difficult, and a unique scheme of the disulfur bridge pattern is not available. Enzymatic or other form of experimental evidence concerning the disulfur bridge pattern is also not available now. We will use the FRMD treatment described previously to study the disulfur bonding scheme of these molecules.

Four distinctive disulfur schemes seem compatible with the Circulin B sequence. These patterns are schematically presented in Figure 10. The possible disulfur schemes are selected by simple inspection of the patterns built by permutating all sulfur atoms between three disulfur bridges, and discarding those which result in knots or other crossings unlikely to occur (J. Casas-Finet, personal communication).

We will consider *all* disulfur bridge schemes (Figure 10) *simultaneously*, by using a FRMD restraint scheme. We will then use Eq. (7), considering each of the possible disulfur bridge patterns as a single restraint for the purpose of normalization. In this way, the molecule will be biased toward the *dominant* configuration of the disulfur bond (of the set of proposed topologies) at each step of the MD trajectory.

Calculations are carried out with M (Cachau, 1994e), using the Xplor 3.1 unified atom model force field (Brünger, 1992). After 10ps of heating (from 10 K to 300 K), and 30ps of equilibration (300 K), 600ps of production run are computed. The simulation included 30 water molecules surrounding the central molecule.

The results are schematically represented in Figure 10. One of the structures, resembling a globular aggregate, dominates the FRMD search. However, all of the putative configurations are visited in more than one occasion during the trajectory. The compact form, described in Figure 10 as Scheme 1, however, dominates the search. Scheme 1 is the favored structure (the one with highest signal after normalization) during 71% of the trajectory. The transition between structures is a low energy process. The total potential energy fluctuates within a small range; the highest energy structure lies +37 Kcal/mol over the minimum potential energy detected during the trajectory.

One of the more remarkable aspects of the dominant structure during the FRMD trajectory is that it encapsulates a water molecule. The conformation (Figure 11) shows a characteristic hydrophobic surface, with the exception of a few charge

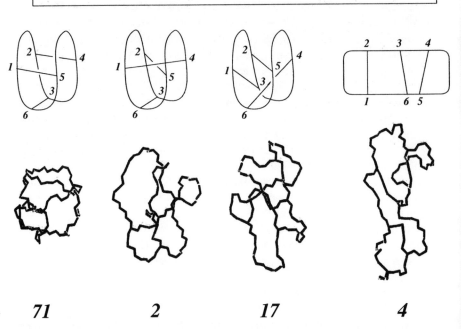

Figure 10. Defensins. From top to bottom: sequence of the modeled defensin, explored schemes, and final conformations that match the scheme showing the percentage of the MD trajectory during which that conformation was the dominant one (only the regions where a dominant conformation was observed are used for this calculation).

residues in opposite faces and some other hydrophilic residues facing toward the inside of the molecule. The overall arrangement resembles a textbook example of what a carrier molecule is expected to be. Because this molecule is expected to function as a carrier, the described conformation constitutes a fascinating example of modeling of function from sequence.

Despite the validity of the determined conformation, which rests on several hypotheses (the a priori selection of a subset of possible disulfur bond schemes is an important one), this example reveals that FRMD has the ability to discern between a set of conformations, all of which are restricted to a small energy gap. Given this result and the validations already discussed in the small peptide and taxol calculations, we will next apply this technique to the description of larger molecular aggregates. Several modifications introduced when doing so will be the subject of the following sections.

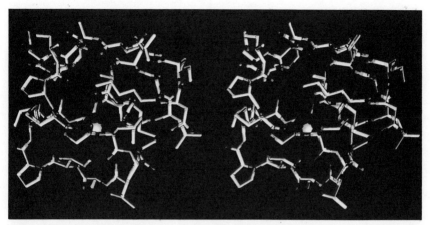

Figure 11. Dominant defensin conformation showing the central water molecule. The outside of the model is clearly hydrophobic. MD calculations are underway to explain the interaction of the outer surface with a lipid bilayer.

IV. MODELING LARGE MOLECULAR AGGREGATES

A. Methods

In the preceding sections we discussed a technique (FRMD) useful for the study of restrained dynamics. In the following sections we will apply this technique to larger molecular aggregates by means of the archetype hypothesis mentioned above. The basic idea is to consider a model protein tethered to its archetype by positional harmonic restraints. These restraints will be treated in an equivalent manner as we did with the NMR-NOE-based restraints. During the MD trajectory the proper scale factors will be determined as a function of the agreement of the modeled structure with the tethering one. We will use this technique to bias the MD trajectory toward conformations which preserve the secondary structure of the parent molecule. This requires tethering restraints between C_α or backbone atoms. A problem not addressed by this procedure is the tendency of proteins, during free MD trajectories, to disorganize the surface amino acid conformations. To deal with this problem and to help preserve the overall shape of the parent molecule in the modeled structure, an X-raylike restraint will also be used.

To describe the details of implementing this technique, we will present it in modeling Glutathione S-Transferases.

B. The Microdomain Description: Modeling of Glutathione S-Transferases

Glutathione S-Transferases (GSTs) are a major class of metabolizing enzymes that catalyze the nucleophilic attack of the sulfur atom of Glutathione (γ-Glu-Cys-Gly; GSH) on a wide variety of xenobiotics (Mannervik and Danielson, 1988).

Their dual roles in eliminating potentially carcinogenic agents and their involvement in the multidrug resistance mechanism (Tsuchida and Sato, 1992) have spurred significant activity aimed at understanding their structure and function.

Multiple forms of GSTs have been discovered in most life forms. In humans, there are four major cytosolic classes: Alpha (A), Mu (M), Pi (P), and Theta (T) (Mannervik et al., 1990). Some of these classes can be further divided into a total of more than nine GSTs isozymes. High sequence homology is found within a class that is maintained even across species, in contrast with the poor sequence similarity across classes even from the same species. All GSTs are globular proteins formed by two identical monomers. These dimer arrangements contain two, twofold related active sites.

The structure of the human isozymes from the A (Sinning et al., 1993) and P (Reinemer et al., 1992) classes are already known. The crystal structure for the Rat 3-3 isozyme (Ji et al., 1992) is the closest structure reported to that of the family of human M-class GSTs (Parker and Reinemer, 1994a and b). The high sequence homology between the rat and human enzymes makes the latter ideal candidates for determining structure via homological modeling, and at the same time, for testing the procedures proposed here. In the following sections we will discuss the models generated for three human GSTs of the M class, M1b-1b, M2-2, and M3-3, from the crystallographic data for the Rat 3-3 isozyme. M1a-1a is a fourth member of the M class that differs by only two residues from M1b-1b. Comparing the structures of these isozymes is worth undertaking as it may provide insights into the reasons for their isomorphism and enhance our knowledge of the requirements for the catalytic process they promote. Furthermore, the resulting structures could be applied in designing selective drugs to potentiate chemotherapy selectively in tumors expressing particular GST isozymes (Flatgaard et al., 1993).

Crystallographic Restraint

The models are refined using a method paralleling the technique of molecular replacement (Brändén and Tooze, 1991) frequently used in crystallography. The molecular replacement method is based on assuming close resemblance of the crystallographic envelopes of two similar proteins. The resemblance among such maps is often enough to find the placement of the protein of unknown structure in the space of the unit cell by searching with the known structure as a surrogate model. Even though this procedure is a time-consuming task, it is one of the most reliable methods for determining initial crystallographic phases.

In the present simulation, the same principles are used but in a reverted fashion (Cachau et al., 1994b). From the structure of the Rat 3-3 template a low-resolution (6Å) electron density map ($F_{calc,t}$) is computed, and the initial coordinates of the structure to be modeled are placed in it. The model structure is then annealed inside this envelope as described below. In this way, the overall shape of the parent enzyme is transferred onto the model structure. The characteristics of the information provided by a very low-resolution electron density map of one protein for the study of another is illustrated in Figure 12. Compared to the high-resolution maps normally used in experimental determinations, the low-resolution map provides only the details of the secondary structure. The high homology among all of the GSTs being considered here made the system ideally suited for this procedure, because the low-resolution crystallographic envelopes of highly homologous enzymes will be similar in nature. In this sense, a low-resolution synthetic X-ray map of the Rat 3-3 structure is used as a close mimic of the envelope of the unknown structure of the M1b-1b enzyme.

The X-ray $F_{calc,t}$ are generated using Xplor v3.1 (Brünger, 1992), by placing the crystallographic coordinates for the Rat 3-3 in a unit cell (space group P_1; unit cell dimensions 50Å, 50Å, 50Å) and by calculating the F_{calc} values for reflections with a resolution of 6Å or less. As mentioned, a map at 6Å resolution does not contain detailed information on the position of the side-chain atoms although the main secondary motifs can be easily recognized (Hendrickson, 1987; Perutz, 1992, Figure 12). A restraint energy function using the crystallographic $F_{calc,t}$ is defined as

$$E_{X-ray} = W_A/N_A \sum_{h,k,l} [|F_{calc,t}(h,k,l)| - |F_{calc,m}(h,k,l)|]^2 \qquad (8)$$

where h,k,l are the Miller indices of the selected reflections, $F_{calc,t}$ are the observed structure factors (in our case those of the synthetic system), and $F_{calc,m}$ the computed factors for the structure being modeled by homology to the template. N_A is a normalization factor, and W_A is an overall weighting factor, following the usual protocol for the refinement of crystallographic structures (Brünger, 1991a, 1991b). No phase restraints are used during our simulations.

Feedback-Restrained Molecular Dynamics in Tethered MD Calculations

Use of an X-raylike restraint mimicking a low-resolution map could hold the model protein within a restricted region of the conformation space during a simulated annealing computer experiment. However, due to the lack of high-resolution information (unknown for the model) the secondary structure of the model would significantly degrade during the annealing procedure if the restraints mentioned above are used *alone*. This problem could be overcome by a template force scheme. In our case, harmonic restraints are applied from the C_α atoms of the model to the positions of the C_α atoms in the Rat 3-3 structure. However, realistic conformations might not have been properly sampled with a priori assigned restraints. To overcome these limitations, the FRMD procedure will be used to scale the *template* restraints during the molecular dynamics trajectory.

In what follows, we will describe the FRMD technique applied to the homology modeling of M1b-1b from Rat 3-3. The modeling of M2-2 and M3-3 proceeded in a similar way starting from the M1b-1b optimized model. Time-dependent (τ)

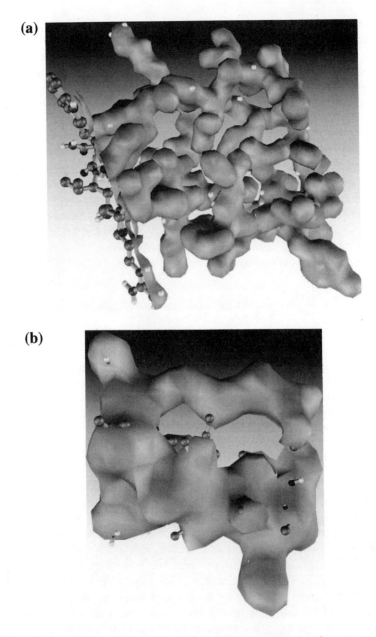

Figure 12. a) High- and b) Low-resolution electron density maps for a segment of the Rat 3-3 structure. Note that in the 6Å resolution map, the details of the structure are smeared, but the secondary structure can be readily recognized.

distance restraints are incorporated as an energy restraint, E_{Rest}, into the force field of Xplor v3.1 (Brünger et al., 1992) using the following relationship:

$$E_{Rest}(\tau) = 1/2 \sum S_{ij}(\tau) K_{ij} R_{ij}(\tau)^2 \qquad (9)$$

where i and j are the running indexes of the C_α atoms on the model and Rat 3:3 proteins; $S_{ij}(\tau)$ are scale factors for the interatomic distance restraints; K_{ij} is a force constant ($K_{ij} = 15.0$ Kcal/mol-Å² for all i,j). Note that $R_{ij}(\tau)$ is the *distance* between atom $C_{\alpha i}$ from the model structure (at time τ), and the coordinate of the nearest C_α atom from the Rat 3-3 structure ($C_{\alpha j}$. $S_{ij}(\tau)$ are actually calculated at discrete intervals, τ_n, using Eq. (2).

As previously discussed, Eqs. (9) and (2) revealed that, if only these functions are used, the model would be trapped in a local minimum after a short time, maximizing only a few restraints. A *quenching* function, similar to the one used in Eq. (4), is introduced to avoid this behavior. The form of this function could be better described in this case as a scale factor on E_{Rest} [Eq. (9)].

$$E^*_{Rest}(\tau) = \omega_{Rest}(\tau) E_{Rest}(\tau). \qquad (10)$$

The functional form used in this work for $\omega_{Rest}(\tau)$ is

$$\omega_{Rest}(\tau) = 1/2 [1 + \sin(\delta + \kappa\tau)] \qquad (11)$$

Eq. (11) is analogous to Eq. (3), however, κ is now chosen so that the period of the function is 2.5ps and δ so that E_{Rest} is out of phase with the envelope restraint energy term that would be affected by a similar quenching function:

$$E^*_{X-ray}(\tau) = \omega_{X-ray}(\tau) E_{X-ray}(\tau) \qquad (12)$$

and

$$\omega_{X-ray}(\tau) = 1/2[1 + \sin(\delta' + \kappa'\tau)]. \qquad (13)$$

As previously mentioned, the quench functions Eqs. (11) and (13) actually *pump* energy into the system. Therefore, the Molecular Dynamics (MD) had to be thermalized at very short intervals (10Fs). The thermalization used is a Gaussian renormalization. Even though the average temperature is set to T = 250 K the highest peaks reach 379 K during the first 40ps of the MD trajectory (<T> = 305 K).

Initial Model

The sequences of the Rat 3-3 and the human M isozymes are aligned using the MACAW program (Schuler et al., 1991). The template and the M1b-1b and M2-2 isozymes contain the same number of residues. No insertions or deletions are required for a successful alignment. The C- and N-termini of the M3-3 isozyme are three residues longer on each side. The percentage identity between the rat and the human isozymes ranged from 60–79%. The human isozyme most closely related to the template is the M1b-1b, while the M3-3 is the most distinct. Therefore, the M1b-1b structure is built from the Rat 3-3 crystallographic data. After the modeling protocol is completed, the resulting M1b-1b structure is then used as a template to build the initial models of the M2-2 and M3-3 isozymes, which are more closely related to M1b-1b than to Rat 3-3.

The coordinates of the three-dimensional structure of the Rat 3-3 GST determined at 2.2 Å resolution (Ji et al., 1992) are obtained from the Brookhaven Protein Databank (Brookhaven code: 1GST). The Insight II package (Biosym Technologies, San Diego, CA) is used to replace the residues that differed in the rat and human M1b-1b structures. The program aligned the replacement residues to the backbone of the original residue and maintained the dihedral angles for the side chains in common between the two residues. For some of the residues, the values of some mutated side-chain torsions must be altered to avoid severe van der Waals collisions.

Initial models for the M2-2 and M3-3 structures are generated from the refined M1b-1b model in an identical manner.

Simulation

After generating an initial structure from the Rat 3-3 template, the M1b-1b structure is refined using the CHARMm 19 set of parameters in Xplor v3.1 (Brünger, 1991), until the norm of the gradient is smaller than 0.5 Kcal/Å. The united atom approximation and reduced charges are used ($\omega_{ele} = 0.33$). The use of reduced charges and the cancellation of the hydrogen bond terms are common practices in the simulated annealing modeling of proteins in X-ray crystallography. The logic behind this technique is to increase the mobility of the different groups to scan a larger portion of the conformation space during the simulated annealing. A cutoff of 10Å is used throughout. Robson's expression for the $1/R_{ij}$ term is used (Robson and Platt, 1986). This function is quick to calculate and reduces the electrostatic interactions even further at extremely short distances. The minimized structure is used as a starting point in a low-temperature simulated annealing. The molecule is heated over 5ps from 10 K to 250 K followed by 10ps equilibration. After equilibration, E_{Rest} and E_{X-ray} are turned on in a ramp of 15ps. The molecular dynamics production step is carried for 100 ps after which the structure is cooled over 20 ps. As the system cools, the restraints are removed and the system is fully charged. The final structure is then fully minimized. Finally, to collect statistics for the comparison of the structures, an MD calculation is performed starting from the minimized model (heating 5ps, equilibration 10ps, production 50ps, T = 250 K).

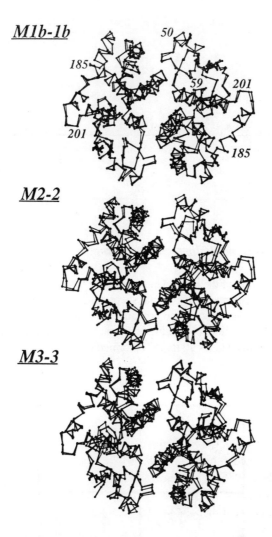

Figure 13. From top to bottom, views of the backbone of the M1b-1b, M2-2 and M3-3 structures generated after refinement superimposed on the Rat 3-3 template shown in bold.

Table 5. Root-Mean-Squared Deviations for the Different GST Structures Using the Trace and the Entire Backbone of the Enzymes

C_α				
	Rat3-3	M1b-1b	M2-2	M3-3
Rat3-3	—	0.567	1.227	1.040
M1b-1b	0.567	—	1.252	1.134
M2-2	1.227	1.252	—	1.493
M3-3	1.040	1.134	1.493	—
Backbone				
	Rat3-3	M1b-1b	M2-2	M3-3
Rat3-3	—	0.667	1.252	1.073
M1b-1b	0.667	—	1.327	1.221
M2-2	1.252	1.327	—	1.532
M3-3	1.073	1.221	1.532	—

Structural Analysis and Refinement Procedure

The crystal structure of the Rat 3-3 enzyme is used as a template (Ji et al., 1992). Rat 3-3 is a dimeric enzyme with globular shape and C_2 point group symmetry. The two identical monomers contain two domains each. Domain I from residues 1–82 has four β-strands and three α-helical segments, forming a βαβαββα motif. The second domain, from residues 90 to 217 is formed by five α-helices. There are two

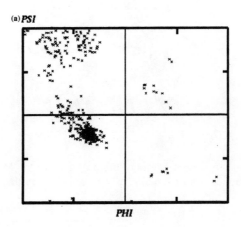

(continued)

Figure 14. Ramachandran plots for the template Rat 3-3 and the three human isozymes modeled. a) Rat 3-3; b) Mu1b-1b; c) Mu2-2; d) Mu3-3.

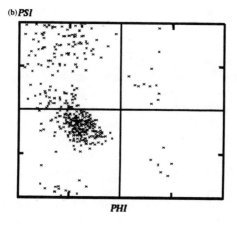

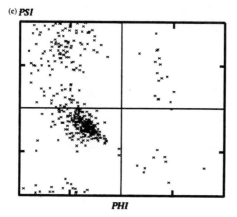

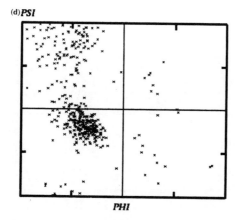

Figure 14. (Continued)

active sites in the dimer (Figure 18). Each is a single, deep cavity, where GSH and the substrate to be conjugated bind. The Rat 3-3 crystal structure is determined in the presence of GSH, which occupies part of the active site. The residues that contact the GSH molecule define the G-site. The H-site or the cosubstrate site is formed by nonadjacent residues from three structural elements with relative flexibility including the *cis*-Pro bend from residues 39–42, the α-β-turn-α from residues 90–127 and the C-terminus β-turn from residues 200–212.

Figure 13 shows the structures that resulted from the homology modeling and refinement steps. The structures preserve the overall folding topology observed in the template, as can be seen by the RMS values reported in Table 1 and by the Ramachandran plots in Figure 14. Overall, M1b-1b and M3-3 appear the most consistently conserved.

Nevertheless, the degree of flexibility permitted by the imposed restraints is sufficient to allow displacements in some localized areas of the molecule. In this sense, the restraints fulfilled their function of preserving the overall structure while allowing for structural rearrangements in certain areas.

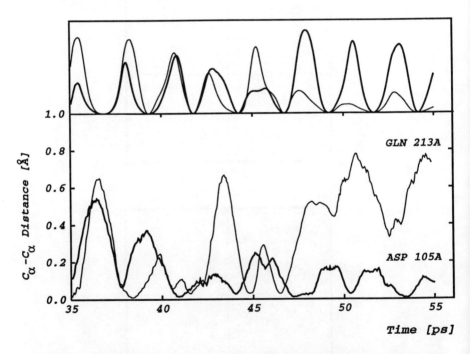

Figure 15. Distance between the position of the C_α Gln213 (thin line) or Asp 105 (bold line) from their original position in the template as a function of the time elapsed from the initiation of the annealing experiment. The upper portion of the plot shows the relative strength of the total restraint imposed on the atom at that time $[S_{ij}(\tau_n) \cdot \omega_{Rest}(\tau)]$.

Protein Model Building

An analysis of the effect of the restraints on the structure can be done by analyzing the trajectories of the C_α atoms relative to the position in the reference structure. Figure 15 illustrates two extreme cases for C_α 215 and C_α 105. In the figure, the resulting $[S_{ij}(\tau_n) \cdot \omega_{Rest}(\tau)]$ for each of the C_α atoms is presented as a function of time, together with the distance of the restrained atom to the tethering atom in the template (R_{ij}). In both cases, the atoms at $\tau = 35$ps are under significant stress, but the trajectories differentiated from that point. C_α 215 lost synchronism with the

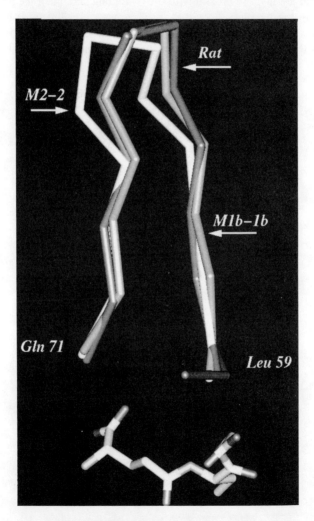

Figure 16. Illustration of the displacements observed for the microdomain ranging from residues 50 to 71. The M1b-1b and M2-2 structures are overlapped on the original Rat 3-3 template. The GSH molecule is given as a reference.

restraint function, and its contribution to E_{Ref} is consequently quenched, while C_α 105 is able to accommodate to the restraint force and gained synchronism with the restraint function. In this manner, by quenching some restraints while reinforcing others, the FRMD protocol allowed for moderate local changes in the structure.

The largest displacements found are confined to three regions of the monomer, residues 59–71, 112–150 and 185–201, which belong to solvent-exposed areas or overlap, in part, with the residues that form the H-site. The residues in the displaced regions that are in close proximity to the G-binding site retain the positions they occupied in the template. Localized displacements are illustrated in Figure 16, which shows the backbone in the microdomain formed by residues 50–71 for Rat 3-3, M1b-1b, and M2-2. The changes are very pronounced at the tip of the loop that

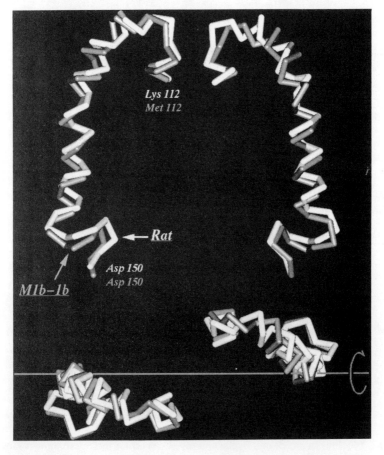

Figure 17. Rearrangement of the α-helix segment formed by residues 112 to 150. The lower middle portion of the helix is part of the GSH binding site.

Protein Model Building 105

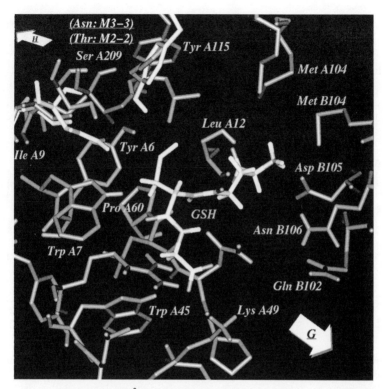

Figure 18. Residues within 6Å of the GSH molecule. The upper part of the figure is next to the H-site, while the lower part is near the G-site. The approximate disposition of the sites is indicated by the arrows. At the center is the GSH molecule. Only the residues of M1b-1b are shown in their entirety. For M2-2 and M3-3 only the residues that varied in any one molecule are shown. The Tyr 115 is shown in the three molecules to provide an insight into the consequences that the displacements observed in some microdomains could have on binding.

is entirely solvent-exposed while the residues closer to the GSH molecule occupy equivalent positions in all modeled isozymes.

The helices from residues 112 to 150 also show displacements illustrated in Figure 17. The structure of this region is shown for M2-2 compared to the template. The lower middle portion of the helix, that is part of the G-site, is a close match to the initial structure. Interestingly, the distortions in this region are more important for the residues that contribute to the H-site, such as 115 at the beginning of the helix, or for the solvent-exposed residues. The residues 181–201 form part of a helix that is solvent-exposed.

Therefore, all the structural changes in the refined structures for these enzymes are confined to solvent-exposed regions or to the H-site. The structure of the G region of the active site is maintained for all of these enzymes (Cachau et al., 1994c).

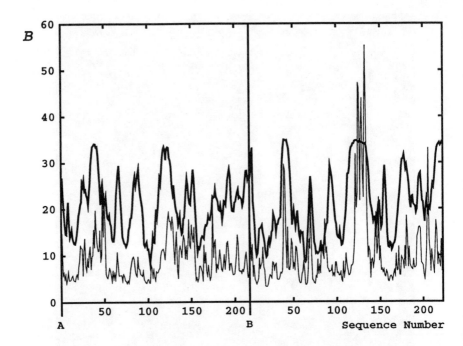

Figure 19. Computed *B* factors for the M2-2 isozyme (thin line) compared to the experimental values for the template (bold line). A and B indicate the two subunits.

Dynamic Properties

The MD simulations carried out reveal that the M1b-1b has the most rigid backbone, while M2-2 appears to be the most flexible. In all cases the regions of higher mobility are preserved corresponding, for the most part, to the residues involved in the H-site around residue 40, from residues 100 to 150, and toward the C-terminus. The M2-2 isozyme displays a very large motion in the region around residues 112–150, which corresponds to one of the areas where significant displacements occurred relative to the template structure, particularly for one of the subunits.

Based on the RMS observed during the MD trajectory, B-factors ($B_{calc} = 8\pi^2/3<RMS>$) for the M2-2 structure are computed, and compared with the experimental B-factors for the Rat 3-3 isozyme (Figure 19). As expected, the computed B values are typically smaller than the experimental ones. The higher values of B_{calc} are for the region formed by residues 112–150 for the second subunit. There is a remarkable asymmetry in this region for both units. The same asymmetry can be found in the crystal structure, for which the experimental temperature factors (Rat 3-3) are clearly not symmetric. The reason for this difference lies in the

Protein Model Building

crystallographic packing. In the crystal, one of the subunits shows more exposure to water for the region 112–150 than the other, which shows contacts with symmetry-related copies in the crystal packing. Thus, the difference in the dynamic behavior could be traced to the different environments of the helix tips. This asymmetry in the structure of the two subunits in Rat 3-3 propagates into the modeled structures through the tethering C_α coordinates. This may explain the largely asymmetric motions observed during the MD simulations for that region in the M2-2 structure.

Validation Experiments

Three computer experiments are carried out to characterize the procedure adopted for refinement. The detailed implications of each will be further discussed in the modeling of Heat Shock Proteins described below. The experiments are as follows: In a first experiment the initial Rat 3-3 is fully minimized and a simulated annealing is performed without restraints. After annealing, the structure is fully minimized. A full modeling calculation on Rat 3-3 is carried out as a second test. This time the restraint protocols applied to the other models are used. After the modeling, the structure is minimized and a MD calculation performed. Finally, a homology modeling of the Rat 3-3 is carried out using a procedure identical to that described for the human enzymes. The modeled Rat 3-3 structure is developed using the M1b-1b model as a template and compared to the experimentally determined structure.

The first experiment resulted in a model with RMS = 1.57Å (only C_α atoms) which is higher than any of the values obtained from the models for the human enzymes. As expected, the second experiment resulted in a model with only a 0.34Å RMS deviation suggesting that the combination of restraints reduces the overall distortion, whenever an appropriate match is possible between the sequence and the template. The reverse modeling of the Rat 3-3 structure gave an RMS = 1.15Å when compared to the experimental structure. The value falls within the range of the RMS values for the M2-2 and M3-3 models (1.22 and 1.04 Å, respectively). The only region of the model (in all three cases) where large reorganizations are observed is at the tip of the central α-helix on the side where the larger B-factors in the Rat 3-3 structure are found.

Two independent criteria, recently introduced by Colovos and Yates (1993) and Sippl (1993), are used to check the model structures. In slightly different ways, both procedures use an analysis of the statistics of nonbonded atomic interactions. Except for the tips of the central α-helices, all other portions of these proteins resemble, in this statistical sense, the core of a regular protein.

At the conclusion of this study, we obtained access to the yet unpublished coordinates for a structure of a M2-2 enzyme from human muscle, resolved at 1.85 Å (Rule, G.S., personal communication). Our modeled M2-2 structure agrees remarkably well in folding topology with the crystallographic structure. The most significant disagreement is localized toward the C-terminus. However, in the resolved enzyme the Trp 214, the largest disagreement observed, has been replaced by a Phe. Moreover, the electron density is poorly resolved at the C-terminus. Visual inspection of our M2-2 model structure, the crystallographically determined one, and the Rat 3-3 structure indicates that the displacements observed in the model, relative to the Rat 3-3 can also be found in the human isozyme. The similarities found increased our confidence in the models developed.

C. The Validation Problem: Modeling of Heat Shock Proteins-70 from HLA Motifs

In the preceding sections the application of FRMD techniques was validated using experimental information or high-level computational techniques. The solutions compatible with the NMR information for GRGDS were known beforehand; the taxol solutions were validated by a) comparing them with the known taxotere structure, b) modeling taxotere with a higher level technique (MOPAC) and, after this technique was validated applying it to the taxol structures, and c) the FRMD showed the expected behavior by rejecting false NOE signals; in the GST family of structures a model has become recently available for M1b-1b. In addition, independent criteria (Colovos and Yates, 1993; Sippl, 1993) were used to prove the validity of the packing on the final models of GST.

All of these validations are of greater value in the development of a new technique, however, they lack the generality required to model a molecule of completely *unknown* structure, where only the parent structure is known. The semiempirical calculations described for the taxol problem are unfeasible for treating larger molecules. In addition, the semiempirical calculations should be verified in analogous systems before they can be considered trustworthy. Verification through knowledge of the final solution is not an option for an unknown structure, and the computation of alternative techniques (e.g., Colovos and Yates, 1993; Sippl, 1993) may only show consistency in the results. However, unless the methods are proven to be free of similar bias, validating the final structure by these criteria is only partial.

Validation of a Model through Statistical Analysis

A homology modeling technique must show some properties concerning sensitivity to sequence and structure arrangement.

One must expect a homology modeling technique to be sensitive to the amino acid sequence of the model. Randomization of the sequence must result in noise in the model structure. There are two simple ways of proving this hypothesis: A FRMD self-modeling (model of a protein from its own sequence) experiment with the sequence of the parent protein partially randomized will show the sensitivity of the technique. Alternatively, the model sequence can be partially randomized in a series of FRMDs to accomplish this same goal.

Protein Model Building

The procedure should converge to the correct conformation from a distorted one. We can imagine the folding of two similar proteins as the distorted image of one into the other. The modeling technique must be able to overcome this distortion and reach the proper final conformation. The ability of the technique to do this can be proven by adding noise to the three-dimensional coordinates of the parent molecule. If a FRMD procedure is too sensitive to the initial set of coordinates, it may prove unsuccessful.

The amount of information lost during a modeling experiment must be measured. For the final model to be useful, an error estimate of the coordinates must be obtained. During a FRMD trajectory, information contained in the initial set of coordinates is irreversibly lost. The information lost will result in errors in the coordinates of the final model. A simple form to determine an upper limit for this error consists of a back-modeling experiment. In such an experiment the parent structure is modeled from the model. The differences between the initial and the back-modeled structure describe an upper bound to the information lost during the FRMD trajectory.

One of the simplest problems in homology modeling is the description of the initial set of rotamer coordinates for the amino acid side chains for the mutations and insertions required to build an initial model from the parent coordinates. Because there is no *ab initio* procedure for assigning these parameters, the sensitivity of the FRMD trajectory to the rotamer library used must be tested. As in the preceding proposed testing protocols, a statistical approach can be also used to deal with this problem; the FRMD calculation must be repeated with different levels of randomization in the rotamers, and the effect of this noise must be measured in the resulting FRMD trajectory. A robust technique must be insensitive to the initial rotamer library used. Following this same logic, a self-modeling protocol can be used to measure the *minimal* amount of noise incorporated in a FRMD trajectory by mutating the amino acids of the parent structure to the model sequence and back to the original with different rotamer libraries.

These statistical exercises are simple to conceive and of general use, however, they are seldom employed in a homology modeling protocol. Procedures of evaluation of the final structure which will not require the recomputing of a MD trajectory are favored instead. The main reason for this treatment is the computational cost of a purely MD self-consistent analysis. This type of study requires multiplying the number of MD trajectories only for the purpose of collecting statistical data. However, given the growing speed of modern supercomputers, we can expect that self-tests will also grow in use in the coming years.

As in previous sections, we will apply these techniques to a working example, the homological modeling of a Heat Shock Protein.

Heat Shock Proteins

The expression of heat shock or stress proteins (HSP) is a common cellular response to a variety of stresses. HSPs are balanced to interact with polypeptides that become denatured as a consequence of stress. Also, expression of HSPs has been correlated with the induction of cellular tolerance to stress. The most studied of the HSPs is the hsp-70 family of polypeptides composed of four highly homologous polypeptides, hsp-72, hsp-73, grp-78, and grp-75, all of which can bind ATP. Some of these proteins (hap-73, grp-78 and grp-75) are present in cells under normal conditions, and they have been named molecular *chaperonins*. These proteins are apparently involved in translocating polypeptides across membranes and in protein folding. hsp-72 and hsp-73 are both localized in the cytosol and nucleus; hsp-73 is constitutive and hsp-72 is inducible following stress. The amino acid sequence of swine hsp-72 is compared with the constitutive form of hsp-70 (named hsp-73 or hsp-70c). Their putative structures are also compared using a three-dimensional model of the protein. We have found a high degree of homology in the putative peptide binding site of both hsp-70 proteins suggesting that, in spite of some differences in the amino acid sequence of both molecules, they are functionally identical. Thus, a unidimensional comparison of amino acid sequences between proteins cannot give information about the homology at the functional biologic site. Consequently, we simulated a three-dimensional model of the peptide binding site domain of hsp-70. The initial peptide binding domain model constructed is based on the one proposed by (Rippman, 1991), claiming similarity between hsp-70 and the HLA I molecule. We then compared the proposed region of hsp-70 that interacts with denatured polypeptides with pig hsp-72 (Figure 20). This region is composed of two α-helices that enclose an area of β-sheets.

A high degree of homology between pig hsp-72 and its human counterpart (78% identity without insertions) can be found in the β-sheet region. Most amino acid changes are localized on one of the α-helices. This alignment of mutations agrees with the consensus mechanism of recognition involved in HLA motif-peptide complexes as demonstrated recently in the structure of MHC class II peptide complex structures. This consensus mechanism describes the specificity of the complex formation interaction as the result of a hydrogen bond network between the peptide backbone and the backbone and side chains of the α-helices, while the thermodynamics of the recognition process would be driven by the hydrophobic interactions between the peptide side chains and the hydrophobic bottom of the recognition grove.

Given the overall agreement in sequence alignment between the HLA motif and the hsp-70 which rationalizes most aspects of the function-specificity between different sequences of HSP, the modeling of HSP on the HLA framework seems a feasible project. We will apply a protocol similar to the one described in Eqs. (8) to (14) to this example.

A protocol similar to the one described for the GSTs was used for the treatment of hsp-70. At the end of the FRMD trajectory, a model which resembles the original HLA motif (RMS: 1.6Å) was obtained. Work is under way in our laboratory to carry this simulation procedure to the study of the HSP-peptide interaction, from where a study similar to the one described for the GSTs would be possible. However, at

Protein Model Building

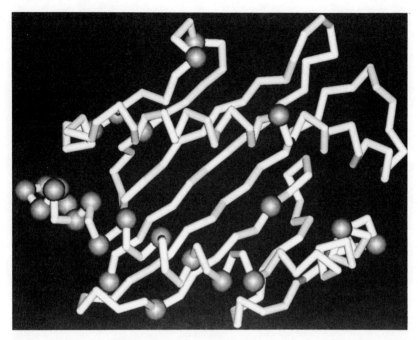

Figure 20. Mutations from the pig-68 to hsp-70 aligned with the three-dimensional structure of HLA. The mutations are displayed as spheres. Note the uneven distribution of the mutations which tend to concentrate in one face of one of the α-helices.

the current stage of the development of this model, the most important result is the validation of the HLA-like folding as a putative folding for HSP sequences, because there are no known solved structures of this domain for HSPs.

The Validation Protocol

We will consider three modeling exercises, in addition to the initial one, to ensure the quality of the model structure. The complete procedure is schematically described in Figure 21.

a. Self-modeling. After the sequence alignment, a list of mutated amino acids is built. The modeling exercise consists in mutating these amino acids to Ala and back to the original amino acids in the known structure. The usual FRMD protocol is then applied to this structure. The typical RMSD after 100ps for this trajectory is 0.6Å. The C_α trace of the resulting model, compared with the initial structure is presented in Figure 21(a). This result is our highest limit of quality. In fact, what we are measuring with this technique is the minimal amount of noise that we will incorporate in the structure due to the FRMD protocol. In large part, the noise is

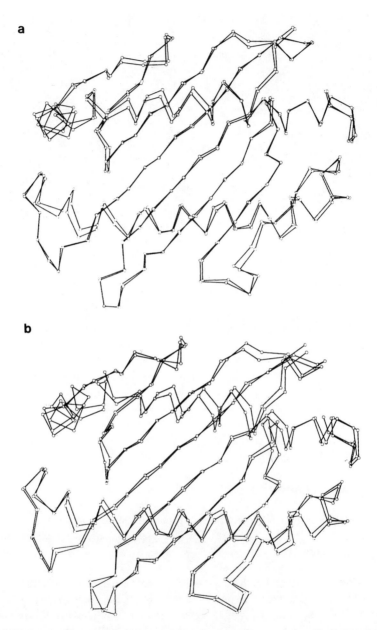

Figure 21. Validation mechanisms in homological modeling. Comparison of the final model with the HLA initial C_α trace. a) self-modeling; b) back modeling; c) best agreement out of 7 trajectories using random mutations.

c

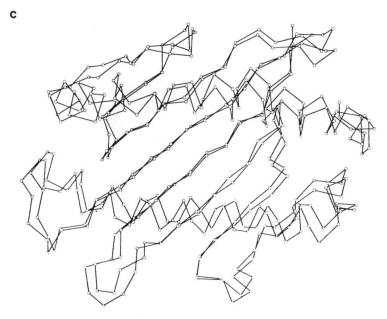

Figure 21. (Continued)

the result of the unlikely placement of the side chains which take place as part of the initial sequence-homology modeling exercise. If only the FRMD is used, starting from the experimental coordinates (no mutation exercise is done), the RMSD is reduced to 0.45Å.

b. Back modeling. During this simulation the initial model is rebuilt starting from the final model. The RMSD for the rebuilt model vs. the experimental one is 1.5Å. The result of this simulation is presented in Figure 21(b). This modeling exercise measures the maximum loss of information during the FRMD process. This is our lower limit of quality. We must do better than that for a one-way simulation.

c. Random mutation modeling. The results reported in a and b indicate that the sequences used in those simulations tend to fold in the HLA-like pattern. This result is expected for simulation b because we are using a sequence that we know from the experimental structure, folds with this pattern. Thus, we must evaluate the tendency of the technique to freeze the folding pattern irrespective of the protein sequence. In other words, we could envision a homology modeling procedure which will hold the secondary structure of the experimental model in place irrespective of the protein sequence being modeled. Such a technique will not be appropriated. Therefore, before we accept the modeled structure, we must prove that the technique used is sensitive to the sequence composition of the protein.

A way of probing the sensitivity of the FRMD calculation consists of introducing random mutations in the sequence and repeating the FRMD experiment. These random mutation experiments measure the sensitivity of the FRMD procedure to the sequence specificity of the folding pattern. The typical RMSD for 7 mutations along the preserved α-helix is 2.5Å (over 5 experiments of 80ps trajectory), showing a high degree of sensitivity to randomized sequence alignment.

V. REMARKS

Homology modeling of proteins is a growing field, which today encompasses many disciplines with diverse approaches to a common goal: predicting the tertiary structure of a protein from its one-dimensional amino acid sequence. The rapid development of some areas of molecular biology is changing the basis for performing these simulation exercises. In fact, the massive amount of information that will be soon available in sequence data banks will dwarf the very large data banks of today. On the structural side of the problem, the picture is clearly less optimistic. The complexity of force field fitting for biomolecules has proven an extremely difficult task. To add to the difficulty, one must realize that many variables critical to the folding of a protein (e.g., pH) are not even considered in our current force fields.

Thus, the state of our knowledge today clearly favors the use of statistical techniques. The main drawback with these techniques, however, is the difficulty in combining different qualities of information as variable as the information contained in the known sequences and three-dimensional structures. The problem become more remarkable if we consider that sequence segments do not carry much three-dimensional information unless they are considered part of a domain, which tends to confuse the sequence alignment in protein prediction approaches.

Even though we have recognized the limitations of the force fields in use, the mathematical nature of the approximations make them well-suited for porting information from the three-dimensional structures to the unidimensional sequences to help in the homology modeling process. Because of the intrinsic deficiencies of these force fields, the information can be used to *bias* them for a given set of problems, so that, within a restricted region of conformations, these corrected force fields have a much better representation of the protein. Two main approaches are used to accomplish this goal: retrofitted force fields and bootstrapping techniques. In the first case the parameters of the force field are modified to incorporate the three-dimensional information. In the second approach, auxiliary functions are used to bias the trajectory of a MD calculation. The technique discussed here belongs to this second family of approaches. One of the main advantages of a bootstrapping algorithm is that it can be used to match one sequence in one three-dimensional structure.

The technique proposed here for homology modeling offers a restraint for the folding topology of the model in the form of a template or parent structure as a rigid reference frame, allowing for some significant changes to take place. The strategy can certainly be applied in situations where the molecular replacement technique would be expected to succeed in crystallographic determinations. In addition, it could be easily implemented to refine models based on the protein core volume concept (Bordo and Argos, 1990; Lim and Sauer, 1989; Hubbare and Blundell, 1987). Because the central idea in this approach is that related proteins will have similar core volumes, the X-ray map of a related enzyme applied to the core could provide a valuable restraint during the refinement. Two main areas of research are currently being explored in our laboratory: 1) the treatment of insertions and deletions and 2) the treatment of domain movements. The first problem has proven the most difficult. This is due, in part, to memory effects during the FRMD trajectory and to sequence mismatching difficulties which, in turn, become model misfittings. The second problem is currently being analyzed by studying the movement of Fab domains. The position of the domains with respect to each other is usually described by an *elbow* angle. The technique that best describes the flexibility of this molecule consists of a self-consistent protocol which is used on top of the FRMD technique. In these simulations the domains are considered independent molecules. During the MD trajectory, the *initial* structure of each domain is superimposed by a rigid body analysis over the current position of the modeled protein. From this new position, a new set of F_{calcs} are computed which, in turn, are used in the ensuing step of the MD trajectory. The main difficulty, resides in the large trajectories required for this protocol to reach the proper elbow angle.

ACKNOWLEDGMENTS

Research is supported in part by the National Cancer Institute, DHHS, under contract NO1CO-74102 with Program Resources, Incorporated. The contents of this publication do not necessarily reflect the views or policies of the DHHS, nor does mention of trade names, commercial products, or organizations imply endorsement by the U.S. Government.

REFERENCES

Argos, P., Vingron, M., & Vogt, G. (1991). Protein sequence comparisons: Methods and significance. Protein Eng. 4, 375–383.

Baker, J. K. (1992). Nuclear Overhauser effect spectroscopy (noesy) and dihedral angle measurements in the determination of the conformation of taxol in solution. Spectroscopy Lett. 25(1), 31–48.

Benner, S. (1992). Predicting de novo the folded structure of proteins. Curr. Opinion Struct. Biol. 2, 402–412.

Blundell, T. L. & Johnson, M. S. (1993). Catching a common fold. Protein Sci. 2, 877–883.

Blundell, T. L., Sibanda, B. L., Sternberg, M. J., & Thornton, J. M. (1987). Knowledge-based prediction of protein structures and the design of novel molecules. Nature 326, 347–352.

Bordo, D. & Argos, P. (1991). Suggestions for safe residue substitutions in site directed mutagenesis. J. Mol. Biol. 217, 721–729.

Bork, P., Ouzounis, C., & Sander, C. (1994). From genome sequence to protein function. Curr. Opinion Struct. Biol. 4(3), 393–403.

Brändén, C. & Tooze, J. (1991). Introduction to Protein Structure. Chapter 17, Garland, London.

Brünger, A. T. (1990). Refinement of Three-dimensional structures of proteins and nucleic acids. In: Molecular Dynamics. Applications in Molecular Biology (Goodfellow, J. M., Ed.), pp. 137–178, CRC, Boca Raton, FL.

Brünger, A. T. (1991a). Crystallographic phasing and refinement of macromolecules. Curr. Opinion Struct. Biol. 1, 1016–1022.

Brünger, A. T. (1991b). Simulated annealing in crystallography. Annu. Rev. Phys. Chem. 42, 197–223.

Brünger, A. T. (1992). XPLOR v3.0 "A System for Crystallography and NMR," The Howard Hughes Medical Institute CT, USA.

Cachau, R. E., Gussio, R., Beutler, J. A., Chmurny, G. N., Hilton, B. D., Muschnik, G. M., & Erickson, J. W. (1994a). Solution structure of taxol determined using a novel feedback-scaling procedure for noe-restrained molecular dynamics. Int. J. Supercomputer Appl. 8(1), 24–34.

Cachau, R. E., Erickson, J. W., & Villar, H. O. (1994b). Novel procedure for structure refinement in homology modeling and its application to the human class Mu glutathione S-transferases. Prot. Eng. 7(7), 831–839.

Cachau, R. E. (1994c). Molecular simulations of water using a floating polynomial force field and an interpolating electrostatic field representation. Biophys. Chem. 51, 1–17.

Cachau, R. E., Ysern, X., Erickson, J. W., & De Maio, A. (1994d). Homology modeling of HSP-72 from an HLA motif. (submitted).

Cachau, R. E. (1994e). Calibration of the FRMD parameters using small peptides of known structure. (submitted).

Cachau, R. E., Serpersu, E. H., Mildvan, A. S., August, J. T., & Amzel, L. M. (1989). Recognition in cell adhesion. A comparative study of the conformations of RGD-containing peptides by Monte Carlo and NMR methods. J. Mol. Recog. 2, 179–186.

Chelvanayagam, G., Heringa, J., & Argos, P. (1992). Anatomy and evolution of proteins displaying the viral capside jellyroll topology. J. Mol. Biol. 228, 220–242.

Chmurny, G. N., Hilton, B. D., Brobst, S., Look, S. A., Whiterup, K.M., & Beutler, J.A. (1992). ^{1}H- and ^{13}C-NMR assignments for taxol, 7-epi-taxol, and cephalomannine. J. Nat. Prod. 55(8), 1157–116.

Chothia, C. (1992). One thousand protein families for the molecular biologist. Nature 357, 543–544.

Colovos, C. & Yates, T. O. (1993). Verification of protein structures: patterns of nonbonded atomic interactions. Protein Sci. 2, 1511–1519.

Crippen, G. (1991). Prediction of protein folding from amino acid sequence over discrete conformation spaces. Biochemistry 30, 4232–4237.

Dandekar, T. & Argos, P. (1992). Potential of genetic algorithms in protein folding and protein engineering simulations. Prot. Eng. 5, 637–645.

DeGrado, W. F. & Matthews, B. W., Eds. (1993). Engineering and Design. In: Current Opinion in Structural Biology, Vol. 3, Number 4, Current Biology, London.

Doolittle, R. F. & Blundell, T. L., Eds. (1993). Sequences and Topology. In: Current Opinion in Structural Biology, Vol. 3, Number 3, Current Biology, London.

Eisenhauer, P. B., Harwig, S. S. L., Szklarek, D., Ganz, T., Selsted, M. E., & Lehrer, R. I. (1989). Purification and antimicrobial properties of three defensins from rat neutrophils. Infect. Immunol. 57, 2021–2027.

Falzone, J., Benesik, A. J., & Lecomte, J. T. J. (1992). Characterization of taxol in methylene chloride by NMR spectroscopy. Tetrahed. Lett. 33(9), 1169–1172.

Farber, G. K. & Petsko, G. A. (1990). The evolution of α/β-barrel enzymes. Trends Biochem. Sci. 15, 228–234.

Finkelstein, A. V. & Ptitsyn, O. B. (1987). Why do globular proteins fit the limited set of folding patterns? Prog. Bioph. Mol. Biol. 50, 171–190.

Flatgaard, J. E., Bauer, K. E., & Kauvar, L. M. (1993). Isozyme specificity of novel glutathione S-transferase inhibitors. Cancer, Chemother. and Pharmacol. 33, 63–70.

Flores, T. P. & Moss, D. S. (1990). Simulating the Dynamics of Macromolecules. In: Molecular Dynamics. Applications in Molecular Biology (Goodfellow, J. M., Ed.), pp. 1–26, CRC, Boca Raton, FL.

Friedrichs, M. S. & Wolynes, P. G. (1989). Toward protein tertiary structure recognition by means of associative memory Hamiltonians. Science 246, 371–373.

Friedrichs, M. S., Goldstein, R. A., & Wolynes, P. G. (1991). Generalized protein tertiary structure recognition by using associative memory Hamiltonians. J. Mol. Biol. 222, 1013–1034.

Fujii, G., Selsted, M. E., & Eisenberg, D. (1993). Defensins promote fusion and lysis of negatively charged membranes. Protein Sci. 2, 1301–1312.

Fujinaga, M., Delbaere, L. T. J., Brayer, G. D., & James, M. N. G. (1985). Refined structure of alpha-lytic protease at 1.7 Å resolution. Analysis of hydrogen bonding and solvent structure. J. Mol. Biol. 184, 479–491.

Ganz, T., Selsted, M. E., & Lehrer, R. I. (1988). Defensins: Antimicrobial/cytotoxic peptides of phagocytes. In: Bacteria-Host Cell Interaction, pp. 3–14, Alan R. Liss, New York.

Goldstein, R. A., Luthey-Schulten, Z. A., & Wolynes, P. G. (1992a). Protein tertiary structure recognition using optimized hamiltonians with local interactions. Proc. Natl. Acad. Sci. USA 89, 9092–9093.

Goldstein, R. A., Luthey-Schulten, Z. A., & Wolynes, P. G. (1992b). Optimal protein-folding codes from spin-glass theory. Proc. Natl. Acad. Sci. USA 89, 4918–4922.

Gueritte-Voegelein, F., Guenard, D., Mangatal, L., Potier, P., Guilhem, J., & Cesario, M. (1990). Structure of a synthetic taxol precursor, n-tert-butoxycarbonyl-10-deacetyl-n-debenzoyltaxol. Acta Cryst. C46, 781–784.

Gueritte-Voegelein, F., Guenard, D., Lavelle, F., Le Goff, M.-T., Mangatal, L., & Potier, P. (1991). Relationships between the structure of taxol analogues and their antimitotic activity. J. Med. Chem. 34, 992–998.

Gustavson, K. R., Sowder, R. C., Henderson, L. E., Parsons, I. C., Kashman, Y., Cardellina, J. H., McMahon, J. B., Buckheit, R. W., Pannel, L. K., & Boyd, M. R. (1994). Circulin A and B, Novel HIV-Inhibitory Macrocyclic Peptides From the Tropical Tree Chassalia Parvifolia (submitted).

Hagler, A. T., Lifson, S., & Dauber, P. (1979). Consistent force field studies of intermolecular forces in hydrogen-bonded crystals. 2. A benchmark for the objective comparison of alternative force fields. J. Am. Chem. Soc. 101, 5122–5130, DISCOVER version 2.7.0 (March 1991), BIOSYM Technologies, 10065 Barnes Canyon Road, San Diego, CA 92121, USA.

Hendrickson, W. A. (1987). X-ray diffraction. In: Protein Engineering (Oxender, D. L., & Fox, C. F., Eds.), pp. 10–11, Alan R. Liss, New York.

Hill, C. P., Yee, J., Selsted, M. E., & Eisemberg, D. (1991). Crystal structure of defensin HNP3, an amphiphilic dimer: Mechanisms of membrane permeabilization. Science 252, 1481–1485.

Hill, T. L. (1968). Introduction to Statistical Thermodynamics, Addison-Wesley, Massachusetts.

Hilton, B. D., Chmurny, G. N., & Muschik, G. M. (1992). Taxol, quantitative internuclear proton-proton Distances in CDCl3 solution from NOE data, 2D NMR ROESY buildup rates at 500MHz. J. Nat. Prod. 55(8), 1157–1161.

Holmes, F. A., Walters, R. S., Theriault, R. L., Forman, A. D., Newton, L. K., Raber, M. N., Buzdar, A. U., Frye, D. K., & Hortobagyi, G. N. (1991). Phase II trial of taxol, an active drug in the treatment of metastatic breast cancer. J. Natl. Cancer Inst. 83(24), 1797–1802.

Hopfield, J. (1982). Neural networks and physical systems with emergent collective computational abilities. Proc. Natl. Acad. Sci. USA 79, 2544–2558.

Horwitz, S. B. (1992). Mechanism of action of taxol. Trends Pharm. Sci. 13, 134–137.

Hubbare, T. J. P. & Blundell, T. L. (1987). Comparison of solvent inaccessible cores of homologous proteins: definitions useful for protein modeling. Prot. Eng. 1, 159–171.

Hue Sun, C. & Dill, K. A. (1991). Polymer principles in protein structure and stability. Annu. Rev. Biophys. Chem. 20, 447–490.

Ji, X., Zhang, P., Armstrong, R. N., & Gilliand, G. L. (1992). Three dimensional structure of a GST from the Mu gene class. Structural analysis of the binary complex of isozyme 3-3 and glutathione at 2.2 Å resolution. Biochemistry 31, 10169–10181.

Kabsch, W., Mannherz, H. G., Suck, D., Pai, E. F., & Holmes, K. C. (1990). Atomic structure of the actin, DNase I Complex. Nature 347, 37–44.

Kingston, D. G. I. (1991). The chemistry of taxol. Pharmac. Ther. 52, 1–34.

Kingston, D. G. I., Samaranayake, G. C., & Ivey, A. (1990). The chemistry of taxol, a clinically useful anticancer agent. J. Nat. Prod. 53(1), 1–12.

Leahy, D. J., Hendrickson, W. A., Aukhil, I., & Erickson, H. P. (1992). Structure of a fibronectin type III domain from tenasein phased by MAD analysis of the seleinomethionyl protein. Science 258, 987–990.

Lim, W. A. & Sauer, R. T. (1989). Alternative packing arrangements in the hydrophobic core of the lambda repressor. Nature 339, 31–36.

Lipshutz, R. J. & Fodor, S. P. A. (1994). Advanced DNA sequencing technologies. Curr. Opinion Struct. Biol. 4(3), 376–380.

Maiorov, V. N. & Crippen, G. M. (1992). A contact potential that recognizes the correct folding of globular proteins. J. Mol. Biol. 227, 876–888.

Mannervik, B. & Danielson, U. H. (1988). Glutathione transferases structures and catalytic activity. CRC Critical Rev. Biochem. 23, 283–337.

Mannervik, B., Aisthi, Y. C., Board, P. G., Hayes, J. D., DiIlio, C., Ketterer, B., Listowsky, B., Morgenstern, R., Muramatsu, M., Pearson, W. R., Pickett, C. B., Sato, K., Widersten, M., & Wolf, C. R. (1990). Nomenclature for human glutathione transferases. Biochem. J. 282, 305–306.

McGuire, W. P., Rowinsky, E. K., Rosenshein, N. B., Grumbine, F. C., Ettinger, D. S., Armstrong, D. K., & Donehower, R. C. (1989). Taxol, a unique antineoplasic agent with significant activity in advance ovarian epithelial neoplasms. Ann. Intern. Med. 111, 273–279.

Monzigo, A. F. & Matthews, B. W. (1982). Structure of a mercaptan-thermolysin complex illustrates mode of inhibition of zinc proteases by substrate-analogue mercaptans. Biochemistry 21, 3390–3397.

Pardi, A., Zhang, X. L., Selsted, M. E., Skalicky, J. J., & Yip, P. F. (1992). NMR studies of defensin antimicrobial peptides. 2. Three dimensional structures of rabbit NP-2 and human HNP-1. Biochemistry 31, 11357–11364.

Parker, M. & Reinemer, P. (1994a). Biochim. Biophys. Acta 1205, 1–18.

Parker, M. & Reinemer, P. (1994b). Eur. J. Biochem. 220, 645–661.

Perutz, M. (1992). Protein Structure: New Approaches to Disease and Therapy, p. 14–17, W. H. Freeman, New York.

Reinemer, P., Dirr, H. W., Ladenstein, R., Huber, R., LoBello, M., Federici, G., & Parker, M. W. (1992). Three-dimensional structure of class Pi glutathione S-transferase from human placenta in complex with S-hexylglutathione at 2.8 Å resolution. J. Mol. Biol. 227, 214–226.

Rippman, F., Taylor, W. R., Rothbard, J. B., & Green, N. M. (1991). A hypothetical model for the peptide binding domain of hsp-70 based on the peptide binding domain of HLA. EMBO 10, 1053–1059.

Robson, B. & Platt, E. (1986). Refined models for computer calculations in protein engineering. calibration and testing of atomic potential functions compatible with more efficient calculations. J. Mol. Biol. 188, 259–281.

Ruhoslahti, E. (1988). Fibronectin and its receptors. Annu. Rev. Biochem. 57, 375–413.

Sander, C. (1991). de novo design of proteins. Curr. Opinion Struct. Biol. 1, 630–638.

Scheek, R. M., van Gunsteren, W. F., & Kaptein, R. (1989). Molecular dynamics simulation techniques for determination of molecular structures from nuclear magnetic resonance data. In: Methods in Enzymology, Vol. 177, Part B, Nuclear Magnetic Resonance (Abelson, J. N., & Simon, M. I., Eds.), pp. 204–282, Academic, San Diego.

Schiff, P. B., Fant, J., & Horwitz, S. B. (1979). Promotion of microtubule assembly in vitro by taxol. Nature 277, 665–667.

Schiff, P. B. & Horwitz, S. B. (1980). Taxol stabilizes microtubules in mouse fibroblast Cells. Proc. Natl. Acad. Sci. USA 77, 1561–1565.

Schuler, G. D., Altschul, S. F., & Lipman, D. J. (1991). A workbench for multiple alignment construction and analysis. Proteins 9, 180–190.

Selsted, M. E., Brown, D. M., DeLange, R. J., Harwig, S. S. L., & Lehrer, R. I. (1985a). Primary structure of six antimicrobial peptides of rabbit peritoneal neutrophils. J. Biol. Chem. 260, 4576–4584.

Selsted, M. E., Harwig, S. S. L., Ganz, T., Schilling, J. W., & Lehrer, R. I. (1985b). Primary structures of three human neutrophil defensins. J. Clin. Invest. 76, 1436–1439.

Selsted, M. E. & Harwig, S. S. L. (1987). Purification, primary structure, and antimicrobial activities of a guinea pig neutrophil defensin. Infect. Immunol. 55, 2281–2286.

Shakhnovich, E. I. & Gutin, A. M. (1990). Implication of thermodynamics of protein folding for evolution of primary sequences. Nature 346, 773–775.

Shulze–Kremer, S. (1992). Genetic algorithms for protein tertiary structure prediction. In: Proceedings of the 2nd Conference on Parallel Problem Solving from Nature. (Manner, R. & Manderick, B., Eds.), p. 391–400, Elsevier, B.V. Amsterdam.

Sinning, I., Kleywegt, G. J., Cowan, S. W., Reinemer, P., Dirr, H. W., Huer, R., Gilliland, G. L., Armstrong, R. N., Ji, X., Board, P. G., Olin, B., Mannervik, B., & Jones, A. T. (1993). Structure determination and refinement of human alpha class glutathione transferase A1-1, and comparison with the Mu and Pi class enzymes. J. Mol. Biol. 232, 192–212.

Sippl, M. J. (1993). Recognition of errors in three-dimensional structures of proteins. Proteins 17, 355–362.

Sondek, J. & Shortle, D. (1990). Accommodation of single amino acid insertions by the native state of staphylococcal nuclease. Proteins 7, 229–305.

Stewart, J. J. P. (1990). MNDO A semiempirical molecular orbital program. J. Comput.-Aided Mol. Design 4(1), 1–101.

Summers, L., Wistow, G., Narebor, M., Moss, D., Lindley, P., Slingsby, C., Blundell, T., & Bartels, K. (1984). X-Ray studies of the lens and specific proteins. Pept. Protein Rev. 3, 147–167.

Swindell, C. S., Krauss, N. E., Horwitz, S. B., & Ringel, I. (1991). Biologically active taxol analogues with deleted a-ring side chain substituents and variable C-2' configurations. J. Med. Chem. 34, 1176–1184.

Tsuchida, S. & Sato, K. (1992). Glutathione transferases and cancer. Crit. Rev. Biochem. and Mol. Biol. 27, 337–384.

Wani, M. C., Taylor, H. L., Wall, M. E., Coggon, P., & McPhail, A. T. (1971). Plant antitumor agents. VI. The isolation and structure of taxol, a novel antileukemic and antitumor agent from *Taxus brevifolia*. J. Am. Chem. Soc. 93, 2325–2327.

Weber, G. (1992). Protein Interactions. Chapman and Hall, London.

Wilde, C. G., Griffith, J. E., Marra, M. N., Snable, J. L., & Scott, R. W. (1989). Purification and characterization of human neutrophil peptide 4, a novel member of the defensin family. J. Biol. Chem. 264, 11200–11203.

Withka, J. M., Swaminathan, S., Srinivasan, J., Beveridge, D. L., & Bolton, P. H. (1992). Toward a dynamical structure of DNA: Comparison of theoretical and experimental noe intensities. Science 255, 597–601.

Wodak, S. J. & Rooman, M. J. (1993). Generating and testing protein folds. Curr. Opinion Struct. Biol. 3, 247–259.

STATISTICAL ANALYSIS OF PROTEIN SEQUENCES

Volker Brendel

	Abstract	122
I.	Introduction	122
II.	Single Sequence Statistics	123
	A. Global Sequence Features	124
	B. Local Sequence Features	126
	C. Repetitive Structures	128
	D. Analysis of Spacings	129
	E. Score-Based Sequence Analysis	131
III.	Pairwise Sequence Comparisons	136
	A. Statistical Theory	137
	B. Amino Acid Substitution Scoring Matrices	138
	C. Applications to Database Searches	143
	D. Applications to Alignments	144
IV.	Multiple Sequence Comparisons	150
	A. Description of Motifs as Profiles	151
	B. Identification of Motifs	152
V.	Perspective	157
	Acknowledgments	157
	References	157

ABSTRACT

Statistical considerations play an important role in many aspects of the analysis of protein sequence and structural data. This chapter presents a review of methods that find application in the analysis of protein primary sequences. The methods are organized according to whether they require as input single sequences, pairs of sequences, or multiple sequences. Statistically significant sequence features in terms of global and local composition, spatial distribution of residues along the primary sequences, and pairwise or multiple similarities are viewed as starting points for further interaction with the data. It is posited that, due to the continued accumulation of molecular sequence data, statistical screening of the data has become a virtual necessity in the interpretation and design of biomolecular experimental studies, as well as a source of novel molecular hypotheses.

I. INTRODUCTION

Protein primary sequences as well as protein three-dimensional structural data are now abundantly available. Distinct primary sequences number in the ten thousands, and distinct structures with atomic coordinates resolved by crystallography or nuclear magnetic resonance range in the hundreds. These data are a rich source for studies concerned with protein structure, function, and evolution. Intensive research is devoted to elucidating reliable associations between residues and particular configurations of residues with, for example, secondary structure, active sites, transmembrane domains, DNA-binding domains, and so forth. Many of the primary sequences derive from translation of open reading frames or cDNAs accumulated in large-scale DNA sequencing projects. For such sequences most of our initial hypotheses about their possible functions and classifications are based upon sequence comparisons with catalogues of known proteins and protein motifs. Not infrequently it occurs that these comparisons are inconclusive, and more general sequence properties have to be investigated to reveal any significant similarities and associations. What is common, however, in all of these cases is that the initial analysis tends to be theoretical rather than experimental. Perhaps not the best denotation in this context, "theoretical" nonetheless serves to make an important distinction: experimental approaches are increasingly preceded by extensive database work on the computer to narrow down hypotheses about the sequences at hand to those giving best prospects for experimental verification (Gilbert, 1991).

Statistical considerations play an important role in these studies by affording a standard for distinguishing what is likely from what is unlikely to occur by chance. Statistical and biological significance are obviously not synonymous. This is partly due to the fact that there are no realistic models known that are obeyed by naturally occurring protein sequences. Simplified statistical models may nonetheless be useful in setting benchmarks for what kind of sequence features and similarities to focus on for more detailed investigation (cf. Karlin & Brendel, 1992). For example, although protein residues are clearly not realizations of independent letter generators, if such a model stated that three leucines in a row would occur in about every fifth protein of typical length and composition, we might not be too much interested in studying experimentally the role of such a leucine run in a given protein. This, on the other hand, does not deny the possible biological significance of these residues, for example, as part of a signal peptide or transmembrane domain. On a case by case basis then, further studies may be motivated by a combination of statistical and biological considerations, with one or the other taking a lead role.

Areas of applying statistical methods to analyzing protein sequence data may be divided according to (i) statistics derived on a single sequence, including sequence length, global composition, local composition and patterns (for example, clusters, runs, and periodic arrangements of particular amino acids), and spacings between sequence markers, (ii) statistics of pairwise sequence comparisons, (iii) statistics of multiple sequence comparisons, including identification of motifs shared by a group of proteins, (iv) statistics of phylogenetic reconstructions based upon sequence similarities, (v) statistical structure prediction methods, and (vi) statistical evaluation of known protein structures. This chapter is limited to the first three topics. For (iv) see Felsenstein (1988). Secondary structure prediction methods have recently been reviewed by Rost, Sander & Schneider (1994); for global structure prediction see Cohen & Kuntz (1989). References to (vi) include Karlin, Zuker & Brocchieri (1994) and references cited therein.

The emphasis in this review will be on biological motivations and methodological approaches. Applications will be given by way of examples, and no attempt is being made to be exhaustive in pursuing possible applications. References to the current research literature will give sufficient pointers for the interested reader to find more detailed information. Similar approaches applied to DNA sequence analysis have recently been reviewed by Karlin and Cardon (1994).

II. SINGLE SEQUENCE STATISTICS

In this section I shall review issues and methods for statistically evaluating sequence features pertinent to single primary sequences. In other words, the focus is on sequence features that can be calculated by programs that take as input one sequence at a time. To establish the underlying probability distributions, or to characterize particular groups of proteins, the statistics are, of course, applied many times over. This immediately raises the very important issue of multiple comparisons. For example, if the statistical threshold is set at the 1% significance level when applied to a given sequence, but then the test is applied to a database of 10,000 sequences, one would expect some 100 positive test results by chance alone. A conservative approach would be to determine the number of tests to be performed beforehand and to adjust the required significance level relative to a given sequence accordingly (i.e., in the above example one would require a threshold of 0.0001% at the single sequence level for an overall significance of 1%).

The simplest sequence features to be discussed in this section are determined globally over an entire sequence, such as its length and amino acid composition. Local sequence features concern clusters, runs, and periodic patterns of particular residues. The methods are illustrated with analysis of the distribution of charged residues, but they apply equally to residues identified by other attributes. After subsections on evaluation of repeats and spacings between sequence markers (e.g., particular residues or oligopeptides), I review a very general and versatile statistical theory of sequence analysis based upon scores associated with the letters of the sequence.

A. Global Sequence Features

Length

The distribution of protein lengths has recently been the subject of analysis and speculation by White & Jacobs (1994), White (1994a,b), and by Berman, Kolker & Trifonov (1994), Trifonov (1994), and Kolker & Trifonov (1995). These two groups of authors observed empirical length distributions consistent with their respective models of protein evolution. Their studies and proposals are intriguing. Further statistical analysis and increased data sets would seem necessary to distinguish between alternative explanations for the observations.

Composition

For a given sequence, determine its amino acid frequencies. Relevant questions are whether the sequence is particularly rich or poor in certain residue types in comparison to standard sets of proteins, grouped, for example, by species, size, structure, function, cellular localization, time of expression, or evolutionary criteria (Karlin et al., 1991). Reliable answers independent of chance fluctuations require some minimal length of the sequence, generally of the order of 200 or more residues.

Quantile distributions of amino acid usage. A detailed analysis of amino acid usage in various protein sets is given in Karlin, Blaisdell & Bucher (1992). It was found that, independent of amino acid type, the median and mean usage are generally very close and often nearly invariant across species. Much greater variance was observed in the tails of the amino acid usage distributions. This variance can be assessed by way of quantile tables (ibid.). By definition, the $x\%$-quantile $Q(x)$ of a residue type for a given set of proteins is the fraction of proteins for which that residue type occurs with frequency at most $x\%$. Thus, x_m yielding $Q(x_m) = 0.5$ indicates the median usage value, and the interquartile range $[x_{0.25}, x_{0.75}]$ is defined by the usage frequency points $Q(x_{0.25}) = 0.25$ and $Q(x_{0.75}) = 0.75$. A quantile distribution $\tilde{Q}(\cdot)$ is said to be stochastically larger than the quantile distribution $Q(\cdot)$ if $\tilde{Q}(x) < Q(x)$ for all x. This relation implies that the mean usage corresponding to the quantile distribution $\tilde{Q}(x)$ exceeds the mean usage corresponding to the quantile distribution $Q(x)$ and, more generally, each monotone transformation on levels of usage is similarly ranked.

Applications of this type of refined compositional analysis range from evolutionary interpretations to implications for database searches. For example, Karlin et al. (ibid.) observe a large number of stochastic orderings between *E. coli* and human protein amino acid usages and suggest that this attests to the ancient divergence of these species. Differences in amino acid usage between species may have to be taken into account in estimating molecular evolutionary distances. Specific biases may also be incorporated into amino acid substitution matrices used in sequence comparisons (see Section IV).

Correlations of amino acid usage. A thorough investigation of correlations between amino acid frequencies was made by Karlin & Bucher (1992). Following their method, correlations of amino acid usage may be evaluated by means of an adjusted Kendall Tau correlation coefficient. For each residue type pair (X, Y) let $p_i(X)$ and $p_i(Y)$ be the corresponding occurrence frequencies in the ith member of a set S of n protein sequences. For each pair of sequences, indexed i and j, determine

$$\tau_{ij} = \begin{cases} +1 \text{ if} & [p_i(X) - p_j(X)][p_i(Y) - p_j(Y)] > 0 \\ -1 \text{ if} & [p_i(X) - p_j(X)][p_i(Y) - p_j(Y)] < 0 \\ 0 \text{ if either} & p_i(X) = p_j(X) \text{ or } p_i(Y) = p_j(Y). \end{cases} \quad (1)$$

The Kendall Tau correlation coefficient is then calculated as

$$\tau(X,Y) = \frac{\sum_{i \neq j} \tau_{ij}}{\sqrt{n(n-1) - 2t} \sqrt{n(n-1) - 2s}} \quad (2)$$

where t and s are the numbers of ties among pairs of $\{p_i(X)\}$ and $\{p_i(Y)\}$, respectively. Clearly $-1 \leq \tau \leq 1$, and $\tau = 1$ or -1 if, and only if, the values of $\{p_i(X)\}$ and $\{p_i(Y)\}$ induce a completely concordant or a completely discordant ordering, respectively. A value $|\tau| \geq 0.25$ for two independent random orderings (with $n \geq 250$) has a probability < 0.01 of occurring. To avoid the effect of compensation due to the constraint that the residue frequencies within each protein add up to 1, Karlin & Bucher calculate an adjusted $\tau(X, Y)$ defined as the average of two separate determinations of $\tau(X, Y)$ by first marginalizing with respect to one or the other residue. Precisely, in the first determination, the $p_i(X)$ remain unchanged, but the $p_i(Y)$ are replaced by the marginal fractions $p_i(Y)/(1 - p_i(X))$, and in the second determination the $p_i(X)$ are replaced by the marginal fractions $p_i(X)/(1 - p_i(Y))$ and the $p_i(Y)$ remain unchanged.

Analyzing a wide range of protein sets, Karlin & Bucher (ibid.) demonstrated some general facts of amino acid usage. Thus, basic and acidic residues were found to be highly positively correlated; structurally and functionally similar amino acids are for the most part positively correlated (but not always, e.g., K and R are uncorrelated or negatively correlated); and there is generally a negative correlation between amino acids encoded from codons of the SSN type versus amino acids encoded from codons of the WWN type, where S and W stand for G or C and A or T in the DNA.

B. Local Sequence Features

Whereas in the previous section we considered variables evaluated over an entire protein sequence, comparing against values derived from some standard set of proteins, here we shall consider the protein sequence and its composition fixed. The variability to be studied relates to the location of particular residues or other sequence markers. For example, for the observed number of charged residues in the sequence, are these charged residues anomalously distributed? Is there a significant amount of clustering, or are there any unusually long runs or periodic patterns of charge? The same questions can be addressed relative to other natural groupings of residues, such as hydrophobics, serine/threonine, proline/glutamine. A general statistical framework suitable for investigating some of these questions will be reviewed in Section III.E. The following two subsections present alternative and complementary approaches to those more general methods.

Clusters of Particular Residue Types

Clustering of certain residues may be evaluated by a simple binomial model. The sequence is rewritten as a succession of successes and failures depending on whether the particular residue type is observed or not. A segment of length W entailing x successes is considered a significant cluster if x exceeds $Wf + \rho\sqrt{Wf(1-f)}$, where the value of ρ is determined by the significance level (using the standard normal distribution as reference). To accommodate the problem of multiple comparisons, the value for ρ must adjusted according to the length of the protein.

The binomial statistic has been used extensively for studying clustering of charged amino acids in proteins (Karlin et al., 1989; Brendel & Karlin, 1989a; Karlin & Brendel, 1990; Karlin, 1993). It was shown that significant charge clusters often occur in nuclear transcription and replication factors and developmental control proteins, but are generally lacking in cytoplasmic enzymes. Interestingly, there are virtually no significant charge clusters in the proteins comprising the current structural database. Thus, charge clusters provide an example of a statistically defined sequence feature that can be reliably associated with biologically natural protein classifications, preceding and suggesting experimental elaborations.

Runs and Periodic Patterns of Particular Amino Acid Types

Benchmarks for evaluating the significance of runs and periodic patterns in sequences may be obtained from comparison of random sequence models, for which in the simplest case the letters are generated independently with probabilities set equal to the observed residues frequencies in the sequence at hand. The relevant formulas and mathematical references are presented in Karlin, Ost & Blaisdell (1989), and further examples and discussion involving charge residues are given in Karlin, Blaisdell & Brendel (1990). The most practical formulas are given below.

For a sequence of length N and a letter occurring with frequency λ, the probability of observing a run of this letter of length L is asymptotically given by the conservative estimate

$$\text{Prob}\{L \geq \frac{\ln(N)}{-\ln(\lambda)} + z\} \approx 1 - e^{-(1-\lambda)\lambda^z}, \tag{3}$$

where $z > 0$ is a variable. Setting the right-hand side equal to .01 we obtain z and L corresponding to the length of runs significant at the 1% level.

A more general formula is available if the success run is interrupted by a small number of insertions of nonmatching letters. Precisely, let there be at most n_1 insertions of length 1, at most n_2 insertions of length 2, ..., and at most n_k insertions of length k. Define $n = \Sigma_{i=1}^{k} n_i$ (number of insertions) and let

$$L_0 = \frac{\ln(N) + n[\ln(\ln(N)) - \ln(-\ln(\lambda))]}{-\ln(\lambda)}.$$

Then for positive z

$$\text{Prob}\{L \geq L_0 + z\} \approx 1 - e^{-\frac{\rho_1 \lambda^z}{n_1! \ldots n_k!} \left(\frac{1-\lambda}{\lambda}\right)^s} \tag{4}$$

where $\rho_1 = 1 - \lambda$, and $s = \Sigma_{i=1}^{k} i n_i$ is the total length of the insertions.

For periodic patterns consider a sequence of letters A, B, and C occurring with probabilities α, β, and γ, respectively. Let $\lambda = \sqrt{\alpha\beta}$ and n and s as before. Then an alternating run $ABA...$ or $BAB...$ obeys Eq. (4) with ρ_1 replaced by

$$\rho_2 = \frac{\alpha + \beta - 2\alpha\beta}{\sqrt{\alpha\beta}}$$

For a periodic pattern $AABAAB...$, $ABAABA...$, or $BAABAA...$, Eq. (4) holds for $\lambda = (\alpha^2\beta)^{1/3}$ and

$$\rho_3 = \frac{(1 - \alpha^2\beta)\alpha[\alpha^2 + 2\beta^2 + \gamma(\alpha + 2\beta)]}{\alpha^3 + \alpha\beta + \beta + \gamma(1 + \alpha + \alpha^2)}$$

replacing ρ_1.

Using these statistical criteria, it was shown that autoantigens targeted in systemic lupus erythematosus and other chronic systemic autoimmune diseases are significantly enriched in highly charged regions (surfaces), and it was proposed that this feature is involved in the autoimmune response (Brendel et al., 1991). As with charge clusters, there are no significant charge runs in the current structural database, and it remains an open question whether these particular charge configurations associate with recurrent structural motifs (Karlin, 1995).

C. Repetitive Structures

Internal repetitiveness within a protein sequence may be approached in a variety of ways. First, determine significant repeats, allowing for a few errors. Another measure of repetitiveness is afforded by means of multiplet counts. And thirdly, periodic occurrences of particular residues may be of interest.

Repeats

Following similar formulas as above, the probability of an r-fold repeat exceeding length L is given by

$$\text{Prob}\{L \geq \frac{\ln\binom{N}{r}}{-\ln(\lambda)} + z\} \approx 1 - e^{-(1-\lambda)\lambda^z}, \tag{5}$$

where $\lambda = \Sigma_{i=1}^{m} p_i^r$ is the per position probability of an r-fold match in an m-letter alphabet (Karlin, Ost & Blaisdell, 1989). Extension allowing for errors are available (ibid.), and programmed algorithms implementing these statistics have been published and distributed (Leung et al., 1991).

Multiplets and Altplets

The SAPS program of Brendel et al. (1992) further investigates sequence repetitiveness by means of multiplet counts. Here multiplet refers to any homooligopeptide (e.g., A_2, Q_7). Extremes in these counts are determined as follows. Let f_i be the frequency of residue type i in the sequence. In an infinite random sequence, a given site is the first residue of a multiplet of residue type i with probability $p_i = \Sigma_{k=2}^{\infty} f_i^k (1-f_i)^2 = (1-f_i) f_i^2$. Consequently, the probability of observing a multiplet of any amino acid is given by $p = \Sigma_{i=1}^{20} (1-f_i) f_i^2$. If N denotes the length of the sequence, then the aggregate multiplet count (combined number of all homooligomers) is taken to be significant if it exceeds $Np + 3\sqrt{Np(1-p)}$. A significant multiplet count for a particular amino acid i is required to exceed $Np_i + 4\sqrt{Np_i(1-p_i)}$; here 4 standard deviations above the mean is chosen as a conservative significant threshold to accommodate the problem of testing all 20 amino acids simultaneously. SAPS also determines exceptional counts of specific altplets, referring to runs alternating in amino acids i and j (e.g., KPKPK). In this case p_i is replaced by $p_{ij} = f_i f_j (2 - f_i - f_j)$.

Comparing sets of proteins from different species it was determined that proteins with significant multiplet counts occur much more frequently in the *Drosophila melanogaster* set compared to other eukaryotes, and that they virtually do not occur among the *Escherichia coli* proteins (Sapolsky, Brendel & Karlin, 1993). A more detailed investigation of this observation and possible interpretations are still outstanding.

Periodicities

For a random sequence of length N and specified letter occurring with frequency f, the probability of observing a run of this letter of repeat count r and given period is closely approximated by $1 - (1 - fx)/[(r + 1 - rx)(1 - f)x^{n+1}]$, where x solves $(1 - f)x(1 + fx + \cdots + f^{r-1}x^{r-1}) = 1$ (e.g., Feller, 1968, page 325). With $f = .05$ and r replaced by $r - 1$ in the above formulas, one obtains a lower bound for the probability of observing an r-repeat of any (not predetermined) amino acid, which indicates a minimum value of r that might convey statistical significance. These considerations have helped in critical evaluation of spottings of "leucine zippers" (Brendel & Karlin, 1989b).

D. Analysis of Spacings

A set of n marked residues in a sequence of length N (e.g., all cysteines, or all basic residues) induces $n + 1$ spacings from the N-terminus of the protein to the first marker, from the first marker to the second, and so on. Analysis of these spacings can aid in assessing the degree of homogeneity in the sequence. Unusually long maximal spacings and unusually short minimal spacings suggest an inhomogeneous distribution of the marker. In the same way, unusually short maximal spacings and unusually long minimal spacings indicate excessive evenness in the positioning of the markers.

For statistical evaluation, spacings between amino acids may be interpreted as runs of failures (designated F) between successes (designated S). Then, the probability that the k-th longest spacing is greater than or equal to s is given by the probability that a sequence of n S's and $N - n$ F's contains at least k runs of F's greater or equal to s. The latter probabilities are given in (Morris, Schachtel & Karlin, 1993) and are implemented in the SAPS program of Brendel et al. (1992). A different approach using statistical results concerning the spacings between random points on the unit interval is reviewed below. A more general theory allows evaluating the cumulative lengths of r consecutive fragments (r-scans), where the single fragment lengths are the distances between two consecutive marker sites. The lengths of the longest and shortest r-scans are appropriate statistics for detecting cases of significant clumping, significant overdispersion, or excessive regularity in the spacings of the marker. These ideas have proved useful in genomic DNA sequence analysis (e.g., Karlin & Brendel, 1992) but have not been fully tested on protein sequences. In particular, it is unclear to what extent the relatively short lengths of protein sequences prevent applying the approximate formulas and what possible adjustments can be made.

Minimal and Maximal Spacings

Consider a sequence of length N and a specified array of n markers randomly distributed in the sequence. These occurrences induce $n + 1$ spacings ($U_0, U_1, \ldots, U_n$), where U_0 is the distance before the first occurrence, U_i is the distance from the ith occurrence of the marker to the $i + 1^{st}$ occurrence, and U_n is the distance after the last occurrence. Distances are scaled such that the distance between immediately adjacent markers equals $1/N$. Let the extremal spacings be $m^* = \min\{U_0, U_1, \ldots, U_n\}$ and $M^* = \max\{U_0, U_1, \ldots, U_n\}$. Then (see Karlin and Brendel, 1992, for rationale and examples)

$$F(a) = \text{Prob}\{m^* < a\} = 1 - [1 - (n+1)a]^n, \quad \text{for } 0 < a \leq \frac{1}{n+1}, \quad (6)$$

and

$$G(b) = \text{Prob}\{M^* \geq b\} = 1 - \sum_{i=0}^{n+1} \binom{n+1}{i}(-1)^i[\delta_i(1-ib)]^n, \quad \text{for } 1 > b \geq \frac{1}{n+1}, \quad (7)$$

where $\delta_i = 1$ if $ib < 1$ and $\delta_i = 0$ otherwise. More generally, let $P_k(x)$ denote the probability that k of the $n + 1$ fragments are of lengths less than x. Then

$$P_k(x) = \binom{n+1}{k}\sum_{i=0}^{k}\binom{k}{i}(-1)^i[\delta_i(1-(n+1+i-k)x)]^n \quad (8)$$

where $\delta_i = 1$ if $(n + 1 + i - k)x < 1$ and $\delta_i = 0$ otherwise. Eq. (8) allows analyzing the data in terms of spacings other than the extremes $k = 0$ (m^*) and $k = n + 1$ (M^*).

The evaluation of an extremal minimum at the 1% significance level rests on determining a^* so that $F(a^*) = 0.01$. For an observed m^* smaller than a^*, the minimum spacing is considered significantly small. Similarly, the largest gap is considered statistically significant if the observed M^* exceeds b^*, where b^* satisfies $G(b^*) = 0.01$. For an observed m^* too large ($m^* \geq c^*$ where $F(c^*) = 0.99$) or an observed M^* too small ($M^* < d^*$ where $G(d^*) = 0.99$) or both, the spacings are considered to be overly regular. The formulas are practical for n small or of moderate size. For large n, one may use the asymptotic probabilities given below.

r-Scan Statistics

For a given set of single spacings $\{U_0, U_1, \ldots, U_n\}$, r-scans are formed according to $R_i = \sum_{j=i}^{i+r-1} U_j$, $i = 0, 1, \ldots, n - r + 1$. Let $M_k^{(r)}$ denote the length of the k-th largest r-fragment, and let $m_k^{(r)}$ be the length of the k-th smallest r-fragment. If the markers were distributed randomly according to the uniform distribution then asymptotically for large n (Karlin & Macken, 1991; Dembo & Karlin, 1992)

$$\text{Prob}\left\{m_k^{(r)} < \frac{x}{n^{1+1/r}}\right\} \approx 1 - \sum_{i=0}^{k-1} \frac{\lambda^i}{i!} e^{-\lambda}, \quad \lambda = \frac{x^r}{r!}. \tag{9}$$

With x chosen so that the right side of equation (9) is equal to 0.01, the observed $m_k^{(r)}$ would be considered significantly small if it were less than $x/n^{1+1/r}$. Furthermore

$$\text{Prob}\left\{M_k^{(r)} > \frac{1}{n}[\ln(n) + (r-1)\ln(\ln n) + x]\right\} \approx 1 - \sum_{i=0}^{k-1} \frac{\mu^i}{i!} e^{-\mu}, \quad \mu = \frac{e^{-x}}{(r-1)!}. \tag{10}$$

With x chosen so that the right side of Eq. (10) is equal to 0.01, the observed $M_k^{(r)}$ would be considered significantly large if it were greater than $1/n[\ln(n) + (r-1)\ln(\ln n) + x]$.

To detect clustering among markers, examine all r-scans and ascertain whether the minimum is especially small with respect to the postulated uniform distribution of markers. Similarly, in deciding whether some successive markers are excessively dispersed, check the maximum length among the r-scans to see whether it is especially large. Conversely, when the minimum r-scan length is especially large or the maximum r-scan length is especially small or both, then the spacings of the marker are assessed to be excessively regular. The value of r can be varied in applications appropriate to the overall number of markers and the level of clustering one wishes to investigate.

E. Score-Based Sequence Analysis

A versatile statistical theory for analyzing sequences has been developed in the framework of score assignments to the residues in the sequence with the objective of determining segments of high cumulative score (Karlin & Altschul, 1990; Karlin & Brendel, 1992; Karlin, 1994). The usefulness of this approach is due to the fact that many sequence analysis problems (including those discussed above) can be naturally adapted into a scoring problem. The statistical theory provides thresholds of significance that aid in data interpretation. Some examples of scoring schemes for different applications are as follows:

(i) Scores for clusters of charged residues
Positive charge clusters may be identified as high-scoring segments with a scoring scheme in which Lys (K) and Arg (R) score +2, Asp (D) and Glu (E) score –2, and all other amino acids score –1; for negative charge clusters, the scores assigned to {K, R} and {D, E} should be interchanged;

(ii) Scores for runs of some residue a
In this case the score for residue a may be set to +1 and the score for all

other residues set to a sufficiently large negative number; obviously, only a run of letter a can have a positive score;

(iii) Scores for hydrophobic segments

An appropriate scoring scheme might be the Kyte–Doolittle scale or any of many other scales that have been proposed for measuring hydrophobicity (Kyte & Doolittle, 1982; Hopp & Woods, 1981). The section on "target scores" below will give a rationale for choosing different relative magnitudes for the scores.

In general, let the alphabet in use be $\mathcal{A} = \{a_1, a_2, \ldots, a_r\}$ with associated scores $S = \{s_1, s_2, \ldots, s_r\}$. Let $X_1, X_2, \ldots, X_N$ be the successive letter scores in a sequence of length N. In the simplest model, the X_i are independently identically distributed with probability distribution $\text{Prob}\{X_i = s_k\} = p_k$. This may be interpreted to mean that generating the letter a_k occurs with probability p_k and yields the score s_k. Generalizations to models in which successive letters have Markov dependence are available, but these models appear computationally prohibitive at present (Karlin & Dembo, 1992). Two essential restrictions are imposed on the set S of scores:

(i) A positive score is accessible, i.e., $\text{Prob}\{X_i > 0\} > 0$, and
(ii) The expected score per letter is negative, i.e., $E[X_i] = \sum_{j=1}^{r} p_j s_j < 0$.

Both restrictions are natural to the task of distinguishing sufficiently high positively scoring segments in the sequence. It is also of note that in the many cases where log-odds scores of the type $s_i = \ln(q_i/p_i)$ are indicated (q_i being some alternative probabilities, see the section on "target scores" below), such scores always satisfy the two conditions (*i*) and (*ii*). This follows from the fact that, subject to $\sum_{i=1}^{20} p_i = \sum_{i=1}^{20} q_i = 1$, (i) $q_i > p_i$ for at least one i, and (ii) (by elementary calculus) $f(q_1, q_2, \ldots, q_r) = \sum_{i=1}^{r} p_i \ln(q_i)$ is maximal for $q_i = p_i$, $i = 1, \ldots, r$.

High Scoring Segments

The maximal scoring segment in the sequence of scores $X_1 X_2 \ldots X_N$ may easily be defined by way of the partial sums $S_0 = 0$ and $S_k = \sum_{i=1}^{k} X_i$, $k = 1, 2, \ldots, N$. In this notation the maximal aggregate score is $S = \max_{0 \leq k < l \leq N} (S_l - S_k)$ and this score is attained by the segment ranging from k to l. Karlin & Dembo (1992) proved that for positive x

$$\text{Prob}\{S \geq \frac{\ln(N)}{\lambda} + x\} \approx 1 - e^{-Ke^{-\lambda x}}, \quad (11)$$

where λ is the unique positive solution to the equation

$$E[e^{\lambda X_i}] = \sum_{j=1}^{r} p_j e^{\lambda s_j} = 1. \qquad (12)$$

K is a parameter that is given by an explicit series expression that can be readily evaluated numerically. Precisely,

$$K = F\frac{e^{-2(A+B)}}{\lambda C}, \qquad (13)$$

where

$$A = \sum_{k=1}^{\infty} \frac{1}{k} E[e^{\lambda S_k}; S_k < 0] = \sum_{s_i < 0} p_i e^{\lambda s_i} + \frac{1}{2} \sum_{s_i + s_j < 0} p_i p_j e^{\lambda(s_i + s_j)} + \ldots,$$

$$B = \sum_{k=1}^{\infty} \frac{1}{k} \Pr\{S_k \geq 0\} = \sum_{s_i \geq 0} p_i + \frac{1}{2} \sum_{s_i + s_j \geq 0} p_i p_j + \ldots,$$

$$C = E[X_i e^{\lambda X_i}] = \sum_{j=1}^{r} p_j s_j e^{\lambda s_j},$$

$F = 1$ for nonlattice scores, and $F = \lambda\delta/(1 - e^{-\lambda\delta})$ for lattice scores with δ being the smallest span of score values (that is, all scores can be written as multiples of δ, $|s_i| = i\delta$). Computer routines that calculate λ and K are available (Karlin & Altschul, 1990).

In practice, one may set the left-hand side of Eq. (11) equal to some predetermined significance level, for example, $P = 0.01$ or $P = 0.05$, and solve for $x = x_p$. Rearrangement of Eq. (11) gives

$$x_P = \frac{1}{\lambda}[\ln K - \ln(-\ln(1 - P))]. \qquad (14)$$

For random sequences, a maximal segment score exceeding $S_P = \ln(N)/\lambda + x_P$ is significant at the P level. Note that using an estimate of K larger than the correct value increases the x_P determined according to Eq. (14). Accordingly, approximations to K that use only a finite number of terms in the above series will be conservative. Simple expressions for λ and K are available for certain scoring schemes (Karlin et al., 1990; Karlin & Dembo, 1992). For example, let score 1 occur with probability p, score 0 with probability r, and score -1 with probability $q > p$; then $\lambda = \ln(q/p)$ and $K = (q-p)^2/q$. For scores $\{-m, \ldots, -1, 0, 1\}$, $K = (1 - e^{-\lambda})C$, and for scores $\{-1, 0, 1, \ldots, m\}$, one obtains $K = (1 - e^{-\lambda})(E[S_1])^2/C$.

Note that λ may be interpreted as a scaling factor. To see this, consider scores s_i with calculated parameters λ, K, and significance threshold S_P. Then the scores

$s'_i = \alpha s_i$ for a positive factor α have parameters $\lambda' = \lambda/\alpha$ and $K' = K$, whence the significance threshold becomes $S'_P = \alpha S_P$, scaled by the same magnitude as the individual scores.

High scoring segments are easily identified by way of an excursion plot (Karlin & Brendel, 1992). Thus, recursively define the excursion scores E_i according to

$$E_0 = 0, \quad E_i = \max\{E_{i-1} + s_i, 0\}, \quad i \geq 1. \tag{15}$$

A plot of E_i as a function of i takes the form of excursions from the i-axis into the positive domain. The value of each excursion is defined as its peak score. If the peak score exceeds the critical value S_P, then the segment from the beginning of the excursion up to the residue where the peak value is first realized within the excursion is a high-scoring segment, significant at the P level.

For sufficiently high score S the number of distinct high-scoring segments with score at least S is approximately Poisson distributed with parameter $\mu = \mu(S) = NKe^{-\lambda S}$. Another statistic that may be useful in appraising a given set of scores concerns the length $L(S)$ of a segment of aggregate score at least S. An asymptotic confidence interval for $L(S)$, S large, is given (Dembo & Karlin, 1993) by

$$L(S) = \frac{S}{v} \pm \rho_\alpha \frac{1}{v} \sqrt{\frac{Sw}{v}} \tag{16}$$

where $v = \Sigma_{i=1}^r p_i s_i e^{\lambda s_i}$, $w = \Sigma_{i=1}^r p_i s_i^2 e^{\lambda s_i} - v^2$, and ρ_α is the quantile point of the standard normal distribution corresponding to an $\alpha\%$ confidence interval (e.g., $\rho = 1.96$ for a 90% confidence interval).

Target Scores

In many situations there may be natural criteria underlying score assignments. An example would be the experimentally derived Kyte–Doolittle scale for measuring hydropathy strength (Kyte & Doolittle, 1982). In other situations, however, no such criterion may be available. A second theoretical result concerning the composition of high-scoring segments is helpful in this case. The result states that in random sequences (with successive letters independently identically distributed) high-scoring segments have an intrinsic biased composition such that letter type a_i occurs with frequency

$$q_i \approx p_i e^{\lambda s_i} \tag{17}$$

(Karlin & Altschul, 1990; Karlin, Dembo & Kawabata, 1990; Dembo & Karlin, 1991). Since λ is positive, this bias is consistent with intuition demanding that, in high-scoring segments, positively scoring letters should be overrepresented and negatively scoring letters should be underrepresented. Turning expression (17) around, it follows that scores defined by

$$s_i = \frac{1}{\lambda} \ln(q_i/p_i) \tag{18}$$

will identify high-scoring segments of "target" composition q_i. As indicated above, the location and significance of high-scoring segments remain unchanged upon scaling the scores s_i by any positive factor. Thus, result (18) may be interpreted as follows: Let p_i be the frequencies of letters in some reference random sequence, and let q_i be the desirable target frequencies, which are derived from known representatives of the type of region we seek to identify; then the score for letter a_i should be set proportional to the corresponding log-likelihood ratio $\ln(q_i/p_i)$.

Applications

Rogers, Wells & Rechsteiner (1986) hypothesized that the presence of regions rich in prolines (P), glutamic acid (E), serine (S), and threonine (T) may be involved in targeting proteins containing such regions for rapid degradation. While the proof or rejection of this hypothesis will have to come from experimental measurements, it serves nonetheless as a good example for the application of the scoring method as a complementary theoretical approach. Table 1 gives scores derived as discussed above that are honed to identify regions of composition similar to those detailed by

Table 1. Scores for PEST Regions

Residue	q^a	p^b	$\log_2(q/p)$	Score[c]
S	0.133	0.071	0.901	4
E	0.129	0.063	1.029	4
P	0.105	0.052	1.017	4
D	0.068	0.053	0.356	1
Q	0.047	0.041	0.183	1
V	0.062	0.065	−0.074	0
T	0.055	0.059	−0.110	0
F	0.039	0.040	−0.019	0
Y	0.031	0.032	−0.028	0
H	0.021	0.023	−0.114	0
C	0.018	0.017	0.100	0
L	0.066	0.091	−0.468	−2
A	0.058	0.076	−0.398	−2
M	0.016	0.023	−0.506	−2
G	0.046	0.071	−0.640	−3
N	0.028	0.045	−0.667	−3
I	0.025	0.055	−1.120	−4
R	0.025	0.052	−1.039	−4
K	0.025	0.057	−1.172	−5
W	0.003	0.013	−2.098	−8

Notes: [a]Residue frequencies in the aggregate of the annotated PEST regions (Rogers, Wells and Rechsteiner, 1986).
[b]Standard residue frequencies (average over samples from the database).
[c]Values of the previous column multiplied by the scale factor 4 and rounded to the nearest integer.

```
ID   FML1_HUMAN        STANDARD;      PRT;    353 AA.
DE   FMLP-RELATED RECEPTOR I (FMLP-R-I).

number of residues:   353;    molecular weight:   40.0 kdal

    1   METNFSIPLN ETEEVLPEPA GHTVLWIFSL LVHGVTFVFG VLGNGLVIWV AGFRMTRTVN
   61   TICYLNLALA DFSFSAILPF RMVSVAMREK WPFASFLCKL VHVMIDINLF VSVYLITIIA
  121   LDRCICVLHP AWAQNHRTMS LAKRVMTGLW IFTIVLTLPN FIFWTTISTT NGDTYCIFNF
  181   AFWGDTAVER LNVFITMAKV FLILHFIIGF TVPMSIITVC YGIIAAKIHR NHMIKSSRPL
  241   RVFAAVVASF FICWFPYELI GILMAVWLKE MLLNGKYKII LVLINPTSSL AFFNSCLNPI
  301   LYVFMGRNFQ ERLIRSLPTS LERALTEVPD SAQTSNTHTT SASPPEETEL QAM

A : 25( 7.1%);  C :  8( 2.3%);  D :  6( 1.7%);  E : 15( 4.2%);  F : 29( 8.2%)
G : 14( 4.0%);  H :  9( 2.5%);  I : 34( 9.6%);  K :  9( 2.5%);  L : 40(11.3%)
M : 14( 4.0%);  N : 18( 5.1%);  P : 16( 4.5%);  Q :  4( 1.1%);  R : 15( 4.2%)
S : 22( 6.2%);  T : 29( 8.2%);  V : 30( 8.5%);  W :  9( 2.5%);  Y :  7( 2.0%)

KR    :   24 (  6.8%);   ED    :  21 (  5.9%);   AGP    :  55 ( 15.6%);
KRED  :   45 ( 12.7%);   KR-ED :   3 (  0.8%);   FIKMNY : 111 ( 31.4%);
LVIFM :  147 ( 41.6%);   ST    :  51 ( 14.4%).

High scoring PEST regions:

score=  4.00 frequency=  0.150   ( SEP )
score=  1.00 frequency=  0.028   ( DQ )
score=  0.00 frequency=  0.317   ( VTFYHC )
score= -2.00 frequency=  0.224   ( LAM )
score= -3.00 frequency=  0.091   ( GN )
score= -4.00 frequency=  0.139   ( IR )
score= -5.00 frequency=  0.025   ( K )
score= -8.00 frequency=  0.025   ( W )

Expected score/letter:  -0.977
M_0.01=  33.71;  M_0.05=  27.13

1) From  316 to  349:  length= 34, score=43.00  **
   316  SLPTSLERAL TEVPDSAQTS NTHTTSASPP EETE
   S:  6(17.6%);   E:  5(14.7%);   T:  7(20.6%);   P:  4(11.8%);
```

Figure 1. Identification of a PEST region in the FMLP-related receptor I sequence (taken from SWISS-PROT, Bairoch & Boeckmann, 1994). Scores are as defined in Table 1; the corresponding frequencies are pertinent to the FML1 sequence. The significant aggregate scores at the 1% and 5% are 34 and 28, respectively. The segment comprising residues 316 to 349 is significant at the 1% level.

Rogers et al. It is of note that threonine is actually not more abundant in the PEST regions than it is in average proteins (thus T is scored 0). The statistical theory can be used to identify significantly high-scoring segments as illustrated in Figure 1. A reasonable strategy to substantiate the PEST hypothesis would involve screening the entire database for hits in this scoring scheme and attempting to interpret the results in terms of tendencies of certain types of proteins toward being among those that have hits.

III. PAIRWISE SEQUENCE COMPARISONS

The statistical theory of score-based sequence analysis extends to the problem of pairwise sequence comparisons when alignment blocks are restricted to possibly contain mismatches but no gaps. Following a brief outline of this generalization we shall discuss how this theory bears on the choice of amino acid substitution scoring matrices.

A. Statistical Theory

Consider two independent random sequences $X_1 X_2 \ldots X_M$ and $Y_1 Y_2 \ldots Y_N$ of respective lengths M and N consisting of letters from the alphabet $\mathcal{A} = \{a_1, a_2, \ldots, a_r\}$ drawn with probabilities $\{p_1, p_2, \ldots, p_r\}$ and $\{p'_1, p'_2, \ldots, p'_r\}$, respectively. For a given alignment and position, some letter a_i in the first sequence will be paired with some possibly different letter a_j in the second sequence. Let any such pairing be associated with a score $s_{ij} = s(a_i, a_j)$, where the s_{ij} are obtained from some $r \times r$ (usually symmetrical) scoring matrix. The maximal segment pair score over all alignment offsets is given by $S = \max \Sigma_{l=1}^{L} s(X_{i+l}, Y_{j+l})$, where the maximum is taken over all triplets (L, i, j) satisfying $1 \leq L \leq \min(M, N)$, $0 \leq i \leq M - L$, and $0 \leq j \leq N - L$. Under certain restrictions (Karlin, 1994; Dembo, Karlin & Zeitouni, 1994a,b), Eq. (11) holds with N replaced by MN, λ replaced by the unique positive root of the equation

$$\sum_{i=1}^{r} \sum_{j=1}^{r} p_i p'_j e^{\lambda s_{ij}} = 1, \qquad (19)$$

and K computed as in Eq. (13) (with appropriate substitutions for the p_i and s_i in the right-hand sides of the expressions following Eq. (13)). Analogously to Eq. (17), high-scoring segment pairs will have a composition biased towards the substitution frequencies

$$q_{ij} = p_i p'_j e^{\lambda s_{ij}} \qquad (20)$$

The foregoing statistical theory is used in protein data library searches performed with the BLASTP program of Altschul et al. (1990). Input to a search consists of the query sequence, a data library typically composed of a large number of individual library sequences, and a particular choice of scoring matrix. BLASTP identifies significant high-scoring segment pairs (not allowing gaps) corresponding to similar regions shared by the query sequence and any member of the data library. Each such high-scoring segment pair will be referred to as a "hit." The p_i and p'_j in Eq. (19) are the residue frequencies in the query sequence and in a standard set of proteins, respectively, and the s_{ij} are the scores associated with matching residues a_i and a_j as specified by the input scoring matrix. The standard residue frequencies as coded in BLASTP are A 8.1%, R 5.7%, N 4.5%, D 5.4%, C 1.5%, Q 3.9%, E 6.1%, G 6.8%, H 2.2%, I 5.7%, L 9.3%, K 5.6.%, M 2.5%, F 4.0%, P 4.9%, S 6.8%, T 5.8%, W 1.3%, Y 3.2%, and V 6.7%. Such a choice of standard frequencies (rather than choosing the p'_j to be the residue frequencies of the individual library sequences) is motivated by the practical need to avoid computer-intensive recalculation of the constants λ and K for each pair of sequences being compared. Related issues in searching molecular sequence databases are reviewed in Altschul et al. (1994).

Statistical evaluation in the case of database searches must accommodate the problem of multiple comparisons. In BLASTP this is done in the following way. Let M and N be the lengths of the query sequence and of a particular library sequence, respectively. Under a random sequence model, the query sequence is thought of as generated by independent sampling of residues with frequencies p'_i, and the library sequence is thought of as generated by independent sampling of residues with frequencies p_j. For such sequences, the number of distinct high-scoring segment pairs with score at least S is approximately Poisson distributed with parameter

$$\mu = \mu(S) = MNKe^{-\lambda S} \tag{21}$$

(Karlin & Altschul, 1990). Thus, the expected number of hits of score at least S in the particular library sequence is μ, and the probability of at least h hits in the sequence at that score level is

$$p_h = p_h(S) = 1 - \sum_{k=0}^{h-1} \frac{\mu^k}{k!} e^{-\mu}. \tag{22}$$

For small μ, $p_h \simeq \frac{\mu^h}{h!}$.

Let L be the total number of residues in the data library. Then the expected number of library sequences with at least h hits of score equal to or exceeding S in a search of the query sequence against the entire library is approximately

$$E_h(S) = \frac{L}{N} p_h(S), \tag{23}$$

and the probability that the library contains at least one such sequence is given by the Poisson approximation

$$P_h(S) = 1 - e^{-E_h(S)}. \tag{24}$$

If a library sequence shares one or more high-scoring segments with the query, then for the k-th hit (with score S_k) the BLASTP program reports $E_k(S_k)$ and $P_k(S_k)$. One may set $E_1(S) = 0.01$ and solve for S to obtain the threshold for hits at the 1% overall significance level. Some refinements to this general scheme are implemented in the program (Altschul et al., 1990) but are not discussed here.

B. Amino Acid Substitution Scoring Matrices

The statistical theory review in the last section gives a rationale for choosing amino acid substitution scores. Different approaches will be discussed. Common to these approaches are substitution data k_{ij}, $i = 1, \ldots, 20, j = 1, \ldots, 20$, where k_{ij} is the count of residue i to residue j substitutions in some set of data. Generally, there is no reason to treat i to j substitutions any differently from j to i substitutions,

Statistical Analysis of Protein Sequences

and consequently the k_{ij} counts are symmetrized to give $c_{ij} = c_{ji} = k_{ij} + k_{ji}$. For the following discussion, let $C_{i.} = \Sigma_{j=1}^{20} c_{ij}$, $C_{.j} = \Sigma_{i=1}^{20} c_{ij}$, and $C_{..} = \Sigma_{i=1}^{20} \Sigma_{j=1}^{20} c_{ij}$. Note that $C_{i.} = C_{.i}$ gives the number of i residues in the data, and $C_{..}$ gives the total number of residues in the data. Thus, residue and substitution frequencies are given by $f_i = C_{i.}/C_{..}$ and $f_{ij} = c_{ij}/C_{..}$, respectively.

PAM Matrices

For the PAM matrices of Dayhoff et al. (1978), the substitution counts c_{ij} are inferred substitution counts from hypothetical phylogenetic reconstructions of ancestral sequences for a small number of highly related sequences (more than 85% identical; in total, Dayhoff et al. counted 1572 substitutions among proteins from 71 groups). Substitution scores were calculated on the basis of a Markov model for the propagation of substitutions through evolutionary time. Precisely, the PAM matrices are derived as follows.

First define the "relative mutability" of residue j as the proportion of j residues involved in substitutions, i.e.,

$$m_j = \frac{C_j - c_{jj}}{C_j}. \qquad (25)$$

Let the one-step transition probabilities for the Markov model be denoted by

$$M_{ij} = \text{Prob}\{\text{residue } j \to \text{residue } i \text{ change during one period of time}\}. \qquad (26)$$

Thus, in this model a sequence of composition $\vec{p}$, where $\vec{p}$ is the column vector of frequencies $p_1, p_2, \ldots, p_{20}$ is thought to undergo random changes during time so that the expected composition of the sequence after one time period is given by $\mathcal{M}\vec{p}$, where $\mathcal{M}$ is the 20 × 20 transition matrix (M_{ij}). Dayhoff et al. set

$$M_{ij} = \rho m_j \frac{c_{ij}}{C_j - c_{jj}}, \quad i \neq j \quad M_{jj} = 1 - \rho m_j, \qquad (27)$$

where ρ is a proportionality constant. The value of ρ is assigned by setting $\Sigma_{i=1}^{20} f_i M_{ii} = 0.99$, which amounts to defining the unit of time as the period that will produce, on average, 1% observable residue changes in the evolution of a given protein sequence (1 **A**ccepted **P**oint **M**utation per 100 residues). $\mathcal{M}$ as defined is a stochastic matrix (the column sums equal 1), and the n-step transition probabilities

$$M_{ij}^{(n)} = \text{Prob}\{\text{residue } j \to \text{residue } i \text{ change during } n \text{ periods of time}\} \qquad (28)$$

are given as the elements of n-th power of $\mathcal{M}$. The stationary distribution of this Markov chain is given by the column vector of residue frequencies $f_1, \ldots, f_{20}$. It is easy to show that

$$f_i M_{ji}^{(n)} = f_j M_{ij}^{(n)}, \qquad (29)$$

and thus substitution scores for the PAMn matrix can be assigned symmetrically as log-odds ratios

$$s_{ij}^{(n)} = \ln\left(\frac{M_{ij}^{(n)}}{f_i}\right). \quad (30)$$

The interpretation of these scores is as follows: For each residue type j, a positive score is assigned to the substitution $j \to i$ if the probability of a $j \to i$ transition exceeds the probability f_i of observing residue type i in a randomly generated sequence; otherwise a negative score is assigned. For small n, $M_{ij}^{(n)}$ will be large for $M_{jj}^{(n)}$ relative to the elements $i \neq j$; in this case, positive scores are mostly confined to identity pairs in an alignment, appropriate for comparing sequences deemed to have undergone little evolutionary divergence. Since $M_{ij}^{(n)}$ approaches f_i for n large, high-numbered PAM matrices, on the other hand, tend to the zero matrix, consistent with the underlying model which implicitly assumes that after long times very little distinction can be assigned to different substitutions. Note that $\Sigma_{j=1}^{20} f_j \Sigma_{i=1}^{20} f_i s_{ij}^{(n)} < 0$, so that the statistical theory of high-scoring segment pairs applies for sequences generated with letter probabilities coinciding with the residue frequencies f_i. This may not be true for sequences generated with very different letter probabilities. In conjunction with the fact that the PAM scores imply the unique stationary distribution $f_1, \ldots, f_{20}$, it would be appropriate to derive PAM matrices specific to groups of proteins with shared compositional biases (this argument has been pursued by I. Ladunga, personal communication).

BLOSUM Matrices

The BLOSUM series of matrices were derived from multiple alignments of functional motifs (Henikoff & Henikoff, 1992). In this approach substitution counts are tallied over all possible pairwise comparisons of residues in each column of an alignment block. For example, assume a column consisting of 5 A's, 3 G's, and 2 P's. Then we would tally 10 AA's, 3 GG's, 1 PP, 15 AG's, 10 AP's, and 6 GP's. Log-odds are obtained by treating the count matrix (c_{ij}) similar to contingency tables, with observed counts c_{ij} compared to expected counts $C_{i.}C_{.j}/C_{..}$, or

$$s_{ij} = \ln\frac{f_{ij}}{f_i f_j} \quad (31)$$

Note that, for sequences generated with residue probabilities coinciding with the frequencies f_i, the λ associated with the above scores equals 1. Furthermore, in this case the target frequencies q_{ij} will be equal to the substitution frequencies f_{ij}. Thus, substitution scoring matrices defined by Eq. (31) will tend to assign significance to segment pairs with compositions similar to that of the data used to generate the scores. An approach to overcoming this limitation (different from the Markov model discussed above) will be presented in Section III.C.

Other Amino Acid Substitution Scoring Matrices

Normalization of substitution counts according to Eq. (31) are also applicable for counts derived in ways different from the blocks approach. For example, the counts may be derived from matching positions of structure (rather than sequence) alignments (Risler et al., 1988; Miyazawa & Jernigan, 1993; Johnson & Overington, 1993). Another alternative is to start from statistically significant block identity pairs among all sequence pairs within a given set of sequences (Brendel, Ladunga & Karlin, unpublished). In this case the selection of trusted substitutions is statistically motivated, thus avoiding the difficult issue of multiple alignment prerequisite to the blocks approach. The selection of block identity pairs can be justified by reference to the theory of longest common words for a pair of random sequences. Following Karlin, Ost & Blaisdell (1989), for two sequences of length M and N and compositions $\{p_i\}$ and $\{q_i\}$, respectively, the probability of an extended match consisting of L identities, at most n_1 single pair mismatches (or insertions), at most n_2 mismatches or insertions of length 2, . . ., and at most n_k mismatches or insertions of length k, is closely approximated by

$$p \approx 1 - e^{-\frac{(1-\lambda)\lambda^{L-L_0}}{n_1!\ldots n_k!}\left(\frac{1-\lambda}{\lambda}\right)^s}, \qquad (32)$$

where $\lambda = \Sigma_{i=1}^{20} p_i q_i$ is the probability of a local match and

$$L_0 = \frac{\ln(MN) + n \ln \ln(MN) - \ln(-\ln \lambda)}{-\ln \lambda}, \qquad (33)$$

with $n = \Sigma_{i=1}^{k} n_i$ and $s = \Sigma_{i=1}^{k} i n_i$. Setting p to some small probability, a minimal significant match length $L > L_0$ can easily be determined, and the substitution counts can be tallied from all significant pairwise matches.

Scaling

For practical purposes it is expedient to use integer scores in sequence comparisons. The conversion of a log-odds matrix with scores as defined in Eq. (31) to an integer scoring matrices is afforded by multiplying the fractional scores by a common scale factor and subsequent rounding to the nearest integer values. Insofar as the score statistics discussed above are concerned, differently scaled variants of a given matrix give the same statistically significant hits, apart from possible small changes due to rounding. Thus, the scaling factor is a free parameter. The minimal significant score at the P significance level is given by

$$S_P = \frac{1}{\lambda}\{\ln MN + \ln K - \ln[-\ln(1-P)]\}. \qquad (34)$$

For purposes of comparisons among different substitution scoring matrices, the matrices should be scaled so that their associated λ values are approximately equal relative to some standard protein residue frequencies p_i, i.e., setting $p'_j = p_i$ in Eq. (19). A convenient choice is $\lambda = \ln(2)/2$; in this case an increase in one sequence length by a factor of 2 requires an added 2 units for a significant score. In Eq. (34) the term $\ln(K)$ is typically small compared to the other terms and not sensitive to the particular scoring scheme. For example, for sequences of lengths 300–700 residues, $\ln(MN) \approx 12$; typically $\ln(K) \approx -1$, and thus $S_{0.01} \approx 46$ and $S_{0.001} \approx 52$.

Characteristics of Substitution Scoring Matrices

Relative to some standard residue frequencies p_i, several quantities can be calculated for comparison of substitution scoring matrices. The expected score per substitution is calculated as

$$\mu = \sum_{i=1}^{20} \sum_{j=1}^{20} p_i p_j s_{ij}, \tag{35}$$

with variance

$$\delta = \sum_{i=1}^{20} \sum_{j=1}^{20} p_i p_j s_{ij}^2 - \mu^2. \tag{36}$$

The substitution frequencies expected in high-scoring segment pairs are given by the q_{ij} of Eq. (20), whence the expected fraction of identities in high-scoring matching segment pairs is calculated as

$$F_I = \sum_{i=1}^{20} q_{ii}. \tag{37}$$

More generally, the expected degree of conservation in high-scoring matching segment pairs may be defined as the expected fraction of positively scoring substitutions,

$$F_P = \sum\sum_{i,j:\, s_{ij}>0} q_{ij}. \tag{38}$$

The relative entropy,

$$H = \lambda \sum_{i=1}^{20} \sum_{j=1}^{20} q_{ij} s_{ij}, \tag{39}$$

is a quantity favored by Altschul (1991, 1993). Its meaning may be seen best by writing the expected length of a high-scoring segment pair at the significance level P as (Karlin, Dembo & Kawabata, 1990; Dembo & Karlin, 1991)

$$L_P = \frac{\ln MN + \ln K - \ln[-\ln(1-P)]}{H} = \frac{\lambda S_P}{H}, \tag{40}$$

where S_P is as defined in Eq. (34). Thus, matrices with the same relative entropy will tend to produce hits of the same length distribution.

C. Applications to Database Searches

From the foregoing considerations it is clear that the choice of amino acid substitution scoring matrix has a direct bearing on what kinds of hits one is to expect in a database search with a BLASTP type method. A matrix with large values for F_I and F_P and concomitant large H (e.g., low n PAMn matrices) will tend to identify only short segment pairs with high degree of similarity, whereas matrices with low values for F_I, F_P, and H (e.g., BLOSUM62) will tend to include long segment pairs with a much higher proportion of unfavorable substitutions. Vogt, Etzold & Argos (1995) give a comprehensive review of the performance of different amino acid substitution scoring matrices.

The statistical theory outlined above suggests a natural way of tuning a substitution scoring matrix to a specified value of $F_I = F$, for example. To this end, consider again substitution counts c_{ij} as defined in Section III.B. Let $g_{ii} = c_{ii}/\Sigma_{i=1}^{20} c_{ii}$ be the fraction of ii identities among the identities in the data collection, and let $h_{ij} = c_{ij}/C_{..} - \Sigma_{i=1}^{20} c_{ii}$ be the frequency of ij substitutions among the mismatches in the data collection. We now assume that these fractions represent the equilibrium frequencies of the various identity and mismatch types in a given alignment position, conditional upon the position being an identity or a mismatch, respectively. In other words, if an alignment position is known to be a match, the probability of it being an ii match will be equal to g_{ii}, whereas if the position is known to be a mismatch, the probability of it being an ij mismatch will be equal to h_{ij}. In this context it becomes clear that for a substitution scoring matrix to yield an expected fraction of identities in high-scoring segment pairs equal to F, the desired value of F may be interpreted as the a priori probability of an identity match in a high-scoring segment pair. Thus, the scores should be set to

$$s_{ii} = \ln\frac{Fg_{ii}}{f_i f_i}, \quad s_{ij} = \ln\frac{(1-F)h_{ij}}{f_i f_j} \quad (i \neq j). \tag{41}$$

Indeed, for such scores

$$\sum_{i=1}^{20}\sum_{j=1}^{20} f_i f_j e^{s_{ij}} = 1, \tag{42}$$

i.e., the scale parameter λ is 1, and

$$F_I = \sum_{i=1}^{20} f_i^2 e^{s_{ii}} = F. \tag{43}$$

Substitution scoring matrices derived according to above protocol have been termed **PIM** (**P**ercent **I**dentity **M**atrices), with a number label indicating the targeted F_I value (Brendel, Ladunga & Karlin, unpublished).

D. Applications to Alignments

For protein sequences that are considered homologous, as well as for protein segments deemed to represent the same structural motif, it is generally desirable to produce an alignment of the sequences. Such alignments will display the amino acids thought to derive from the same ancestral gene positions or believed to occupy equivalent positions in the three-dimensional structures. For alignments of two sequences at a time, the most commonly used methods are dynamic programming methods that find alignments with maximal score. The alignment score is calculated as the sum of substitution scores appropriate to alignment positions in which a residue from one sequence is paired with a residue from the second sequence, plus the sum of gap scores associated with insertions or deletions in one sequence with respect to the other (Needleman & Wunsch, 1970; Smith & Waterman, 1981). For example, consider the task of globally aligning the two sequences $\mathcal{A} = a_1 a_2 \ldots a_M$ and $\mathcal{B} = b_1 b_2 \ldots b_N$, where pairings $a_i b_j$ are assigned the substitution score $\sigma(a_i, b_j)$, internal gaps of length k are levied a negative gap penalty $w_k = -\alpha - \beta k$ ($\alpha, \beta \geq 0$), and end gaps are not weighted. Then the dynamic programming algorithm can be formulated by setting $S_{i0} = 0$, $i = 1, \ldots, M$, $S_{0j} = 0$, $j = 1, \ldots, N$, and

$$S_{ij} = \max \begin{cases} S_{i-1,j-1} + \sigma(a_i, b_j) \\ S_{i-k,j} + w_k & k = 1, \ldots, i \\ S_{i,j-l} + w_l & l = 1, \ldots, j \end{cases}. \tag{44}$$

The optimal alignment score is given by $S = \max\{S_{iM}, i = 1, \ldots, N-1, S_{Nj}, j = 1, \ldots, M\}$ (Smith & Waterman, 1981).

Whereas the assignment of substitution scores is well understood in terms of any of the approaches discussed above, the assignment of gap scores is problematic. The relative magnitude of a gap score compared to a mismatch bears heavily on which alignments score optimally. However, there is no statistical theory for the assignment of significance of scores obtained with allowance for variable numbers of gaps, and neither is there a generally accepted rationale for the choice of gap scores (e.g., Argos, Vingron & Vogt, 1991; Vingron & Waterman, 1994; Vogt, Etzold & Argos, 1995).

In the following I shall review a different method that circumvents the problem of gap score assignment by deriving alignments entirely from high-scoring segment pairs (HSSPs) that are identified according to the statistical theory discussed above. This method, termed SSPA for **S**ignificant **S**egment **P**air **A**lignment (Karlin, Weinstock & Brendel, 1995; see also Karlin, Mocarski & Schachtel, 1994), also produces an overall score for each alignment which can be used as a distance

Statistical Analysis of Protein Sequences

measure suitable for studying (evolutionary) relationships within protein families. There are three distinct steps in the method: identification of HSSPs, consistent maximal scoring ordering of the HSSPs, and calculation of SSPA values.

Identification of HSSPs

Two aligned segments corresponding to residues a_i to a_{i+L-1} in the first sequence and residues b_j to b_{j+L-1} in the second sequence are assigned the score $S = S(i, j, L) = \Sigma_{l=0}^{L-1} s(a_{i+l}, b_{j+l})$, where the $s(a_{i+l}, b_{j+l})$ are the appropriate substitution scores taken from a particular amino acid substitution scoring matrix. Segment pairs defined by all permissible starting points i and j and lengths L are evaluated. HSSPs are identified as distinct, nonredundant segment pairs exceeding the threshold score S_P given in Eq. (34) relative to the lengths M and N and residue frequencies p_i and p'_j of the given proteins. In most applications $P = 0.01$ is a good choice.

Algorithmically, the HSSPs are determined by means of a recursion that differs somewhat from the recursion underlying Eqs. (15). Let $l = 1, 2, \ldots$ be the index of successive residue pairs for a particular alignment offset of the two sequences (for example, pairing residue a_1 with residue b_{15} and continuing until one of the sequences terminates), and let s_l be the associated substitution score. In the algorithm l is increased and the substitution scores are cumulated until either the cumulative score becomes negative or it decreases by more than S_P from the current maximal cumulative score. In the latter case an HSSP is identified extending from the starting point of the recursion up to the point where the maximal cumulative score is attained. In either case cumulation of substitution scores starts over (recursion). Formally, let $E_0 = 0$, $\overline{E}_0 = S_P$; then the recursion is given by looping over l from $l = 1$ according to the script

$E_l = E_{l-1} + s_l$
if $\quad\quad E_l < \max\{\overline{E}_{l-1} - S_P, 0\}$ then
$\quad\quad$ if $\quad \overline{E}_{l-1} \geq S_P$: found HSSP with score $\overline{E}_{l-1}$ $\quad\quad\quad\quad$ (45)
$\quad\quad E_l = 0, \quad \overline{E}_l = S_P$
else
$\quad\quad \overline{E}_l = \max\{\overline{E}_{l-1}, E_l\}$
endif
next l

Note that this modified recursion does not necessarily identify the maximal scoring segment pairs. To wit, if two HSSPs with the same alignment offset are separated by a region scoring less than $-S_P$, then they are considered distinct HSSPs, although their combination including the negative scoring region may have a score exceeding either HSSP value. The rationale for this definition of HSSPs is that, for alignments, we are interested in highlighting the highly similar regions from the background of dissimilar (negatively scoring) regions. Current implementations of the algorithm for database searches (BLASTP, Altschul et al., 1990), on the other hand, rely on the statistics of the overall maximal scoring HSSP and thus employ the break criterion $E_l < 0$ instead of the criterion in line 2 of above script.

Consistent Maximal Scoring Ordering of High-Scoring Segment Pairs

The next step in the algorithm involves editing the list of HSSPs generated in the previous step. The goal is to identify a maximal subset of residue pairs (a_i, b_j) from the union of the HSSPs with the following properties: (i) each residue is part of at most one pair, and, if (a_i, b_j) is in the set, then a_{i+k}, $k > 0$, can only be matched with some residue $b_{j+k'}$ where $k' > 0$ (alignment consistency), (ii) if (a_i, b_j) and (a_{i+k}, b_{j+k}) are part of the same HSSP and both are in the set, then (a_{i+l}, b_{j+l}), $1, \ldots, k-1$, are also in the set (exception: an HSSP may be split into two pieces by another HSSP that is entirely contained in the former) (continuity), and (iii) the sum of the substitution scores corresponding to the residue pairs in the set is maximal with respect to all other sets satisfying (i) and (ii) (optimality). In case of overlaps between HSSPs of the original list, this editing will generate shortened pieces of HSSPs (Figure 2). If such shortened pieces of HSSPs end in negatively scoring substitutions, they should be shortened further to end in nonnegatively scoring substitutions. The resulting alignment score S_{AB} for two sequences A and B can be normalized in a variety of ways, as discussed in the next section.

Determination of SSPA Values

Let S_{AA} and S_{BB} be the self-scores comparing each protein sequence with itself. The **global** comparison values are obtained by normalizing with respect to the maximal and minimal self-scores, i.e.,

$$\Gamma_M = 100 \times \frac{S_{AB}}{\max\{S_{AA}, S_{BB}\}}, \quad \Gamma_m = 100 \times \frac{S_{AB}}{\min\{S_{AA}, S_{BB}\}}. \quad (46)$$

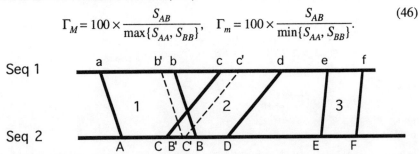

Figure 2. Illustration of significant segment pair alignment (SSPA). Sequences 1 and 2 have three high scoring segment pairs (HSSPs) as defined by the method of Karlin & Altschul (1990): residues a to b of Seq 1 align with residues A to B of Seq 2, and, similarly, residues c to d and e to f match up with residues C to D and E to F, respectively. The overlap of the first two HSSPs is resolved by shortening the segments to new end- and beginning points b', B' and c', C', respectively. The positioning of the new endpoints is determined by maximizing the sum of substitution scores for the two HSSPs combined.

G6PI_ECOLI versus G6PI_ZYMMO

```
 52  DYSKNRITEE TLAKLQDLAK ECDLAGAIKS MFSGEKINRT ENRAVLHVAL RNRSNTPILV
     D+SKN +  +  L   + L +   CD         K+ +F+GEKIN T E+RAV H+A  R +      +
 49  DFSKNHLDSQ KLTAFKKLLE ACDFDARRKA LFAGEKINIT EDRAVEHMAE RGQGAPASVA

112  DGKDVMPEVN AVLE       125
     K+ +       ++E
109  RAKEYHARMR TLIE       122

147  ITDVVNIGIG GSDLGPYMVT EALRPYKNHL NMHFVSNVDG THIAEVLKKV NPETTLFLVA
     +   +++IGIG GS LGP ++  +AL        ++ VSNVDG  + EV KK   NP   TL  VA
132  VKHLLHIGIG GSALGPKLLI DALTRESGRY DVAVVSNVDG QALEEVFKKF NPHKTLIAVA

207  SKTFTTQETM TNAHSARDWF LK         228
     SKTFTT ETM NA SA +W   K
192  SKTFTTAETM LNAESAMEWM KK         213

242  ALSTNAKAVG EFGIDTANMF EFWDWVGGRY SLWSAIGLSI VLSIGFDNFV ELLSGAHAMD
     AL+ N      E GID   +  F + +GGRY SLWS+IG    L++G++ F  +LL G  AMD
225  ALTANPAKAS EMGIDDTRIL PFAESIGGRY SLWSSIGFPA ALALGWEGFQ QLLEGGAAMD

302  KHFSTTPAEK NLPVLVALIG IWYNNFFGAE TEAILPYDQY MHRFAAYFQQ GNMESNGKYV
     +HF        EK N P+L A    +Y+    GA+ T  I   YD+    +   Y QQ    MESNGK V
285  RHFLEAAPEK NAPILAAFAD QYYSAVRGAQ THGIFAYDER LQLLPFYLQQ LEMESNGKRV

362  DRNGNVVDYQ TGPIIWGEPG TNGQHAFYQL IHQGTKMVPC DFIAPAITHN PLS        414
     D +GN++D+   +   I WG   G T+ QHA +QL +HQGT++VP  +FIA   +  L+
345  DLDGNLIDHP SAFITWGGVG TDAQHAVFQL LHQGTRLVPI EFIAAIKADD TLN        397

416  HHQKLLSNFF AQTEALAFGK SRE        438
     HH+ LL+N F AQ  AL G+    +
400  HHKTLLTNAF AQGAALMSGR DNK        422

459  PFKVFEGNRP TNSILLREIT PFSLGALIAL YEHKIFTQGV ILNIFTFDQW GVELGKQLAN
     P + +  G+RP + +IL+ E+   P   LGALIA  YEH+ FT GV +L I +FDQ+ GVELGK++A+
424  PARSYPGDRP STTILMEELR PAQLGALIAF YEHRTFTNGV LLGINSFDQF GVELGKEMAH

519  RI         520
     I
484  AI         485

Segment Pair 1:      52 to  125 vs    49 to  122; length:    74; score:    119
Segment Pair 2:     147 to  228 vs   132 to  213; length:    82; score:    199
Segment Pair 3:     242 to  414 vs   225 to  397; length:   173; score:    386
Segment Pair 4:     416 to  438 vs   400 to  422; length:    23; score:     53
Segment Pair 5:     459 to  520 vs   424 to  485; length:    62; score:    180
                                                                 ‾‾‾         ‾‾‾
                                                                 414         937

G6PI_ECOLI:    length=  549, selfscore=   2821
   matching region:   52 to   520 ( 469 aa, selfscore 2407)
G6PI_ZYMMO:    length=  507, selfscore=   2541
   matching region:   49 to   485 ( 437 aa, selfscore 2199)

SSPA values:    0.33/0.37    0.39/0.43    0.44/0.45
```

Figure 3. High-scoring segment pairs (HSSPs) and SSPA value determination for the glucose-6-phosphate-isomerase sequences of *Escherichia coli* and *Zymomonas mobilis*. The five HSSPs (after removal of overlaps) are displayed in their order from N- to C-terminus in the proteins. The aggregate score is 937. The SSPA values displayed are, in order, the global scores $\Gamma_M = 937/2821 = 0.33$, $\Gamma_m = 937/2541 = 0.37$, the local scores $\Lambda_M = 937/2407 = 0.39$, $\Lambda_m = 937/2199 = 0.43$, and the quality scores $\Phi_M = 0.44$ and $\Phi_m = 0.45$ (all values to be multiplied by 100; Eq. 46). The amino acid substitution scoring matrix used was BLOSUM62 (Henikoff & Henikoff, 1992).

Table 2. SSPA Γ_M/Γ_m Values Comparing Glucose-6-Phosphate-Isomerase (Meuroleukin) Sequences*

	ZYMMO	ECOLI	HAEIN	BACST G6PA	BACST G6PB	MYCGE	PLAFA	TRYBB	LEIME	HUMAN	YEAST	ARATH	ORYSA G6PA	CLALE G6P1	CLAUN chl
Length	507	549	563	449	445	433	591	607	605	558	553	560	567	569	548
Self-Score	2541	2821	2893	2261	2262	2193	3024	3068	3076	2876	2814	2821	2851	2881	2806
ZYMMO		33/37	32/36	6/6	6/6	4/5	23/28	29/35	28/34	31/35	33/36	24/27	24/27	24/27	33/36
ECOLI			77/79	5/6	6/7	4/6	35/37	54/59	55/60	63/64	57/57	42/42	40/41	40/41	89/89
HAEIN				4/5	5/7	4/5	35/37	52/55	55/59	64/64	56/57	42/43	40/41	41/42	76/78
BACST A					72/72	28/29	9/12	4/5	4/5	6/7	5/6	8/10	6/8	6/7	5/6
BACST B						28/29	9/12	5/7	4/5	4/5	7/8	8/10	8/10	8/10	6/8
MYCGE							2/3	3/4	3/5	3/4	6/8	7/10	8/10	6/8	4/5
PLAFA								32/33	32/33	32/34	32/35	45/48	43/46	44/46	35/38
TRYBB									71/72	53/57	49/53	38/41	38/41	38/41	54/59
LEIME										53/57	51/55	39/42	38/41	39/42	55/60
HUMAN											57/58	44/45	42/42	43/43	62/64
YEAST												41/41	39/39	38/39	58/59
ARATH													80/81	84/85	45/45
ORYSA A														79/80	41/41
CLALE 1															42/44

Note: *Sequences were taken from SWISS-PROT (Bairoch & Boeckmann, 1994). Only one representative is displayed from groups of very close sequences: human also represents mouse and pig; *Saccharomyces cerevisiae* (YEAST) also represents *Kluyveromyces lactis* (KLULA), and for both rice (ORYSA) and *Clarkia lewisii* (CLALE) only one isozyme each is included (cf. Fig. 4). The amino acid substitution scoring matrix used was BLOSUM62 (Henikoff & Henikoff, 1992).

Statistical Analysis of Protein Sequences

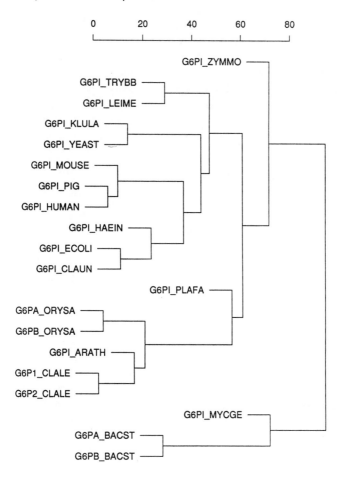

Figure 4. Hierarchical clustering (average linkage) of glucose-6-phosphate-isomerase sequences based on SSPA distances (defined as $100 - \Gamma_M$; e.g., the two *Bacillus stearothermophilus* (BACST) sequences have a Γ_M value of 72 and are clustered at distance 28).

The **local** comparison values Λ_M and Λ_m are obtained in the same way, but confining S_{AA} and S_{BB} to the sum of substitution scores from the first residue in the most N-terminal HSSP to the last residue in the most C-terminal HSSP (residues a to f in the upper sequence and residue A to F in the lower sequence of Figure 2). Lastly, the **quality** of the HSSPs alone can be assessed by the comparison values Φ_M and Φ_m, again obtained in the same way as the global comparison values, but now confining the S_{AA} and S_{BB} to the sum of substitution scores within the HSSPs only (residues a to b', c' to d, and e to f in the upper sequence, and residues A to B', C' to D, and E to F in the lower sequence of Figure 2). The comparison values (ranging from 0 to 100) may be interpreted as percentage similarities over the respective regions. Alternatively, the complement 100 minus the values defined above may be used as a distance measure. In some cases it is also useful to compare values of $S_{AB}/\min\{M, N\}$ (Karlin, Mocarski & Schachtel, 1994). These values depend more strongly on residue biases in the sequences (e.g., highly similar sequences rich in rare amino acids would generally score higher than highly similar sequences rich in abundant amino acids).

Applications

Figure 3 illustrates the SSPA method by comparing the glucose-6-phosphate-isomerase sequences from *Escherichia coli* and *Zymomonas mobilis*. The third HSSP shows some matching beyond the displayed endpoints, however the overlapping part of HSSP 4 scores better and thus is retained instead of the former. The different normalization schemes give rise to the indicated SSPA values, as explained above and in the legend. Table 2 gives the pairwise comparisons among glucose-6-phosphate-isomerase sequences from a wide spectrum of species. The relationships between the sequences are summarized in Figure 4. Of interest is the anomalous positioning of the *E. coli* and *Haemophilus* influenzae bacteria and the *Clarkia* chloroplast (CLAUN) sequences close to each other and close to the mammalian sequences, but clearly distinct from the other bacteria (ZYMMO, BACST, MYCGE). Smith & Doolittle (1992) interpreted these sequence similarities as a possible instance of horizontal transfer from a plant to an enteric bacterium, with the chloroplast enzyme having descended from a nuclear source. The positioning of the recently sequenced cytoplasmic enzymes of the plants *Arabidopsis*, *Clarkia*, and rice quite distant from all the other sequences confounds the picture. It is possible that the data reflect multiple transfers or a multigene family.

IV. MULTIPLE SEQUENCE COMPARISONS

Common measures of pairwise sequence similarity are not transitive: similarity between sequences $\mathcal{A}$ and $\mathcal{B}$ and between sequences $\mathcal{B}$ and C does not necessarily imply similarity of $\mathcal{A}$ and C. In particular, this nontransitivity holds with respect to detectable or pairwise significant similarity. On the other hand, while the pairwise similarities may all be weak, it may also be that these weak similarities all occur within the same protein segments. In that case, one would naturally assign greater significance to the observation than if many segments were involved depending on which two sequences are pairwise compared. Thus, the simultaneous comparison of multiple sequences affords a potentially more powerful means of detecting sequence relationships.

Generalizations of the alignment algorithms discussed above for two sequences are mathematically straightforward but generally computationally prohibitively expensive. For example, the dynamic programming algorithm of Eq. (44) is easily extended to the case of three sequences but would require operations numbering at least of the order of the third power of the average sequence length. Also, the assignment of gap penalties as well as the resolution of multiple alignments of optimal or near-optimal score are even more daunting problems than in the pairwise case. More practical are progressive alignment methods that build up a multiple alignment through successive pairwise alignments of groups of sequences (Feng & Doolittle, 1987; Corpet, 1988; Thompson, Higgins & Gibson, 1994). Pairwise alignment of groups of sequences is afforded by replacing the weight $\sigma(a_i, b_j)$ in recursions of the type Eq. (44) by a weight

$$\sigma(\vec{a}_i, \vec{b}_j) = \sum_{k=1}^{K} \sum_{l=1}^{L} \gamma_{kl}\, \sigma(a_{ik}, b_{jl}), \tag{47}$$

where a_{ik} is the residue of sequence k in the i-th position of the first group, b_{jl} is the residue of sequence l in the j-th position of the second group, and γ_{kl} is a suitable weight, e.g., $\gamma_{kl} = 1/(KL)$ for equal weighting of all sequences. The multiple alignment is obtained by first aligning two sequences, replacing any gaps in their alignment by a gap symbol (say, X), and then continuing according to some prespecified order (scoring 0 for any pairwise letter matching involving X). Published algorithms differ in how the order of progressive alignment is established.

In the following I shall discuss in greater detail approaches related to the identification and application of motifs that are characteristic for specific protein families. In the simplest case a motif may be described by a consensus sequence. This type of description may be appropriate for highly conserved motifs. A more general type of description may be by way of regular expressions that indicate the tolerated variability in a given position (e.g., [LIV] indicating the occurrence of either leucine, isoleucine, or valine at the given site). Still more general are weight matrices or profiles.

A. Description of Motifs as Profiles

A profile description of a motif is a simple generalization of a consensus. Assume the motif occupies W (not necessarily consecutive) positions in the aligned sequences. Then its **probability profile** is defined by the probabilities $\{p_{ij}\}$ where i varies from 1 to W, j indexes all possible letters, and p_{ij} gives the probability of observing residue j in position i of the motif. In the case where a multiple alignment of a training set of sequences over the motif segments is available, the probability profile is simply obtained by listing the observed letter frequencies in each motif position. The probability profile can be turned into a **scoring profile** by replacing the probabilities p_{ij} by suitable scores s_{ij}, for example, log-odds ratio scores $s_{ij} = \ln(p_{ij}/f_j)$ where the f_j are a standard set of residue frequencies. In this formulation special care needs to be exercised in the (to be expected) case of some zero p_{ij}'s. One solution would be to add a small increment to all probabilities and renormalize. An alternative approach is to use scores defined as $s_{ij} = \sum_{k=1}^{20} p_{ik}\sigma_{kj}$ where the σ_{kj} are the entries of some general log-odds ratio substitution scoring matrix.

B. Identification of Motifs

In many applications one is confronted with the task of finding a motif that is characteristic for a given set of sequences known or presumed to be related. Let the sequences be denoted by $S_1, S_2, \ldots, S_n$ of respective lengths $L_1, L_2, \ldots, L_n$. Recently statistical algorithms have been introduced in the field of sequence analysis for testing the hypothesis that several of the sequences (possibly all) contain a common motif that can be described by a profile of fixed size W. Because the presence and location of the motif is unknown, the problem is essentially one of selection and alignment: we wish to extract from the set of sequences those which contain a motif site, align these sequences so that equivalent motif positions match up, and then describe the motif as a profile. I discuss first an expectation maximization (EM) algorithm and then a more recent method based upon Gibbs sampling. Approaches based on hidden Markov models are similar, allowing more flexibility in terms of insertions or deletions at the cost of increased numbers of parameters (Krogh et al., 1993; Baldi et al., 1994).

Expectation Maximization

This algorithm was proposed by Lawrence and Reilly (1990) in the context of identification and characterization of protein binding sites in a set of unaligned DNA fragments. Here I follow a more general model (Brendel & Kleffe, unpublished). The following assumptions are made:

(i) The motif may or may not be present in a given sequence.
(ii) If present, the motif is embedded in a context with generally different compositional bias in the left context compared to the right-context.
(iii) Successive sequence positions are assumed to represent independent random variables following specified distributions.

Let the parameters of the model be denoted by P_{Lj} (probability of observing residue j in a position of the left context), P_{Rj} (probability of observing residue j in a position of the right context), P_{ij} (probability of observing residue j in the i-th motif position), and P_σ (probability of a sequence containing a motif). For the 20-letter amino acid alphabet and a motif spanning W positions, this amounts to $1 + (W + 2) \times 20$ parameters. Let the symbol θ represent a particular choice for all these parameters, and let S denote the set of sequences $S_1, \ldots, S_n$. Writing $g_\theta(S)$ for the likelihood of the observations S under the model specified by parameters θ, the goal is to determine parameters θ^* that maximize $g_\theta(S)$ over all possible choices of θ. The corresponding parameters P_{ij}^* form a probability profile for the most characteristic motif. The strength of the motif can be assessed by the increase in the likelihood after replacing an initial random choice of parameters by θ^*.

The problem, as formulated, is one of maximizing a function of many variables. This is a very difficult problem, and generally there are no practical algorithms to arrive deterministically at the global maximum. Application of the EM algorithm (Dempster, Laird & Rubin, 1977) in this context, however, allows finding at least local maxima of the likelihood function in an iterative procedure in which new parameters are estimated from fitting the sequences S to the model based on current parameters. For an initial choice of parameters $\theta = (\{P_{Lj}\}, \{P_{Rj}\}, \{P_{ij}\}, P_\sigma)$, the algorithm determines parameters $\theta' = (\{P'_{Lj}\}, \{P'_{Rj}\}, \{P'_{ij}\}, P'_\sigma)$ giving an equal or increased likelihood of the observations, i.e., $g_{\theta'}(S) \geq g_\theta(S)$. Subsequently replacing θ by θ' and iterating this cycle until the increase in likelihood becomes marginal determines a local maximum of the likelihood function. To locate possible different local maxima, this procedure is repeated in practice for a large number of initial choices of θ.

First consider a single sequence S_s of length L_s. This sequence may contain the motif starting in position k, where k is any of $1, 2, \ldots,$ or $L_s - W + 1$, or it may not contain the motif. In the latter case the sequence may be all right context (this possibility represented by $k = 0$) or it may be all left context (represented by $k = L_s - W + 2$). If the motif position k were known, the probability of S_s (conditional on k) would be obtained as

$$f_\theta(S_s|k) = \prod_j (P_{Lj}^{n(Lskj)} P_{Rj}^{n(Rskj)}) \prod_i P_{ij}^{n(skij)}, \qquad (48)$$

where $j = 1, \ldots, 20$, $i = 1, \ldots, W$, and $n(Lskj)$, $n(Rskj)$, and $n(skij)$ are the counts of residue j in left context, right context, and motif position i, respectively, for the given choice of k (note that in this formulation $n(skij)$ is 1 if the residue in the i-th motif position is of type j, and 0 otherwise). The joint probability for S_s and k is given by

$$h_\theta(S_s, k) = f_\theta(S_s|k) w_{s\theta}(k), \qquad (49)$$

where $w_{s\theta}(k)$ is the a priori probability of the motif occurring at position k in sequence S_s. For unaligned sequences we would typically specify

$$w_{s\theta}(k) = P_\sigma \frac{1}{L_s - W + 1} \quad (k = 1, \ldots, L_s - W + 1),$$

$$w_{s\theta}(0) = w_{s\theta}(L_s - W + 2) = (1 - P_\sigma)/2, \qquad (50)$$

where P_σ is the probability of the sequence containing a motif. With these specifications

$$g_\theta(S_s) = \sum_k h_\theta(S_s, k) \tag{51}$$

and

$$w_\theta(k|S_s) = \frac{h_\theta(S_s,k)}{g_\theta(S_s)} \tag{52}$$

are accessible. The appropriate choice of θ' will become obvious after writing Eq. (52) in the form

$$\log g_\theta(S_s) = \log h_\theta(S_s, k) - \log w_\theta(k|S_s) \tag{53}$$

and taking the expectation with respect to the $w_\theta(k|S_s)$:

$$\log g_\theta(S_s) = \sum_k w_\theta(k|S_s) \log h_\theta(S_s, k) - \sum_k w_\theta(k|S_s) \log w_\theta(k|S_s) \tag{54}$$

(the left-hand side of Eq. (53) is independent of k and therefore remains unchanged). Replacing θ with θ' in Eq. (53) and repeating the previous step, it is seen that

$$\log g_{\theta'}(S_s) - \log g_\theta(S_s) = \sum_k w_\theta(k|S_s) \log \frac{h_{\theta'}(S_s, k)}{h_\theta(S_s, k)} - \sum_k w_\theta(k|S_s) \log \frac{w_{\theta'}(k|S_s)}{w_\theta(k|S_s)}. \tag{55}$$

Here the term following the minus sign is nonpositive for all choices of θ' (subject to the constraint $\Sigma_k w_{\theta'}(k|S_s) = 1$). Thus the desired increase in the (log-)likelihood is attained if we can maximize $H(\theta') = \Sigma_k w_\theta(k|S_s) \log h_{\theta'}(S_s, k)$ with respect to θ'. This, however, is easily achieved by employing Eqs. (48) and (49) and collecting terms:

$$H(\theta') = \sum_j \sum_k w_\theta(k|S_s) n(Lskj) \log P'_{Lj} + \sum_j \sum_k w_\theta(k|S_s) n(Rskj) \log P'_{Rj}$$

$$+ \sum_i \sum_j \sum_k w_\theta(k|S_s) n(skij) \log P'_{ij} + \sum_k w_\theta(k|S_s) \log w_{s\theta'}(k). \tag{56}$$

Subject to the constraints $\Sigma_j P'_{Lj} = 1$, $\Sigma_j P'_{Rj} = 1$, $\Sigma_j P'_{ij} = 1$, and $\Sigma_k w_{s\theta'}(k) = 1$, $H(\theta')$ is maximized for the choice

$$P'_{Lj} = \frac{\sum_k w_\theta(k|S_s) n(Lskj)}{\sum_j \sum_k w_\theta(k|S_s) n(Lskj)}, \quad P'_{Rj} = \frac{\sum_k w_\theta(k|S_s) n(Rskj)}{\sum_j \sum_k w_\theta(k|S_s) n(Rskj)},$$

$$P'_{ij} = \frac{\sum_k w_\theta(k|S_s)n(skij)}{\sum_j \sum_k w_\theta(k|S_s)n(skij)}, \quad w_{s\theta'}(k) = w_\theta(k|S_s). \tag{57}$$

The generalization to $n > 1$ independent sequences follows immediately from

$$g_\theta(S) = \prod_s g_\theta(S_s), \tag{58}$$

which merely leads to replacing the sums over k in the previous equation by sums over s and k. For unaligned sequences with no prior information on motif occurrence and location, we would restrict $w_{s\theta'}(k)$ to be of the form Eq. (50) with parameter P'_σ and maximum likelihood estimate $P'_\sigma = 1 - \Sigma_s[w_\theta(0|S_s) + w_\theta(L_s - W + 2|S_s)]/n$.

Gibbs Sampling

The Gibbs sampling method of Lawrence et al. (1993) for the motif identification problem is viewed as a stochastic analog of the expectation maximization method. These authors assume a uniform motif context, i.e., $P_{Lj} = P_{Rj} = P_j$ for all j, and seek to maximize the loglikelihood ratio

$$\sum_{s=1}^{n} \sum_{i=1}^{W} \sum_{j=1}^{20} n(k_s ij) \log \frac{P_{ij}}{P_j} \tag{59}$$

over all sets of motif positions k_s, where $n(k_s ij)$ is the count of residue j in motif position i assuming the motif has starting position k_s in sequence S_s. Similarly to the EM algorithm, the P_{ij} and P_j are estimated from the observed residue frequencies in motif and background positions assuming the validity of the current motif positions. The recursion in the algorithm involves (1) selecting one of the sequences at random, (2) estimating the parameters from the current motif positions in the remaining sequences, and (3) selecting at random a new motif position in the sequence selected in step 1 with sampling probabilities proportional to the loglikelihood ratios of motif over background probability calculated for all possible motif positions. The algorithm is seeded with randomly selected motif positions in all sequences. After some of the motif positions have been selected "correctly" (in the sense of representing a meaningful local alignment) by chance, their corresponding pattern will be reflected in the parameter estimates and will thus tend to pick out matching positions in the other sequences.

Applications

Figure 5 gives the best local matching for a window of size 18 in a set of 20 Fos/jun DNA-binding proteins with an added five helix-turn-helix DNA-binding

```
ATF3_HUMAN    131  KRRRERNKIAAAKCRNKK   148
FRA2_CHICK    128  RIRRERNKLAAAKCRNRR   145
FOS_CHICK     140  RIRRERNKMAAAKCRNRR   157
FOSB_MOUSE    159  RVRRERNKLAAAKCRNRR   176
FOS_AVINK      95  RIRRERNKMAAAKCRNRR   112
AP1_COTJA     238  ERKRMRNRIAASKCRKRK   255
FOSX_MSVFR    117  RIRRERNKMAAAKCRNRR   134
BZLF_EBV      176  EIKRYKNRVASRKCRAKF   193
HBPA_WHEAT    256  QKRKLSNRESARRSRLRK   273
FOS_HUMAN     141  RIRRERNKMAAAKCRNRR   158
JUNB_HUMAN    272  ERKRLRNRLAATKCRKRK   289
TJUN_AVIS1    221  ERKRMRNRIAASKSRKRK   238
JUND_HUMAN    272  ERKRLRNRIAASKCRKRK   289
OP2_MAIZE     229  VRKKESNRESARRSRYRK   246
JUND_MOUSE    266  ERKRLRNRIAASKCRKRK   283
AP1_DROME     216  ERKRQRNRVAASKCRKRK   233
AP1_HUMAN     256  ERKRMRNRIAASKCRKRK   273
CYS3_NEUCR    103  EDKRKRNTAASARFRIKK   120
FRA1_RAT      111  RVRRERNKLAAAKCRNRR   128
JUND_CHICK    246  ERKRLRNRIAASKCRKRK   263
RPSB_BACSU    110  RIKELGPRIKMAVDQLTT   127
RPSK_BACSU    112  IEKRALMKMFHEFYRAEK   129
NAHR_PSEPU    162  FQRRLLQNHYVCLCRKDH   179
HMAN_DROME    342  KIWFQNRRMKWKKENKTK   359
NTRC_BRASR     68  RIKKMRPNLPVIVMSAQN    85
```

Figure 5. Local sequence matching found by the one-sided background EM algorithm for a library of 20 Fos/jun DNA-binding proteins and five helix-turn-helix DNA-binding proteins (*cf.* PROSITE database; Bairoch & Bucher, 1994). Sequences were taken from SWISS-PROT (Bairoch & Boeckmann, 1994).

```
ATF3_HUMAN    at pos 131:  KRRRERNKIAAAKCRNKK   35.19
FRA2_CHICK    at pos 128:  RIRRERNKLAAAKCRNRR   35.28
FOS_CHICK     at pos 140:  RIRRERNKMAAAKCRNRR   35.50
FOSB_MOUSE    at pos 159:  RVRRERNKLAAAKCRNRR   35.21
FOS_AVINK     at pos  95:  RIRRERNKMAAAKCRNRR   35.50
AP1_COTJA     at pos 238:  ERKRMRNRIAASKCRKRK   37.58
FOSX_MSVFR    at pos 117:  RIRRERNKMAAAKCRNRR   35.50
BZLF_EBV      at pos 176:  EIKRYKNRVASRKCRAKF   23.19
HBPA_WHEAT    at pos 256:  QKRKLSNRESARRSRLRK   17.25
FOS_HUMAN     at pos 141:  RIRRERNKMAAAKCRNRR   35.50
JUNB_HUMAN    at pos 272:  ERKRLRNRLAATKCRKRK   36.05
TJUN_AVIS1    at pos 221:  ERKRMRNRIAASKSRKRK   33.60
JUND_HUMAN    at pos 272:  ERKRLRNRIAASKCRKRK   37.08
OP2_MAIZE     at pos 229:  VRKKESNRESARRSRYRK   16.54
JUND_MOUSE    at pos 266:  ERKRLRNRIAASKCRKRK   37.08
AP1_DROME     at pos 216:  ERKRQRNRVAASKCRKRK   37.39
AP1_HUMAN     at pos 256:  ERKRMRNRIAASKCRKRK   37.58
CYS3_NEUCR    at pos 103:  EDKRKRNTAASARFRIKK   17.80
FRA1_RAT      at pos 111:  RVRRERNKLAAAKCRNRR   35.21
JUND_CHICK    at pos 246:  ERKRLRNRIAASKCRKRK   37.08

RPSB_BACSU    at pos 117:  RIKMAVDQLTTETQRSPK    1.32
RPSK_BACSU    at pos 112:  IEKRALMKMFHEFYRAEK    0.16
NAHR_PSEPU    at pos 162:  FQRRLLQNHYVCLCRKDH    1.41
HMAN_DROME    at pos 322:  LTRRRRIEIAHALCLTER    6.04
NTRC_BRASR    at pos  68:  RIKKMRPNLPVIVMSAQN    1.81
```

Figure 6. The best matches in a profile scanning using the EM algorithm derived parameters are weighted BLOSUM62 scores as applied to the original library of Fos/jun (first 20) and helix-turn-helix (last five) DNA-binding proteins. It is seen that the latter set is clearly separated from the former.

proteins as determined by the application of the described EM algorithm (left and right context parameters set equal in this case). The region found in the Fos/jun proteins is the well known basic motif preceding the leucine zipper domain (cf. PROSITE database of motifs, Bairoch & Bucher, 1994). The helix-turn-helix proteins are also matched up, but for the given ordering in the display it is quite clear that their matching to the motif is very weak. Figure 6 provides a quantitative verification of this impression. Shown are the results of screening the original sequences with the EM algorithm-derived motif profile, scoring by weighted BLOSUM62 substitution scores as described in Section IV.A. Here the two groups of sequences are clearly separated. This technique has been applied widely in motif matching algorithms (e.g., Gribskov, 1994; Lüthy, Xenarios & Bucher, 1994).

V. PERSPECTIVE

The analysis of biomolecular sequences will remain a very active field of research well into the next century. Statistical and computational methods have a firm place in all aspects of organizing and evaluating the sequence data. Some new methods will become practical only with increased computing power and increased data sets. But, even in the current databases and with current methods, I suspect that there are many more discoveries to be made.

ACKNOWLEDGMENTS

Samuel Karlin has been at the source of most of the ideas, methods, and applications presented in this review. He has been an inspirational mentor, and in many ways, this review reflects what I hope to have learned from him. I am also indebted to my colleagues Edwin Blaisdell, Luciano Brocchieri, Chris Burge, Jürgen Kleffe, and Istvan Ladunga who contributed through collaborations and discussions. This work has been supported in part by NIH Grant 5R01HG00335-07.

REFERENCES

Altschul, S. F. (1991). Amino acid substitution matrices from an information theoretic perspective. J. Mol. Biol. 219, 555–565.
Altschul, S. F. (1993). A protein alignment scoring system sensitive at all evolutionary distances. J. Mol. Evol. 36, 290–300.
Altschul, S. F., Gish, W., Miller, W., Myers, E. W., & Lipman, D. J. (1990). Basic local alignment search tool. J. Mol. Biol. 215, 403–410.
Altschul, S. F., Boguski, M. S., Gish, W., & Wootton, J. C. (1994). Issues in searching molecular sequence databases. Nature Genet. 6, 119–129.
Argos, P., Vingron, M., & Vogt, G. (1991). Protein sequence comparisons: methods and significance. Protein Eng. 4, 375–383.
Bairoch, A. & Boeckmann, B. (1994). The SWISS-PROT protein sequence data bank: current status. Nucl. Acids Res. 22, 3578–3580.
Bairoch, A. & Bucher, P. (1994). PROSITE: recent developments. Nucleic Acids Res. 22, 3583–3589.

Baldi, P., Chauvin, Y., Hunkapiller, T., & McClure, M. A. (1994). Hidden Markov models of biological primary sequence information. Proc. Natl. Acad. Sci. USA 91, 1059–1063.

Berman A. L., Kolker, E., & Trifonov, E. N. (1994). Underlying order in protein sequence organization. Proc. Natl. Acad. Sci. USA 91, 4044–4047.

Brendel, V., P. Bucher, I.R. Nourbakhsh, B.E. Blaisdell & S. Karlin (1992). Methods and algorithms for statistical analysis of protein sequences. Proc. Natl. Acad. Sci. USA 89, 2002–2006.

Brendel, V., Dohlman, J., Blaisdell, B. E., & Karlin, S. (1991). Very long charge runs in systemic lupus erythematosus associated autoantigens. Proc. Natl. Acad. Sci. USA 88, 1536–1540.

Brendel, V. & Karlin, S. (1989a). Association of charge clusters with functional domains of cellular transcription factors. Proc. Natl. Acad. Sci. USA 86, 5698–5702.

Brendel, V. & Karlin, S. (1989b). Too many leucine zippers? Nature 341, 574–575.

Cohen, F. E. & Kuntz, I. D. (1989). Tertiary structure prediction. In: Prediction of Protein Structure and the Principles of Protein Conformation (Fasman, G. D., Ed.), pp. 647–705, Plenum, New York.

Corpet, F. (1988). Multiple sequence alignment with hierarchical clustering. Nucl. Acids Res. 16, 10881–10890.

Dayhoff, M. O., Schwartz, R. M., & Orcutt, B. C. (1978). A model of evolutionary change in proteins. In: Atlas of Protein Sequence and Structure (Dayhoff, M.O., Ed.). Vol. 5, Suppl. 3, pp. 345–352, Nat. Biomed. Res. Found., Washington, DC.

Dembo, A. & Karlin, S. (1991). Strong limit theorems of empirical functionals for large exceedences of partial sums of i.i.d. variables. Ann. Prob. 19, 1737–1755.

Dembo, A. & Karlin, S. (1992). Poisson approximations for r-scan processes. Ann. Appl. Prob. 2, 329–357.

Dembo, A. & Karlin, S. (1993). Central limit theorems of partial sums for large segmental values. Stoch. Process. Appl. 45, 259–271.

Dembo, A., Karlin, S., & Zeitouni, O. (1994a). Critical phenomena for sequence matching with scoring. Ann. Prob. 22, 1993–2021.

Dembo, A., Karlin, S., & Zeitouni, O. (1994b). Limit distribution of maximal nonaligned two-sequence segmental score. Ann. Prob. 22, 2022–2039.

Dempster, A. P., Laird, N. M., & Rubin, D. B. (1977). Maximum likelihood from incomplete data via the EM algorithm. J. Royal Stat. Soc. Series B 39, 1–38.

Feller, W. (1968). An Introduction to Probability Theory and its Applications, 3rd edition, Vol. I, Wiley, New York.

Felsenstein, J. (1988). Phylogenies from molecular sequences: inference and reliability. Ann. Rev. Genet. 22, 521–565.

Feng, D.-F. & Doolittle, R. F. (1987). Progressive sequence alignment as a prerequisite to correct phylogenetic trees. J. Mol. Evol. 25, 351–360.

Gilbert, W. (1991). Towards a paradigm shift in biology. Nature 349, 99.

Gribskov, M. (1994). Profile analysis. Methods Mol. Biol. 25, 247–266.

Henikoff, S. & Henikoff, J. G. (1992). Amino acid substitution matrices from protein blocks. Proc. Natl. Acad. Sci. USA 89, 10915–10919.

Hopp, T. P. & Woods, K. R. (1981). Prediction of antigenic determinants from amino acid sequences. Proc. Natl. Acad. Sci. USA 78, 3824–3828.

Johnson, M. S. & Overington, J. P. (1993). A structural basis for sequence comparisons. An evaluation of scoring methodologies. J. Mol. Biol. 233, 716–738.

Karlin, S. (1993). Unusual charge configurations in transcription factors of the basic RNA polymerase II initiation complex. Proc. Natl. Acad. Sci. USA 90, 5593–5597.

Karlin, S. (1994). Statistical studies of biomolecular sequences: score-based methods. Phil. Trans. R. Soc. Lond. B 344, 391–402.

Karlin, S. (1995). Statistical significance of sequence patterns in proteins. Curr. Opin. Struct. Biol. 5, 360–371.

Karlin, S. & Altschul, S. F. (1990). Methods for assessing the statistical significance of molecular sequence features by using general scoring schemes. Proc. Natl. Acad. Sci. USA 87, 2264–2268.

Karlin, S., Blaisdell, B. E., & Brendel, V. (1990). Identification of significant sequence patterns in proteins. Methods Enzymol. 183, 388–402.

Karlin, S., Blaisdell, B. E., & Bucher, P. (1992). Quantile distributions of amino acid usage in protein classes. Protein Eng. 5, 729

Karlin, S., Blaisdell, B. E., Mocarski, E. S., & Brendel, V. (1989). A method to identify distinctive charge configurations in protein sequences with application to human herpesvirus polypeptides. J. Mol. Biol. 205, 165–177.

Karlin, S. & Brendel, V. (1990). Charge configurations in oncogene products and transforming proteins. Oncogene 5, 85–95.

Karlin, S. & Brendel, V. (1992). Chance and statistical significance in protein and DNA sequence analysis. Science 257, 39–49.

Karlin, S. & Bucher, P. (1992). Correlation analysis of amino acid usage in protein classes. Proc. Natl. Acad. Sci. USA 89, 12165–12169.

Karlin, S., Bucher, P., Brendel, V., & Altschul, S. F. (1991). Statistical methods and insights for protein and DNA sequences. Annu. Rev. Biophys. Biophys. Chem. 20, 175–203.

Karlin, S. & Cardon, L. R. (1994). Computational DNA sequence analysis. Annu. Rev. Microbiol. 48, 619–654.

Karlin, S. & Dembo, A. (1992). Limit distributions of maximal segmental score among Markov-dependent partial sums. Adv. Appl. Prob. 24, 113–140.

Karlin, S., Dembo, A., & Kawabata, T. (1990). Statistical composition of high-scoring segments from molecular sequences. Ann. Stat. 18, 571–581.

Karlin, S. & Macken, C. (1991). Some statistical problems in the assessment of inhomogeneities of DNA sequence data. J. Amer. Statist. Assoc. 86, 27–35.

Karlin, S., Mocarski, E. S., & Schachtel, G. A. (1994). Molecular evolution of herpesviruses: Genomic and protein sequence comparisons. J. Virol. 68, 1886–1902.

Karlin, S., Ost, F., & Blaisdell, B. E. (1989). Patterns in DNA and amino acid sequences and their statistical significance. In: Mathematical Methods for DNA Sequences (Waterman, M., Ed.), pp. 133–157, CRC, Boca Raton, FL.

Karlin, S., Weinstock, G., & Brendel, V. (1995). Bacterial classifications derived from RecA protein sequence comparisons. J. Bacteriol. 177, 6881–6893.

Karlin, S., Zuker, M., & Brocchieri, L. (1994). Measuring residue associations in protein structures. Possible implications for protein folding. J. Mol. Biol. 239, 227–248.

Kolker, E. & Trifonov, E. N. (1995). Periodic recurrence of methionines: fossil of gene fusion? Proc. Natl. Acad. Sci. USA 92, 557–560.

Krogh, A., Brown, M., Mian, I. S., Sjolander, K., & Haussler, D. (1994). Hidden Markov models in computational biology. Applications to protein modeling. J. Mol. Biol. 235, 1501–1531.

Kyte, J. & Doolittle, R. F. (1982). A simple method for displaying the hydropathic character of a protein. J. Mol. Biol. 157, 105–132.

Lawrence, C. E., Altschul, S. F., Boguski, M. S., Liu, J. S., Neuwald, A. F., & Wootton, J. C. (1993). Detecting subtle sequence signals: a Gibbs sampling strategy for multiple alignment. Science 262, 208–214.

Lawrence, C. E. & Reilly, A. A. (1990). An expectation maximization (EM) algorithm for the identification and characterization of common sites in unaligned biopolymer sequences. Proteins 7, 41–51.

Leung, M.-Y., Blaisdell, B. E., Burge, C., & Karlin, S. (1991). An efficient algorithm for identifying matches with errors in multiple long molecular sequences. J. Mol. Biol. 221, 1367–1378.

Lüthy, R., Xenarios, I., & Bucher, P. (1994). Improving the sensitivity of the sequence profile method. Protein Sci. 3. 139–146.

Miyazawa, S. & Jernigan, R. L. (1993). A new substitution matrix for protein sequence searches based on contact frequencies in protein structures. Protein Eng. 6, 267–278.

Morris, M., Schachtel, G., & Karlin, S. (1993). Exact formulas for multitype run statistics in a random ordering. SIAM J. Discrete Math. 6, 70–86.

Needleman, S. B. & Wunsch, C. (1970). A general method applicable to the search for similarities in the amino acid sequences of two proteins. J. Mol. Biol. 48, 443–453.

Risler, J. L., Delorme, M. O., Delacroix, H., & Henaut, A. (1988). Amino acid substitutions in structurally related proteins. A pattern recognition approach. Determination of a new and efficient scoring matrix. J. Mol. Biol. 204, 1019–1029.

Rogers, S., Wells, R., & Rechsteiner, M. (1986). Amino acid sequences common to rapidly degraded proteins: the PEST hypothesis. Science 234, 364–368.

Rost, B., Sander, C., & Schneider, R. (1994). Redefining the goals of protein secondary structure prediction. J. Mol. Biol. 235, 13–26.

Sapolsky, R. J., Brendel, V., & Karlin, S. (1993). A comparative analysis of distinctive features of yeast protein sequences. Yeast 9, 1287–1298.

Smith, M. W. & Doolittle, R. F. (1992). Anomalous phylogeny involving the enzyme glucose-6-phosphate isomerase. J. Mol. Evol. 34, 544–545.

Smith, T. F. & Waterman, M. S. (1981). Identification of common molecular subsequences. J. Mol. Biol. 147, 195–197.

Thompson, J. D., Higgins, D. G., & Gibson, T. J. (1994). CLUSTAL W: improving the sensitivity of progressive multiple sequence alignment through sequence weighting, position-specific gap penalties and weight matrix choice. Nucl. Acids Res. 22, 4673–4680.

Trifonov, E. N. (1994). On the recombinational origin of protein-sequence-subunit structure. J. Mol. Evol. 38, 543–546.

Vingron, M. & Waterman, M.S. (1994). Sequence alignment and penalty choice. Review of concepts, case studies and implications. J. Mol. Biol. 235, 1–12.

Vogt, G., Etzold, T. & Argos, P. (1995). An assessment of amino acid exchange matrices in aligning protein sequences: the twilight zone revisited. J. Mol. Biol. 249, 816–831.

White, S. H. (1994a). The evolution of proteins from random amino acid sequences: II. Evidence from the statistical distributions of the lengths of modern protein sequences. J. Mol. Evol. 38, 383–394.

White, S. H. (1994b). Global statistics of protein sequences: implications for the origin, evolution, and prediction of structure. Annul Rev. Biophys. Biomol. Struct. 23, 407–439.

White, S. H. & Jacobs, R. R. (1994). The evolution of proteins from random amino acid sequences. I. Evidence from the lengthwise distribution of amino acids in modern protein sequences. J. Mol. Evol. 36, 79–95.

PROGRESS IN LARGE-SCALE SEQUENCE ANALYSIS

Jean-Michel Claverie

I.	Introduction	162
	A. Sequences	162
	B. Databases	163
	C. ... and Problems	164
	D. The Query Side	165
	E. The General Constraints of Large-Scale Sequence Analysis	166
II.	The Concept of Query Masking	168
	A. The "Overwhelming Output" Problem	168
	B. Low-Entropy Matches	168
	C. Artifactual Matches	172
	D. Query Masking: A Practical Solution to Output Inflation	174
	E. XNU: Finding and Masking Internal Repeats	175
	F. XBLAST: Finding and Masking "Junk" Sequences	177
	G. Query Masking: Conclusion	179
III.	Identifying Exons in Mammalian Genomic Sequences	180
	A. The Problem	180
	B. Grail: Finding Coding Exons from Their "Word" Usage	181
	C. Database "Look Up": An Alternative Method for Finding Exons	184
	D. The Grail/Look Up Composite Approach	187

Advances in Computational Biology
Volume 2, pages 161–208
Copyright © 1996 by JAI Press Inc.
All rights of reproduction in any form reserved.
ISBN: 1-55938-979-6

IV. Statistical Significance in Sequence Motif Searching 188
   A. The Problem 188
   B. Mathematical Representation of Sequence Motifs 189
   C. C, P and M Modes: Three Solutions to the Missing Data Problem 191
   D. Statistical Significance of Motif Scores 195
   E. Motif Vs. Blast Search: The Whole Vs. the Sum of Its Parts 199
   Acknowledgments 202
   References 202

I. INTRODUCTION

A. Sequences ....

A growing number of laboratories throughout the world are actively engaged in the systematic sequencing of the genomes of a wide range of organisms: gram-negative bacteria *Escherichia coli* (Fujita et al., 1994; Blattner et al., 1993), gram-positive *Bacillus subtilis* (Kunst and Devine, 1991), *Mycobacterium leprae* (Honore et al., 1993; Smith, 1994), yeast *Saccharomyces cerevisiae* (Dujon et al., 1994; Oliver et al., 1992), nematode *Caenorhabditis elegans* (Wilson et al., 1994; Sulston et al., 1992), mammals like humans (Martin-Gallardo et al., 1992; Koop et al., 1994; Slightom et al., 1994; Chissoe et al., 1994), and mouse (Koop et al., 1992). The complete genomic sequences of many viruses have also been determined, notably those of large DNA viruses like cytomegalovirus (Chee et al., 1990), Vaccinia (Goebel et al., 1990) and Variola viruses (Shchelkunov et al., 1993; Massung et al., 1993), Epstein–Barr virus (Baer et al., 1984), other herpesviruses (McGeoch et al., 1988; Telford et al., 1992; Albrecht et al., 1992) and one baculovirus (Ayres et al., 1994). In the meantime, the positional cloning approach (i.e., on the basis of chromosomal location, prior knowledge about a defective protein; see Monaco (1994) for a recent review) to the identification of human disease genes is contributing large human genomic sequences such as a 180-kb region within the retinoblastoma susceptibility gene (Toguchida et al., 1993), a 101-kb region within the neurofibromatosis type 1 gene (Cawthon et al., 1990) or a 62-kb region encompassing the fragile X syndrome gene (Verkerk et al., 1991; Richards et al., 1994).

In a complementary approach, expressed sequences tags or ESTs (Adams et al., 1991), i.e., short (300–400 nucleotide) single-pass partial cDNA sequences from a number of organisms and tissues, have been rapidly accumulated by various laboratories (Sikela and Auffray, 1993). The main projects includes human (Adams et al., 1992; Okubo et al., 1992; Khan et al., 1992; Grausz and Auffray, 1993; Adams et al., 1993 a & b), *Caenorhabditis elegans* (Waterston et al., 1992; McCombie et al., 1992), *Arabidopsis thaliana* (Newman, 1993; Desprez et al., 1993), *Oriza sativa* (rice) (Yuzo and Takuji, 1993), mouse (Davies et al., 1994), and *Plasmodium falciparum* (Chakrabarti et al., 1994). A similar approach has been applied to mung bean nuclease fragments (Reddy et al., 1993) from a *Plasmodium falciparum* genomic library.

B. Databases ....

Altogether, these massive genomic and EST sequencing efforts are contributing approximately 100 millions nucleotides a year to the cooperative EMBL/GenBank/DDBJ database system (Rice et al., 1994). The volume of these databases has been doubling in size every 18 month since 1987. As of December 1994, GenBank (Benson et al., 1994) release 86.0 contained 237,775 entries representing over 230,485,000 nucleotides. Compared to other factual scientific databases, GenBank (as EMBL and DDBJ) is of a respectable size, but is specially unique for its extreme heterogeneity, both in terms of size and quality of the individual entries. The three longest sequences presently found in GenBank are a 685,973-nucleotide long region within the human T-cell receptor locus (accession: L36092; Rowen et al., 1994; Slightom et al., 1994), a 338,534-nucleotide long region (92.8 to 00.1 minutes) from the *E. coli* genome (accession: U14003; Burland et al., 1994) and the 315,338 nucleotides of yeast chromosome III (accession: X59720, Oliver et al., 1992); including those, 32 entries correspond to sequences longer than 100,000 nucleotides, and 302 entries to sequences longer than 30,000 nucleotides. On the other side of the spectrum, three entries in GenBank describe one-nucleotide sequences (accession: M81653, M81649 & I02850), and more than 4100 entries correspond to sequences of less than 20 nucleotides (mostly derived from patent applications). Those anomalies reflect the editorial policy adopted by primary sequence databases that any voluntary submission, no matter how bad, should be accepted. However, some of those very short entries do actually have a biological *raison d'être* such as a 3-nucleotide exon in some NCAM transcripts (Prediger et al., 1988). In quality, the a priori accuracy of the sequences described may also vary tremendously, from the theoretical 10^{-6} obtained through highly redundant genomic sequencing, to about 1% in the case of single-pass automated sequencing of partial cDNAs. According to an independent estimation (Kristensen et al., 1992) the average accuracy of the sequences in EMBL/GenBank might be a low 3%.

GenBank (or the DDBJ and EMBL databases) is also extremely redundant, as multiple overlapping fragments of the same region (or close homologous sequences from various alleles, bacterial strains or virus isolates) coexist in the database (as a consequence of the editorial policy "everything in, nothing out"). Finally, a large number of entries (10% or more) are potentially misleading due to a variety of reasons: sequencing errors (Claverie, 1986; Posfai and Roberts, 1992; Claverie, 1993), uncertainty on the origin of the DNA (Savakis and Doelz, 1993), and erroneous and/or undocumented features in the sequences, such as vector sequences (Lopez et al., 1992; Lamperti et al., 1992), coding regions, introns (Burglin and Barnes, 1992), and splice sites (Senapathy et al., 1990).

C. ... and Problems

The size, heterogeneity and high signal-to-noise ratio of primary nucleotide databases such as GenBank have rendered them progressively useless for sensitive sequence comparison and/or rigorous statistical analysis. Direct scanning of GenBank has become only useful as a reference tool, mainly to look for occurrences of (nearly) identical segments long enough to be of statistical significance (e.g., matching 24 nucleotides from a 25-nucleotide query), a specific task for which the BLASTN program is well suited (Altschul et al., 1990). Even in that limited context, unexpectedly strong matches have to be carefully checked for the eventual involvement of common cloning vector sequences, poly-linkers, or frequent but unnoticed contamination and/or artifactual recombination with mitochondrial DNA or ribosomal RNAs.

Progress toward more usable information resources has been the creation of specialized (secondary) databases, sometimes adding mapping and functional information to the sequences. By construction, those databases are more homogeneous, smaller in size, and usually better annotated and quality-controlled by experts in their respective fields. There are several dozens of them, in various states of maintenance. Among these thematic databases some are specific for an organism, such as *E. coli* (Rudd, 1993; Mount and Schatz, 1994), *Caenorhabditis elegans* (Durbin and Thierry-Mieg, 1991), *Arabidopsis* (Cherry et al., 1992), or yeast (Linder et al., 1993). Others gather information on a given type of molecule such as ribosomal RNAs (16S: Neefs et al., 1993; Gutell, 1993; 23S: Gutell et al., 1993), snRNA (Shumyatsky and Reddy, 1993), a type of sequence segment (human genomic repeats: Jurka et al., 1992; Claverie and Makalowski, 1994; membrane spanning protein segments: Hofmann and Stoffel, 1993), or synthetic oligonucleotides used as molecular probes (Romano et al., 1993). There are also databases of specific biological functions such as transcription (Ghosh, 1993) or G-protein-coupled receptors (Kolakowski, in press). Protein motifs and "functional signatures" are collected in two useful databases: PROSITE (Bairoch and Bucher, 1994) and BLOCKS (Henikoff and Henikoff, 1994). Finally, some specialized databases have been designed to help molecular evolution studies (HOVERGEN, Duret et al., 1994).

A useful compromise between specialized databases and the EMBL/GenBank/DDBJ repository are the protein sequence databases Swiss-Prot (Bairoch and Boeckmann, 1994) and PIR (Barker et al., 1993). Although most of protein sequences are now derived from the conceptual translation of genomic or cDNA sequences, there are many advantages in working at the amino acid level. Protein databases are smaller in size (e.g., 14,147,368 amino acids, for 40,292 sequences in Swiss-Prot 40.0), less redundant and, in particular for Swiss-Prot, much better annotated. Similarity searches will then run faster and provide more readily interpretable outputs. Due to the degeneracy of the genetic code, the function and structure of a protein is more directly related to its sequence of amino acids than to the encoding nucleotide sequence. As a consequence, the conceptual translations of genes encoding structurally or functionally related proteins are much more conserved than are the DNA sequences. Thus, amino acid sequence similarities are still detectable long after any significant trace of nucleotide sequence homology has been lost: for instance, proteins with 50% of identical residues (a strong evidence of similar three-dimensional structure and function) can be encoded by nucleotide sequences less than 30% identical, a percentage that cannot be reliably distinguished from the random resemblance expected between unrelated DNA segments (i.e., 25%). This is the conservation of amino acid sequence motifs that enabled us to recognize clearly the relationships between protein-encoding genes from organisms separated by billions years of evolution such as bacteria, yeast, plants or vertebrates (Green et al., 1993; Claverie, 1993). In section III.C. we will show how to locate coding exons using the conceptual translations of large genomic sequences

Finally, another useful secondary database, dbEST, is the separate collection of all the ESTs (more than 100,000 entries) submitted to the EMBL/GenBank/DDBJ consortium (Boguski et al., 1993). Despite their small size and poor accuracy, ESTs have already proven useful in identifying important human disease genes (Boguski et al., 1994), as well as discovering unexpected genes homologous between organisms as distant as yeast and human (Tugendreich et al., 1993). For the same reasons as above, EST sequences are best used in the form of their six putative translations.

D. The Query Side

Sequencing machines of the current generation are easily capable of delivering 30kb of raw sequences per day. Technological and cost improvements will no doubt contribute to further accelerate this rate. Thus, an active sequencing laboratory (such as those involved in highly competitive gene hunts), might need to query all databases, with 30,000 nucleotides of new sequences every day. In laboratories involved in truly massive sequencing projects and using 3, 5 or 10 sequencers, 100,000 to 300,000 nucleotides have to be analyzed daily. When those laboratories reach the final stage of their positional cloning projects, large contigs are eventually to be assembled and the next crucial problem is rapidly identifying one or more exons within the anonymous contigs. Recent progress in the computational methods developed for this purpose are discussed in Section II and III of this chapter.

Another laboratory might be involved in generating a large number of ESTs. There, the main task will be the detection of "anomalous" or noninformative sequences, the estimation of the coding potential of each individual clone, the sensitive detection of an eventual similarity with already characterized genes, etc.

Finally, molecular evolution studies or works in automated classification will also require exhaustive comparisons of the entire content of databases.

From the point of view of computational biology, these separate scientific endeavors have in common the necessity of performing a very large number of sequence comparisons and obtaining a manageable, rapidly interpretable output. In this output, the contribution of the anomalies, errors and artifacts, that we know are abundant in both the databases and newly determined data (constituting a single large, or multiple query sequences), should be negligible. Throughout this chapter, and in the context of different experimental settings, I will discuss the recent progress that has made this possible.

E. The General Constraints of Large-Scale Sequence Analysis

Commercial or academic software packages typically incorporate one hundred or more different algorithms to compute various properties of nucleotide and amino acid sequences. These programs address all aspects of research (secondary/tertiary structure, functional signatures and motifs, optimal alignments, multialignment/phylogeny) for both types of sequences. The most advanced systems have interactive, mouse-driven interfaces and use color graphics to present the results in full details. In summary, those packages are optimally designed for the detailed analysis of a single (reasonably sized) sequence.

In contrast, large-scale sequence analysis involves the recurrent application of a limited number of different algorithms, mostly to determine which regions of a large contig or which members of a large collection are either protein encoding, and/or can be characterized by similarity with other sequences. The similarity assessment is most often performed by direct sequence-to-sequence alignment, but can also involve the more sophisticated detection of a "motif" (see Section IV).

For instance, the analysis of a newly determined genomic contig of human DNA will involve the localization of putative exons, and the prediction of the function of the corresponding protein from similarity or motif analysis. Similarly, the analysis of ESTs will involve determining their coding potential and the tentative prediction of the function of the corresponding protein, again by sequence alignment or motif analysis.

Because of the volume of data involved (i.e., the size of the genomic sequence or of the EST collection and the size of the databases of reference), the procedure habitually used for single sequence analysis, might suddenly become unadaptable because

- i) the program is much too slow,
- ii) its output requires human judgment (and nobody wants to visually analyze 50kb-wide graphics),
- iii) its usage requires human interactions (and nobody wishes to browse through hundreds of result screens with "yes/no" buttons).

Large-Scale Sequence Analysis 167

Finally, the programs employed to analyze well-validated sequences one-at-a-time are usually not equipped to deal with the high error/artifact rate characterizing large-scale projects.

The constraints of large-scale sequence analysis thus have three main implications:

1. The best, most rigorous available programs and algorithms may have to be replaced by faster, more convenient approximations.
2. The programs have to function with a large data set without human intervention.
3. The output must be of manageable size, and its content mostly biologically relevant (i.e., containing very few "false positive" results).

Faster programs may reduce sensitivity (we may miss something). However, this is the whole point of the "large-scale approach": the few results we might miss become negligible compared to the numerous ones we get with the standard treatment. For instance, there is no point in trying to rescue information from bad EST sequences (e.g., with many ambiguous positions, eventual deletion/insertions), knowing that another sequence, corresponding to the same gene, will eventually appear.

Large data sets disqualify any interactive, "point & click" application. One must return to "batch mode" processing, using files of commands (e.g., scripts) and automated procedures to run combinations of algorithms. With its innate ability to transmit results ("pipe") from one job to the next, the UNIX system (in its various shell flavors: Bourne, C, Korn, TCL/tk) is ideally suited for this task. 2 also implies that one abandons fancy color graphics and returns to sober, computer readable outputs, although retaining human readability is important for debugging or serendipity (lucky browsing).

Finally, 3 demands that the program incorporate a measure of *biological significance*, i.e., a way to determine a priori which results are worth retaining in the final, limited output produced for human interpretation (if any). Short of giving the program an encyclopedic knowledge of biology, this is usually not possible. However, a measure of *statistical significance* (how unlikely the result is) can, and must, be incorporated to serve as an objective relevance filter. When this is in place, human efforts can then be solely devoted to the detailed analyses of already statistically significant results, using expert biological knowledge and/or intuition. Of course, this approach will cause some interesting biological findings (such as a distant evolutionary relationship) to be missed from time to time, but this is a consequence that we already accepted in 1. However, a close (even if not perfect) match between statistical significance ("unlikely result") and biological significance ("interesting finding") is necessary for the whole approach to work with a satisfactory yield. This demands that our automated relevance filter incorporate a minimum of biological knowledge (in particular about possible misleading results).

In the next chapter, we present the concept of query masking as a general way of incorporating this knowledge in large-scale sequence analysis.

II. THE CONCEPT OF QUERY MASKING

A. The "Overwhelming Output" Problem

Although the protein sequence databases are of much better quality than the EMBL/GenBank/DDBJ nucleotide sequence repository, they contain a number of anomalous entries or flaws that might produce dreadful consequences (i.e., gigantic senseless output) in the context of large-scale sequence analysis. Let us say, for instance, that we have just produced a 90-kb long sequence of the human genome in the region of an interesting human disease gene. Our first impulse is to compare the six conceptual translations of this sequence (or all possible ORFs) to all known proteins as stored in Swiss-Prot and PIR (or their nonredundant combination). BLASTX, a local alignment program for nucleic acids vs. proteins (Gish and States, 1993), can do that for us in almost no time. We thus fix a reasonable statistical significance threshold for the alignments we wish to retain (a unique feature of the BLAST algorithm (Karlin and Altschul, 1990; Altschul et al., 1990; Karlin et al., 1990)), run the search and expect a manageable output, eventually locating a number of exons in this region.

Whatever the genomic sequence we started from, I can predict that we will get back a huge listing reporting numerous matches with most of the candidate ORFs we generated by conceptual translation. And this cannot be biologically significant. At most, we should expect 5 to 20 exons to be in a region of this size, about half of them (Claverie, 1993; Green et al., 1993) detectable by homology in the protein databases. What went wrong? A closer look at this output reveals two main types of alignments: those involving stretches of sequence with a biased amino acid composition, and those, with no apparent compositional bias, but always involving the same database entries. In the following we will respectively refer to these alignments as "low entropy" and "artifactual" matches.

B. Low-Entropy Matches

Figure 1 (A to F, G to P, and Q to Y) illustrates the notion of "low-entropy" matches (also called "low-complexity" or "simple sequences"). Such caricatural (albeit real) alignments can be produced by using "homopolymeric" queries against the database. What is happening can be readily explained by the following simple computer experiment. Let us construct a *theoretical* protein with the sequence,

80[L]-64[A]-56[G]-56[V]-56[S]-48[K]-48[T]-48[E]-40[R]- ... etc.,

where the numerical coefficient approximates the abundance of each amino acid found in the whole Swiss-Prot database (Figure 2). Then, let us use this theoretical

```
>sp|P02734|ANP4_PSEAM ANTIFREEZE PEPTIDE 4 PRECURSOR.
    Score= 123, P= 5.7e-12, Identities= 31/38 (81%)

Query:    1 AAAAAAAAAAAAAAAAAAAAAAAAAAAAAAAAAAAAAA 38
            AAAAAAA AA AAAAAAA AA AAAAAAA AA AA AA
Sbjct:   27 AAAAAAATAATAAAAAAATAATAAAAAAATAATAAKAA 64


>sp|P26371|KRUC_HUMAN KERATIN, ULTRA HIGH-SULFUR MATRIX PROTEIN
    Score= 98, P= 4.2e-10, Identities = 14/28 (50%),

Query:    1 CCCCCCCCCCCCCCCCCCCCCCCCCCCC 28
            C C   C CC  CCC   C   CC C CC
Sbjct:   75 CGCSQCSCCKPCCCSSGCGSSCCQCSCC 102


>sp|P31231|CAQS_RANES CALSEQUESTRIN, SKELETAL MUSCLE.
    Score= 246, P= 1.3e-26, Identities = 41/41 (100%)

Query:    1 DDDDDDDDDDDDDDDDDDDDDDDDDDDDDDDDDDDDDDDDD 41
            DDDDDDDDDDDDDDDDDDDDDDDDDDDDDDDDDDDDDDDDD
Sbjct:  377 DDDDDDDDDDDDDDDDDDDDDDDDDDDDDDDDDDDDDDDDD 417


>sp|P19351|TRT_DROME TROPONIN T, SKELETAL MUSCLE
    Score= 169, P= 1.8e-13, Identities= 33/35 (94%),

Query:    1 EEEEEEEEEEEEEEEEEEEEEEEEEEEEEEEEEEE 35
            EE+EE+EEEEEEEEEEEEEEEEEEEEEEEEEEEEE
Sbjct:  361 EEDEEDEEEEEEEEEEEEEEEEEEEEEEEEEEEEE 395


>sp|P15605|YM04_PARTE HYPOTHETICAL 18.8 KD PROTEIN (ORF4).
    Score= 112, P= 4.6e-13, Identities= 21/41 (51%),

Query:    1 FFFFFFFFFFFFFFFFFFFFFFFFFFFFFFFFFFFFFFFFF 41
            FFFF +   FFF FF FF   FF +F +  FFF   F FFF
Sbjct:   83 FFFFNYLSGFFFLFFVFFTSFFVYFSYLLFFFVPVFVLFFF 123
```

Figure 1. (a) Low-entropy database matches: A to F.

sequence to query Swiss-Prot, using a highly statistically significant score threshold. The result is an overwhelming output describing megabytes of matches against more than 2,000 database entries (Figure 2). None of them are of course "biologically significant," because the theoretical query cannot possibly be biologically or evolutionarily related to any real sequence in the databases. Then, is the statistical theory behind the BLAST scoring system wrong? No, because we can easily prove it. If we now randomly shuffle our "funny looking" query (Figure 2) into a "statistically correct sequence" (Figure 3) and again scan the database with it, all the matches disappear. We are now left with a "biologically significant" null output, consistent with the fact that no real sequence is related to such a random query.

>sp|P09026|HXB3_MOUSE HOMEOBOX PROTEIN HOX-B3 (HOX-2.7) (MH-23).
    Score= 161, P= 1.1e-16, Identities= 28/34 (82%),

Query: 1 GGGGGGGGGGGGGGGGGGGGGGGGGGGGGGGGGG 34
            G   G GGGGGGGGGGGGGGGG   GGGGGGGGGG
Sbjct: 148 GTAEGCGGGGGGGGGGGGGGGGSSGGGGGGGGGG 181

>sp|P04929|HRPX_PLALO HISTIDINE-RICH GLYCOPROTEIN PRECURSOR.
    Score= 319, P= 3.8e-41, Identities= 41/46 (89%),

Query: 1 HHHHHHHHHHHHHHHHHHHHHHHHHHHHHHHHHHHHHHHHHHHHHH 46
            HHHHHHHH HHHHHHHHH HHHHHHHHH HHHHHHHH   HHHHHHHH
Sbjct: 215 HHHHHHHHGHHHHHHHHHGHHHHHHHHHGHHHHHHHHHDAHHHHHHH 260

>sp|P29974|CGCC_MOUSE CGMP-GATED CATION CHANNEL PROTEIN
    Score= 137, P= 1.6e-11, Identities= 27/47 (57%),

Query: 1 KKKKKKKKKKKKKKKKKKKKKKKKKKKKKKKKKKKKKKKKKKKKKKK 47
            K+KKKKKK+KK K   K + K   +KKKKK+K+K+KKKK++K K+KK+
Sbjct: 94 KEKKKKKKEKKSKADDKNEIKDPEKKKKKEKEKEKKKKEEKTKEKKE 140

>sp|P32770|ARP_YEAST ARP PROTEIN.
    Score= 198, P= 1.7e-18, Identities= 32/48 (66%),

Query: 1 NNNNNNNNNNNNNNNNNNNNNNNNNNNNNNNNNNNNNNNNNNNNNNNN 48
            N N N NN+NNNNN+NNN+NNN++N + N+N+N NNNNNNNN NN+NN
Sbjct: 517 NGNGNGNNSNNNNNHNNNHNNNHHNGSINSNSNTNNNNNNNNGNNSNN 564

>sp|P12978|EBN2_EBV EBNA-2 NUCLEAR PROTEIN.
    Score= 276, P= 2.3e-32, Identities= 40/42 (95%),

Query: 1 PPPPPPPPPPPPPPPPPPPPPPPPPPPPPPPPPPPPPPPPPP 42
            PPP PPPPPPPPPPPPPPPPPPPPPPPPP PPPPPPPPPP
Sbjct: 59 PPPLPPPPPPPPPPPPPPPPPPPPPPPPPSPPPPPPPPPPP 100

Figure 1. (b) Low-entropy database matches: G to P.

By telling us that a given alignment score is statistically significant, BLAST simply states that such a good match is very unlikely between two evolutionarily unrelated sequences. However, this conclusion is only valid provided that both the query and target sequences are reasonably close to the random model of amino acid composition *and distribution*. If one of these assumptions is not valid, artifactual alignments such as those in Figure 1 will pop up all the time. Our fundamental problem, however, is that there are actual proteins (i.e., valid database entries) exhibiting such troublesome locally biased amino acid compositions. In fact, I found examples (see Figure 1) of real proteins with homopolymeric (or quasi-homopolymeric) stretches for all amino acids except for the most hydrophobic ones: Ile, Leu, Met, Val and Trp. Many storage proteins and also transcription factors, homeobox proteins, and important structural proteins exhibit stretches of polar

```
>sp|P20226|TF2D_HUMAN TRANSCRIPTION INITIATION FACTOR TFIID
         Score= 205, P= 1.5e-22, Identities= 40/43 (93%),

Query:     1  QQQQQQQQQQQQQQQQQQQQQQQQQQQQQQQQQQQQQQQQQQQ  43
              ++QQ+QQQQQQQQQQQQQQQQQQQQQQQQQQQQQQQQQQQQQ
Sbjct:    53  EEQQRQQQQQQQQQQQQQQQQQQQQQQQQQQQQQQQQQQQQQQ  95


>sp|P80001|PRT1_SEPOF SPERMATID-SPECIFIC PROTEIN T1
         Score= 172, P= 5.1e-22, Identities= 37/46 (80%),

Query:     1  RRRRRRRRRRRRRRRRRRRRRRRRRRRRRRRRRRRRRRRRRRRRRR  46
              RRRRRR RRRRRR RRR R   RRR RRRRRRRRR RRRR RRRR
Sbjct:    22  RRRRRRSRRRRRRSRRRSRSPYRRRYRRRRRRRRRSRRRRYRRRR  67


>sp|P02845|VIT2_CHICK VITELLOGENIN II PRECURSOR
         Score= 155, P= 1.1e-09, Identities= 40/45 (88%),

Query:     1 SSSSSSSSSSSSSSSSSSSSSSSSSSSSSSSSSSSSSSSSSSSSS  45
             SSSSSSSSSS+S SSSSSS SSSSSS S SSS SSSSSSSSSSSS
Sbjct: 1206  SSSSSSSSSSNSKSSSSSSKSSSSSSRSRSSSKSSSSSSSSSSSS  1250


>sp|Q05049|MUC1_XENLA INTEGUMENTARY MUCIN C.1 (FIM-C.1)
         Score= 184 , P= 6.7e-17, Identities = 38/46 (82%),
Query:     1 TTTTTTTTTTTTTTTTTTTTTTTTTTTTTTTTTTTTTTTTTTTTTT  46
             TTTT TTTTT TTTTTT   TTTT TTTTTT TTTTTTTTT    TTT
Sbjct:   403 TTTTPTTTTTPTTTTTTKATTTTPTTTTTTPTTTTTTTTTKATTT  448


>sp|P36079|YKI3_YEAST HYPOTHETICAL 23.7 KD PROTEIN
         Score= 101, P= 7.0e-08, Identities= 17/34 (50%),

Query:     1 YYYYYYYYYYYYYYYYYYYYYYYYYYYYYYYYYY  34
             +Y  Y  Y Y Y Y Y Y + Y Y YY   YYY  YY
Sbjct:     3 WYIYRYIYIYMYLYIYVHTYIYIYYCCYYYRDYY  36
```

Figure 1. (c) Low-entropy database matches: Q to Y.

residues. Those databases entries have a higher than average chance of matching any query of random amino acid composition. By using the conceptual translation of a genomic nucleotide sequence as a query, we expose ourselves to even more trouble. The vertebrate genome contains many regions of "simple sequence," i.e., local stretches of single, double or triple base repeats. Once conceptually translated, those simple nucleotide sequences generate low-entropy amino acid sequences responsible for the pathological output.

Other nonhomopolymeric types of short amino acid repeats (such as those found in collagens: "GXYGXYGXY") also fall in the category of "low-entropy" segments and also tend to produce inflated output. For instance, Figure 4 shows the result of using human collagen alpha 1 as a query: a huge output, reporting many matches with totally unrelated proteins.

Amino acid composition

```
Ala (A) 7.65    Gln (Q) 4.07    Leu (L) 9.15    Ser (S) 7.08
Arg (R) 5.23    Glu (E) 6.26    Lys (K) 5.83    Thr (T) 5.84
Asn (N) 4.45    Gly (G) 7.10    Met (M) 2.33    Trp (W) 1.30
Asp (D) 5.24    His (H) 2.27    Phe (F) 3.97    Tyr (Y) 3.22
Cys (C) 1.81    Ile (I) 5.46    Pro (P) 5.08    Val (V) 6.49
```

```
>aa.seq | Low Entropy | Swissprot 21 | 495
AAAAAAAAAAAAAAAAAAAAAAAAAAAAAAAAAAAAAAAAAARRRRRRRRRRRRRRRRRRRRR
RRRRNNNNNNNNNNNNNNNNNNNNNNDDDDDDDDDDDDDDDDDDDDDDDDDDCCCCCCCCC
CQQQQQQQQQQQQQQQQQQQQQEEEEEEEEEEEEEEEEEEEEEEEEEEEEEEGGGGGGGG
GGGGGGGGGGGGGGGGGGGGGGGGGGHHHHHHHHHHHIIIIIIIIIIIIIIIIIIIIIIII
IIIIILLLLLLLLLLLLLLLLLLLLLLLLLLLLLLLLLLLLLLLLLLLLLLLKKKKKKKKK
KKKKKKKKKKKKKKKKKKKMMMMMMMMMMMFFFFFFFFFFFFFFFFFFFPPPPPPPP
PPPPPPPPPPPPPPPPPSSSSSSSSSSSSSSSSSSSSSSSSSSSSSSSSSSSSSTTTTTTTT
TTTTTTTTTTTTTTTTTTTTWWWWWWYYYYYYYYYYYYYYYYYVVVVVVVVVVVVVVVV
VVVVVVVVVVVVVVV
```

blastp swissprot aa.seq S=70 > out

WARNING: Descriptions of 1555 database sequences were not reported

```
sp|P15771|NUCL_CHICK NUCLEOLIN (PROTEIN C23).                    160  3.8e-61
sp|P08199|NUCL_MESAU NUCLEOLIN (PROTEIN C23).                    181  2.1e-50
sp|P19338|NUCL_HUMAN NUCLEOLIN (PROTEIN C23).                    193  7.2e-50
sp|P10495|GRP1_PHAVU GLYCINE-RICH CELL WALL STRUCTURAL            154  3.4e-44
..................................................              ...  ......
..................  500 matches !  ..................            ...  ......
..................................................              ...  ......
sp|P09021|HXA5_MOUSE HOMEOBOX PROTEIN HOX-A5 (HOX-1.3)             72  2.6e-07
sp|P17588|LRP1_HSV1F LATENCY-RELATED PROTEIN 1.                   112  2.7e-07
sp|P19660|BCT5_BOVIN BACTENECIN 5 (PR-42).                         64  2.9e-07
sp|P20806|7LES_DROVI SEVENLESS PROTEIN (EC 2.7.1.112).             90  3.0e-07
sp|P14682|UBC3_YEAST UBIQUITIN-CONJUGATING ENZYME E2-34           104  3.0e-07
..................................................              ...  ......
..................  1555 more !  ..................              ...  ......
```

Figure 2. The low-entropy explosion.

C. Artifactual Matches

Low-entropy segments are not the only source of spurious database matches and inflated search outputs. The ubiquitous presence of repeated elements in the human (and mammalian) genome can lead to many problems when combined with the database errors. Although protein databases are much more reliable than the EMBL/GenBank/DDBJ repository databases, they still contain misleading entries, a large number of frame-shifted (and sometimes recombined) sequences (Claverie, 1993), amino acid sequences (translated) from unrecognized vector sequences, and amino acid sequences derived from common genomic repeats, such as Alu (or the murine B1 and B2 repeats). A list of the misleading protein sequences from various

```
>aa.seq shuffled
YLALVDRIGGIANKIASAQSTHIASDQREPKQTNMNPPAEEEKNVEDNPPRKWTKESAVC
QQTDYVAEYPPLLTKFEPVALEELSGLILWVNYAYGLIRSRLGFARAVLVEGGRAGGKAV
NTAYNEHSKSASLCVKCASAVSNPESGGGKRMDNILINQPFDEGEVGALPLRLQCLQYKK
GLIKVIETFDTTTFSSSVHILEDFTHNLEQCYAHTTKKIEDRMLNIFFMSFPVMPALIRV
RKKTPDLSPYILDTSVDIRVSHLLHRFWGFHSPKLDRAKKGNMGVLLRSAGEAWARTCAF
ALMLVDIELTFDFPFEIYCVIQFMSSPSYEPDSTGVTLQARDLFLPNAYDTIGGDNCMDK
HVSEAMEHGPTNVINYVRTWMQAAGGNAYGNDGKLLTLSRQYLCQIIDGKQGSKAWVGLG
QQDVLAGYSTNHQKSETFDSVDNVDLKTEKPKQMLPRGPVRVIESRSITLRSEATLSGRT
PEGAKILFAQEVIFR
```

blastp swissprot aa.seq S=70 > out

------------> NO HITS

Figure 3. "Statistically correct" sequence.

```
>CA19_HUMAN COLLAGEN ALPHA 1(IX)
MKTCWKIPVFFFVCSFLEPWASAAVKRRPRFPVNSNSNGGNELCPKIRIGQDDLPGFDLISQQ
VDKAASRRAIQRVVGSATLQVAYKLGNNVDFRIPTRNLYPSGLPEEYSFLTTFRMTGSTLKKN
WNIWQIQDSSGKEQVGIKINGQTQSVVFSYKGLDGSLQTAAFSNLSSLFDSQWHKIMIGVERS
SATLFVDCNRIESLPIKPRGPIDIDGFAVLGKLADNPQVSVPFELQWMLIHCDPLRPRRETCH
ELPARITPSQTTDERGPPGEQGPPGASGPPGVPGIDGIDGDRGPKGPPGPPGPAGEPGKPGAP
GKPGTPGADGLTGPDGSPGSIGSKGQKGEPGVPGSRGFPGRGIPGPPGPPGTAGLPGELGRVG
PVGDPGRRGPPGPPGPPGPRGTIGFHDGDPLCPNACPPGRSGYPGLPGMRGHKGAKGEIGEPG
RQGHKGEEGDQGELGEVGAQGPPGAQGLRGITGLVGDKGEKGARGLDGEPGPQGLPGAPGDQG
QRGPPGEAGPKGDRGAEGARGIPGLPGPKGDTGLPGVDGRDGIPGMPGTKGEPGKPGPPGDAG
LQGLPGVPGIPGAKGVAGEKGSTGAPGKPGQMGNSGKPGQQGPPGEVGPRGPQGLPGSRGELP
VGSPGLPGKLGSLGSPGLPGLPGPPGLPGMKGDRGVVGEPGPKGEQGASGEEGEAGERGELGD
IGLPGPKGSAGNPGEPGLRGPEGSRGLPGVEGPRGPPGPRGVQGEQGATGLPGVQGPPGRAPT
DQHIKQVCMRVIQEHFAEMAASLKRPDSGATGLPGRPGPPGPPGPPGENGFPGQMGIRGLPGI
KGPPGALGLRGPKGDLGEKGERGPPGRGPNGLPGAIGLPGDPGPASYGKNGRDGERGPPGLAG
IPGVPGPPGPPGLPGFCEPASCTMQLVSEHLTKGLTLERLTAAWLSA
```

blastp swissprot collagen S=70 > out

```
sp|P20849|CA19_HUMAN COLLAGEN ALPHA 1(IX) CHAIN PRECURSOR.      5141  0.0
sp|P02461|CA13_HUMAN PROCOLLAGEN ALPHA 1(III) CHAIN PRECU..     1032  0.0
sp|P08122|CA24_MOUSE PROCOLLAGEN ALPHA 2(IV) CHAIN PRECUR..      560  4.9e-252
sp|P02452|CA11_HUMAN PROCOLLAGEN ALPHA 1(I) CHAIN PRECURSOR.     965  2.9e-243
sp|P20908|CA15_HUMAN PROCOLLAGEN ALPHA 1(V) CHAIN PRECURSOR.     962  2.7e-238
sp|P02458|CA12_HUMAN PROCOLLAGEN ALPHA 1(II) CHAIN PRECUR...     947  8.6e-234
sp|P20850|CA19_RAT   COLLAGEN ALPHA 1(IX) CHAIN (FRAGMENT).     1662  4.1e-227
..............310 matches ! (mostly with unrelated proteins) ..............
sp|P12948|DH3_HORVU  DEHYDRIN DHN3.                               51  0.40
sp|P10357|V70K_TYMV  69 KD PROTEIN.                               46  0.41
sp|P12980|LYL1_HUMAN LYL-1 PROTEIN.                               58  0.42
sp|P28925|ICP4_HSVEB TRANS-ACTING TRANSCRIPTIONAL PROTEIN...      61  0.42
sp|P17473|ICP4_HSVEK TRANS-ACTING TRANSCRIPTIONAL PROTEIN...      61  0.42
sp|Q01786|GUN5_THEFU ENDOGLUCANASE E-5 PRECURSOR (EC 3.2....      68  0.43
```

Figure 4. Effect of short period repeats: collagen.

Protein	Accession[b]	Span	Element
A4 amyloid peptide	M34875	284-258	Alu-J
Biliary glycoprotein	M76741	annotated	Alu-J
Breast cancer BRCAa	U15595	annotated	Alu-J
c-rel phosphoprotein	A60646[c]	308-339	Alu-Sx
cdc2-related kinase	112668[d]	13-44	Alu
Decay accel. factor	M30142	annotated	Alu-Sc
HLA-DR-b1	X12544	annotated	Alu-Sx
Integrin b1	M84237	15-132	Alu-Sx
Lectin-like protein	L14542	annotated	Alu-J
Neurofibromatosis 2	L27065	1-117	Alu-Sc
PM-Scl-75 autoantigen	U09215	905-954	Alu-S
Pregnancy specific glycoprot.	U04325	1261-1583	Alu-J
Ser/Thr kinase 2	L20321	1684-1544	Alu-J
Trombaxane A2 receptor	U11271	1196-1337	Alu-Sx
Uroporphyrinogen synth.[a]	78430d	211-247	Alu
Cholinesterase [a]	S75201	1-397	Alu-Sb2
Insulin/IGF-I receptor	A18622	3081-3141	Alu
Platelet glycoprt IIb	J02963	annotated	Alu-Sx
B-cell growth factor	M15530	59-133/318-599	Alu-J/-Sx
Interferon α/β receptor	X77722	1102-1238	Alu-J
KRAB Z-finger protein	L11672	3567-3839	Alu-Sb
Anti-lectin epitope	X58236	2-155	Alu-Sx
Mahlavu hepato.carcinoma	X55777	1498-1820	Alu-Sx
Malaria antigen	M63279	1-121	Alu-Sp
RMSA-1	L26953	1276-1612/1866-2130	Alu-Sx
Transformation-related prot.	L24521	1-283/645-719	Alu-Sq
X-linked retinopathy prot.	S58722	1-390	Alu-Sx
MHC class I	X16213	274-329	B1
EpoR gene upstream ORF	S52010	417-618	B2
Na/K-ATPase beta 2 sub.	X56007	873-1039	B2
Somatotropin prot. RDE-.25	A60716[c]	1-38	B2

Notes: [a]genes that are inactivated by the Alu insertion; [b]GenBank accession numbers except: [c]PIR accession number, [d]NCBI backbone accession number

Figure 5. Protein database entries and ORFs with Alu-, B1- or B2-like sequences.

databases is given in Figure 5. It is most likely that these sequences are due to experimental errors or nonfunctional splicing events (Claverie and Makalowski, 1994). However, these database entries represent a major nuisance when analyzing human genomic DNA, as 5 to 10% (even up to 50% in some AT-rich regions (Iris et al., 1993)) of it is composed of Alu-like sequences. Thus, running a conceptual translation of human genomic sequence through the protein databases produces many misleading matches with erroneous entries (which keep on accumulating, Claverie and Makalowski, 1994). In addition, less frequent, but similarly misleading matches are induced by the presence of less common repeats (e.g., Line-1), or vector sequences (a frequent contamination of raw sequence data).

D. Query Masking: A Practical Solution to Output Inflation

Both of the problems caused by low-entropy segments and/or artifacts in the databases could be solved by preprocessing the databases and removing all entries capable of inducing noninformative matches. However, this is not a practical solution because the content of the database keeps changing and novel anomalous entries are introduced every day. Also, what is "junk" to someone (e.g., Alu segments, vector sequences, low-complexity segments) might be interesting to others. Thus, multiple copies of the custom-processed databases would have to be maintained locally.

The practical solution is to address the problem at the level of the query itself: this is the general concept of "query masking." Query masking consists of using our biological knowledge of sequences (e.g., low complexity segments, ubiquitous repeats), and of the expected experimental artifacts (e.g., vector contamination) to remove the a priori troublesome segments from the query, thus minimizing their impact on the database scan. This approach preserves the freedom and individuality of each scientist, who can specify which results are uninformative, while using the same unaltered versions of the databases.

Once the categories of "undesirable" sequence segments are defined, masking the query simply consists of replacing the corresponding residues with a placeholder for which a "neutral" matching score (usually 0) will be granted. The masked part of the query then becomes "transparent" to the databases, with the effect of removing their contribution from the output. Query masking does not involve any manipulation of the statistical significance threshold and, when all sources of anomalous output inflation have been identified, is sufficient to reduce the output to a core of nonambiguous, biologically significant answers. In the following sections, I present the procedures most commonly used to filter out low-entropy segments and reduce the impact of artifactual matches.

E. XNU: Finding and Masking Internal Repeats

To mask the low-entropy segments within a query, we first need an algorithm to delineate them. There are two approaches to this problem. One relies on the computation of a local information or local linguistic complexity index, and the application of an empirical threshold. This approach has been used by Wootton and Federhen (1993) and is implemented in the program SEG. The other, implemented in the program XNU (Claverie and States, 1993), consists of characterizing the low-complexity segments as pseudo periodical (quasi-repeated) sequences. In this context, the homopolymeric segment "AAAAA" has a period of one residue, the collagen repeat GXYGXYGXY a period of three, etc. Figure 6 illustrates the simple algorithm used in XNU. Briefly, the sequence is compared to itself; the local alignments are scored with an amino acid scoring matrix (PAM, BLOSUM, etc.), preferentially the same matrix that will be used for the subsequent database search. The significance of the alignment scores are estimated according to Karlin & Altschul (1990). In a dot-matrix representation, the matched internal repeats appear as off-diagonal segments (Figure 6). Short period repeats will produce local alignments very close to the main diagonal (often as a smear) while "normal" repeats appear at a distance at least equal to their length. This property immediately suggests an algorithm to discriminate between low-complexity regions and regular

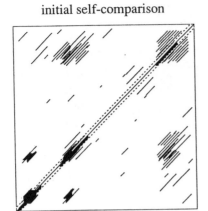 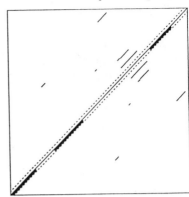

Figure 6. Eliminating short period repeats.

repeats which does not involve any *ad hoc* measure of the local information content. Low-complexity segments will simply be recognized because they repeat themselves with a periodicity of a few positions, while regular internal repeats are associated with much larger periodicity (e.g., 100 residues for an immunoglobulin or a fibronectin domain).

The repeated elements in the sequence are mapped by projecting (vertically and horizontally) the off-diagonal elements onto the main diagonal. A selective mapping (and eventual masking) of the troublesome low-complexity segments is accomplished by only taking into account the off-diagonal elements close to the main diagonal (Figure 6). With a short periodicity threshold (the default value is 10), XNU operates on low-complexity repeats. With a higher threshold (up to half the sequence length), XNU operates on both regular and low-complexity repeats.

In the current implementation, the XNU command line recognizes a variety of parameters controlling the repeat periodicity, the score threshold (or probability) for considering local alignments, and the scoring matrix to be used. Various options allow the output to take alternative forms, one of them is the masking mode, now routinely used with BLAST (with the filter option). In the standard mode (no parameters or options specified) XNU detects and masks the low-complexity repeats, thus ensuring that each reported match will reveal a biologically significant similarity (e.g., common ancestry).

Figure 7 shows the performance of XNU on one of the worst cases: collagen. Processing a collagen sequence with XNU allows a typical BLASTP search output to be reduced from 130 reported matches (most of them without any biological significance and highly redundant) to only 5, all of them with bona fide collagen-related entries.

xnu collagen > collagen.xnu

```
>CA19_HUMAN COLLAGEN ALPHA 1(IX)
MKTCWKIPVFFFVCSFLEPWASAAVKRRPRFPVNSNSNGGNELCPKIRIGQDDLPGFDLISQFQ
VDKAASRRAIQRVVGSATLQVAYKLGNNVDFRIPTRNLYPSGLPEEYSFLTTFRMTGSTLKKNW
NIWQIQDSSGKEQVGIKINGQTQSVVFSYKGLDGSLQTAAFSNLSSLFDSQWHKIMIGVERSSA
TLFVDCNRIESLPIKPRGPIDIDGFAVLGKLADNPQVSVPFELQWMLIHCDPLRPRRETCHELP
ARITPSQTTDERXXXXXXXXXXXXXXXXXXXXXXXXXXXXXXXXXXXXXXXXXXXXXXXXXXXX
XXXXXXXXXXXXXXXXXXXXXXXXXXXXXXXXXXXXXXXXXXXXXXXXXXXXXXXXXXXXXXXX
XXXXXXXXXXXXXXXXXXXXFHDGDPLCPNACPXXXXXXXXXXXXXXXXXXXXXXXXXXXXXXX
XXXXXXXXXXXXXXXXXXXXXXXXXXXXXXXXXXXXXXXXXXXXXXXXXXXXXXXXXXXXXXXX
XXXXXXXXXXXXXXXXXXXXXXXXXXXXXXXXXXXXXXXXXXXXXXXXXXXXXXXXXXXXXXXX
XXXXXXXXXXXXXXXXXXXXXXXXXXXXXXXXXXXXXXXXXXXXXXXXXXXXXXXXXXXXXXXX
XXXXXXXXXXXXXXXXXXXXXXXXXXXXXXXXXXXXXXXXXXXXXXXXXXXXXXXXXXXXXXXX
XXXXXXXXXXXXXXXXXXXXXXXXXXXXXXXXXXXXXXXXXXXXXXXXXXRAPTDQHIKQVCMR
VIQEHFAEMAASLKRPXXXXXXXXXXXXXXXXXXXXXXXXXXXXXXXXXXXXXXXXXXXXXXXX
XXXXXXXXXXXXXXXXXXXXXXXXXXXXXXXXXXXXXXXXXXXXXXXXXXXXXXXXXXXXXXXX
XXXFCEPASCTMQLVSEHLTKGLTLERLTAAWLSA
```

blastp swissprot collagen S=70 > out

```
sp|P20849|CA19_HUMAN COLLAGEN ALPHA 1(IX) CHAIN PRECURSOR.      1418  4.1e-196
sp|P12106|CA19_CHICK COLLAGEN ALPHA 1(IX) CHAIN PRECURSOR...     942  4.4e-129
sp|P13944|CA1C_CHICK COLLAGEN ALPHA 1(XII) CHAIN (FRAGMENT).     153  1.5e-29
sp|P20850|CA19_RAT   COLLAGEN ALPHA 1(IX) CHAIN (FRAGMENT).      155  7.7e-14
sp|P32018|CA1E_CHICK COLLAGEN ALPHA 1(XIV) CHAIN (FRAGMENT).     108  3.1e-10
sp|P12107|CA1B_HUMAN COLLAGEN ALPHA 1(XI) CHAIN PRECURSOR.        79  0.0067
sp|P32017|CA39_CHICK COLLAGEN ALPHA 3(IX) CHAIN PRECURSOR.        70  0.11
sp|P20908|CA15_HUMAN PROCOLLAGEN ALPHA 1(V) CHAIN PRECURSOR.      70  0.12
```

Figure 7. Low-entropy/repeat masking by XNU.

F. XBLAST: Finding and Masking "Junk" Sequences

The purpose of XBLAST is to remove from the search output the voluminous report of a priori unworthy results: matches that we expect to occur, but we do not wish to see. XBLAST works by simply masking in the original query sequence (e.g., translated genomic sequences, putative ORFs, ESTs, etc.) the precise segments bearing the undesirable similarity.

The information used to direct the masking will be supplied by a BLAST search against a database of selected "junk" sequences. For instance, if we are analyzing a translated human genomic sequence, we will include Alu elements, Line-1, MER, etc., in our "junk" database. If we are working with unassembled DNA sequences or ESTs, we may also want to include a selection of vector sequences matching our experimental protocol (Lambda, M13, poly-linker sequences). To process our query (ies), we will then use a command line like:

For (all queries) do
{*blastn junk.db query* |*xblast + query N > query.masked*},

that is, the output of the BLAST search is "piped" through the XBLAST filter, masking the query (e.g., write "N" within) according to the local similarity detected with any of the sequences contained in the "junk" database. Note that all BLAST programs (BLASTN, BLASTP, BLASTX, TBLASTN) can be used with XBLAST. As an example, Alu-like sequences would be filtered from a set of candidate ORFs by

For (all ORFs) do
{blastp   Alu.db ORFs |xblast + ORFs X > ORF.masked}.

In this case, the junk database is a translation of a set of Alu segments (Claverie and Makalowski, 1994), and "X", the ambiguity code for amino acids is used to mask the Alu-like segments in the ORF peptide sequences. The syntax and usages of XBLAST have been described elsewhere (Claverie, 1994).

A) Original query : NF-kappa B p49

```
>NF-kappa B p49
MESCYNPGLDGIIEYDDFKLNSSIVEPKEPAPETADGPYLVIVEQPKQRGFRFRYGCEGPSHGG
LPGASSEKGRKTYPTVKICNYEGPAKIEVDLVTHSDPPRAHAHSLVGKQCSELGICAVSVGPKD
MTAQFNNLGVLHVTKKNMMGTMIQKLQRQRLRSRPQGLTEAEQRELEQEAKELKKVMDLSIVRL
RFSAFLRASDGSFSLPLKPVTSQPIHDSKSPGASNLKISRMDKTAGSVRGGDEVYLLCDKVQKD
DIEVRFYEDDENGWQAFGDFSPTDVHKQYAIVFRTPPYHKMKIERPVTVFLQLKRKRGGDVSDS
KQFTYYPLVEDKEEVQRKRRKALPTFSQPFGGGSHMGGGSGGAAGGYGGAGGGEGVLMEGGVKV
REAVEEKNLGEAGRGLHACNPAFGRPRQADYLRSGVQDQLGQQRETSSLLKIQTLAGHGGRRL
```

B)   Original database search output

```
                                                         Probability
Sequences producing High-scoring Segment Pairs:     Score    P(N)

KBF1_HUMAN  DNA-BINDING FACTOR KBF1                   345   7.3e-132
TREL_MELGA  C-REL PROTO-ONCOGENE PROTEIN              257   5.6e-51
TREL_CHICK  C-REL PROTO-ONCOGENE PROTEIN              253   6.3e-51
TREL_AVIRE  REL TRANSFORMING PROTEIN (P58 V-REL)      257   1.7e-49
DORS_DROME  EMBRYONIC POLARITY DORSAL PROTEIN.        240   1.4e-47
TREL_MOUSE  C-REL PROTO-ONCOGENE PROTEIN              231   5.1e-39
GRP1_PETHY  GLYCINE-RICH CELL WALL STRUCTURAL PROTEIN  80   .7.5e-14
K1CJ_HUMAN  KERATIN, TYPE I CYTOSKELETAL 10            70   1.7e-11
GRP2_PHAVU  GLYCINE-RICH CELL WALL STRUCTURAL PROTEIN  84   1.3e-08
HM27_HUMAN  HOMEOBOX PROTEIN HOX-2.7 (HOX2G).          71   4.4e-07
GRP1_PHAVU  GLYCINE-RICH CELL WALL STRUCTURAL PROTEIN  70   1.0e-06
ALUF_HUMAN  !!!! ALU CLASS F WARNING ENTRY !!!!        99   1.0e-06
ANDR_HUMAN  ANDROGEN RECEPTOR.                         72   3.7e-06
KG3A_RAT    GLYCOGEN SYNTHASE KINASE-3 ALPHA           79   5.4e-05
ALUB_HUMAN  !!!! ALU CLASS B WARNING ENTRY !!!!        85   0.00013
ALUE_HUMAN  !!!! ALU CLASS E WARNING ENTRY !!!!        78   0.0014
K1C6_BOVIN  KERATIN, TYPE I CYTOSKELETAL VIB           73   0.0039
MYSB_ACACA  MYOSIN HEAVY CHAIN IB·.                    71   0.016
```

Figure 8.   (a) Suspicious sequence.

C) xblast (*Alu*) and xnu-masked query

```
>NF-kappa B p49   [xblast and xnu-masked]
MESCYNPGLDGIIEYDDFKLNSSIVEPKEPAPETADGPYLVIVEQPKQRGFRFRYGCEGPSHGG
LPGASSEKGRKTYPTVKICNYEGPAKIEVDLVTHSDPPRAHAHSLVGKQCSELGICAVSVGPKD
MTAQFNNLGVLHVTKKNMMGTMIQKLQRQRLRSRPQGLTEAEQRELEQEAKELKKVMDLSIVRL
RFSAFLRASDGSFSLPLKPVTSQPIHDSKSPGASNLKISRMDKTAGSVRGGDEVYLLCDKVQKD
DIEVRFYEDDENGWQAFGDFSPTDVHKQYAIVFRTPPYHKMKIERPVTVFLQLKRKRGGDVSDS
KQFTYYPLVEDKEEVQRKRRKALPTFSQPXXXXXXXXXXXXXXXXXXXXXXXXXXXXVLMEGGVKV
REAVEEKNLGEAGRxxxxxxxxxxxxxxxxxxxxxxxxxxxxxxxxxxxxxxxxxxxxRRL
```

D) new database search output

```
                                                        Probability
Sequences producing High-scoring Segment Pairs:      Score  P(N)

KBF1_HUMAN  DNA-BINDING FACTOR KBF1                   336   2.1e-111
TREL_MELGA  C-REL PROTO-ONCOGENE PROTEIN              257   1.2e-50
TREL_CHICK  C-REL PROTO-ONCOGENE PROTEIN              253   1.4e-50
TREL_AVIRE  REL TRANSFORMING PROTEIN (P58 V-REL)      257   3.6e-49
DORS_DROME  EMBRYONIC POLARITY DORSAL PROTEIN.        240   3.0e-47
TREL_MOUSE  C-REL PROTO-ONCOGENE PROTEIN              231   1.1e-38
```

Figure 8. (b) Masking of the low-entropy and Alu-like segments.

Because XBLAST and XNU involve the same basic matching algorithm and rely on the same statistical model of alignments (Karlin & Altschul, 1990), they can be used in combination with each other and with subsequent BLAST searches, in a consistent way. Equivalent score/significance thresholds can be used throughout the entire process of i) masking the query for unworthy matches (XBLAST), ii) masking its low-complexity segments (XNU) and, iii) running the final database search. For instance, using a significance threshold of $p \leq 0.01$ for each masking step will warrant that no biologically relevant match of equal or greater significance is at risk of disappearing from the final output.

Figure 8 illustrates the combined use of XBLAST and XNU in analyzing the transcription factor NF-kappaB p49 sequence. The initial database search output is confusing. It features suspicious matches with *Alu*-related entries and similarities with DNA-binding proteins, cell wall structural protein, surface receptor, keratin and myosin! In contrast, the processed entry is solely found to match with its true, biologically relevant, relatives: transcription factors and DNA-binding proteins.

G. Query Masking: Conclusion

The limiting step of large sequencing projects is the scientific analysis and biological interpretation of the data, the generation of which can be totally automated (if highly accurate sequences are not required). Given the present rate of data acquisition (30 kb/day per sequencer), and to meet the challenge of any further increase, there is no choice but to rely mostly on automated computer analysis. The initial stage of data analysis essentially relies on comparison with various databases in search for similarities. We have seen that databases, even the best curated one, do contain anomalous entries most likely to make the result of large-scale searches worthless. In the meantime, sequence data is generated at an increasing pace (currently 300 new entries per day in GenBank) and incorporated into databases with no or very little validation. Anonymous sequence data (i.e., on which very little is known but the sequence itself, a crude genomic location (e.g. "X" chromosome), and/or expression information (e.g. "fetal brain")) now constitute the major (and increasing) fraction of public databases. In the absence of reliable annotations of the entries, the detection of anomaly or "error" cannot rely on anything else but the sequence information. In this context, automatically producing intelligible results from massive database searches involving either large or numerous anonymous sequence queries is the real challenge.

After identifying the two main causes of output inflation (and corruption), we have introduced the general concept of query masking and described two methods to improve the signal-to-noise ratio in database similarity searches. Both methods, XBLAST and XNU, detect regions of the query a priori likely to produce redundant and/or biologically irrelevant matches. Those two programs illustrate a new class of software tools required to compensate for the lack of human expertise in automated, large-scale sequence analysis. In the following, I present the identification of coding exons in mammalian genomic sequences as an application of large-scale database search for which the concept of query masking is central.

III. IDENTIFYING EXONS IN MAMMALIAN GENOMIC SEQUENCES

A. The Problem

The identification of human disease genes by positional cloning typically involves the subcloning of the candidate region, contained in a Yeast Artificial Chromosome (YAC), into a set of λ clones. Each of these λ clones are separately sequenced and assembled into a finished contig of 50 to 100 kb. Computers methods are then used to try to locate the most likely candidate protein-coding exons, which typically represent only a few percent of the whole sequence. Exon finding is a particularly difficult exercise within mammalian genomic sequences, where they are short (~150 nucleotides in average for internal exons) and can be variably and widely interspersed (one exon per 2 to 20 kb). In addition intron and intergenic mammalian sequences abound in misleading repeats (e.g., Alu elements, up to 20% in some regions), patches of simple sequences (homopolymeric, di- or trinucleotide repeats), and transposonlike structures with apparent ORFs (Line-1). Intron/exons

Large-Scale Sequence Analysis

borders (i.e., splice sites) are not conserved enough to become the basis of any useful detection method. Finally, the mammalian genome appears to consist of regions with markedly different overall nucleotide composition (G+C content, the so-called isochores (Bernardi, 1989)), thus reducing the effectiveness of the best available statistically-based exon recognition methods.

B. Grail: Finding Coding Exons from Their "Word" Usage

The most frequently used computer resources for locating protein-coding exons in anonymous vertebrate genomic sequences is Grail (Uberbacher and Mural, 1991; Xu et al., 1994), both available as an E-mail server (send a message with the word "help" on the first text line to Grail@ornl.gov) and a client-server implementation.

The coding recognition module of Grail uses a neural network which combine a series of coding prediction algorithms. Uberbacher and Mural (1991) found most of the necessary discriminating information in three statistical measures: the distribution of hexamers (6-nucleotide overlapping "words" (Claverie and Bougueleret, 1986), the distribution of in-phase hexamers (Claverie et al., 1990), and the codon positional asymmetry (Fickett, 1982). A comprehensive review of these methods (and all others) can be found in Fickett and Tung (1992). There are three versions of Grail simultaneously present on the server. Grail 1, the original program, evaluates coding potential within a fixed size (100 nucleotides) window, and does not rely on any putative splice site information. Grail 1a, still uses a fixed-length window but then evaluates a number of discrete candidates around that window to find the best boundaries. The latest version, Grail 2, uses a variable-length window, defined as an open reading frame bounded by a pair (acceptor and donor) of putative splice sites. All three systems have been trained and optimized to recognize coding exons in human DNA, but can also be used successfully on other vertebrate sequences.

All current Grail versions rank their exon predictions into three classes: "excellent", "good" and "marginal". Overall, the performance of Grail has been rated very high. According to the Grail user's guide (send "help" to Grail@ornl.gov), the Grail 1 program is claimed to find about 90% of coding exons greater than 100 bases (with 100%, 69% and 16% being real for the "excellent", "good" and "marginal" predictions, respectively). The Grail 1a program, reportedly improved, is rated as predicting 95% of exons longer than 100 nucleotides. The most recent Grail 2 program is claimed to find 91% of all coding exons (of any size) with a low false positive rate of 8.6%.

There is no dispute that the Grail system constitutes a useful resource to the genome research community, and is an improvement over previous methods (Fickett, 1982) taken individually, mostly by decreasing the fraction of false positive identification. However, a recent independent evaluation of Grail performance using the 41 human genomic contigs longer than 15 kb (in contrast with the original Grail training set) suggested that it might not be the definitive solution to

Approx. loc.	Strand	GRAIL Version 1	1a	2	Proven	Consistency
600	+	Exc	---	Exc	+	0.66
1850	+	---	Exc	Exc	+	0.66
21250	+	marg	---	---		0.33
27150	+	good	---	---		0.33
27850	+	---	Exc	---		0.33
35250	+	---	---	good		0.33
41050	+	---	marg	---		0.33
42250	+	---	good	---		0.33
50850	+	good	Exc	good		0.66
63850	+	---	---	marg		0.33
3150	−	---	marg	---		0.33
17150	−	---	---	marg		0.33
27700	−	marg	---	---		0.33
29150	−	---	Exc	---		0.33
31700	−	good	Exc	---		0.33
31900	−	marg	Exc	---		0.33
40300	−	---	---	good		0.33
42800	−	marg	---	marg		0.66
43950	−	Exc	Exc	good		0.33

Figure 9. Grail predictions from an unpublished 67-kb contig from human chromosome X (Xp22.3) [A: 30%, T: 31%, C: 19.5%, G: 19.5%].

the problem, contrary to common belief. In the more realistic condition set by Lopez et al. (1994), Grail 1 and Grail 2 only identified 68% and 71% of the coding exons, respectively. In the meantime, the rates of false positive identification were found to be 46% and 30.5%, respectively, much higher than found with the original training set.

The rigorous test of an exon prediction algorithm is difficult for a variety of reasons. First, sequences used in the test set which, of course, must be different from any constituting the training set, could still bear some resemblance. The performance of the neural network might then be influenced by the recognition of the similarity of the encoded exons (very similar proteins can be encoded by very dissimilar DNA sequences). The evaluation of false positive predictions is also difficult, because they could correspond to real exons not yet characterized (alternative mRNA, rarely expressed genes, etc.).

However, a semiquantitative test of Grail performance is possible that does not require any knowledge of the location of the actual exons. It is simply to determine the coherence of the predictions made by the various Grail programs on the same sequence. If, as originally claimed, the predictions ranked as "excellent" are correct ~100% of the time, their location should most often agree, whatever Grail program

Large-Scale Sequence Analysis 183

we use. For a realistic test (e.g., mimicking the situation of a researcher analyzing a brand-new sequence), it is best to use long sequences which have not yet appeared in the database (so no training on them is possible). Figures 9 and 10 present the results of such a consistency test using two unpublished contigs, 67 kb from the human Xp22.3 region (Legouis et al., 1991) and 94 kb from the murine XD region (Simmler, Cunningham, Avner et al., personal communication). For those two sequences, the average concordance for the "excellent" predictions is only 45.3% and 47.7%, respectively. Surprisingly, there is NOT a single case for which the three programs agreed on any prediction. No preferred concordance pattern can be found between any two of the three Grail versions, suggesting that none of them are better (or worse) than any others.

Given those surprising results, we evaluated the concordance of the three Grail programs on another unpublished, long, human genomic contig, encompassing the

Approx. loc.	Strand	GRAIL Version 1	1a	2	Proven	Consistency
150	+	---	---	good		0.33
11000	+	---	good	---		0.33
16400	+	---	Exc	---		0.33
16600	+	marg	---	---		0.33
18150	+	marg	marg	---		0.66
32850	+	---	Exc	good		0.33
33250	+	marg	---	---		0.33
38700	+	good	---	---		0.33
40150	+	marg	---	---		0.33
47000	+	Exc	Exc	good		0.66
82400	+	---	---	Exc		0.33
85000	+	---	good	---		0.33
86200	+	marg	Exc	---		0.33
3800	-	---	marg	---		0.33
18350	-	---	marg	---		0.33
19000	-	---	---	good		0.33
23150	-	---	Exc	---		0.33
24550	-	---	marg	---		0.33
35500	-	Exc	Exc	good		0.66
44700	-	good	Exc	Exc	+	0.66
49500	-	good	---	---		0.33
53650	-	good	---	Exc	+	0.33
56300	-	---	marg	---		0.33
71300	-	marg	---	---		0.33
89300	-	---	---	marg		0.33

Figure 10. Grail predictions from an unpublished 94-kb contig from mouse chromosome X (XD) [A: 30%, T: 27.5%, C: 21%, G: 21.5%].

GRAIL loc.	Strand	GRAIL Version 1	1a	2	Validation
6350	+	good	Exc	---	
8600	+	---	---	good	
9100	+	Exc	Exc	Exc	Line-1
9550	+	Exc	Exc	---	Line-1
10500	+	---	good	---	
18750	+	---	marg	---	
20454-20531	+	---	---	---	exon1, missed
21000	+	---	good	---	
22500	+	Exc	---	marg	exon2
24000	+	Exc	Exc	Exc	exon3
33000	+	Exc	Exc	Exc	exon4
43500	+	good	Exc	Exc	exon5
45100	+	---	---	Exc	exon6
200	-	---	marg	good	
9400	-	marg	---	---	
14100	-	good	Exc	---	
25350	-	---	good	---	
27350	-	Exc	---	---	
32500	-	---	Exc	marg	

Figure 11. Grail prediction from an unpublished 46-kb contig (Y chromosome) encompassing the RPS4 gene [A: 27%, T: 31.5%, C: 20%, G: 21.5%].

RPS4 gene (Zinn et al., 1994; and personal communication), encoding the well characterized 40S ribosomal protein S4, for which there are many homologues in the databases. Figure 11 shows the results of this experiment, where the average agreement between "excellent" prediction is 57.3%. A complete agreement is observed twice, but one of them identifies a clear "false positive", in fact, a Line-1 repeat. However, considering all categories, and all Grail versions, five out the six proven exons in this regions are successfully identified by at least one "excellent" prediction.

Finally, the last test involved the well-characterized human serum albumin 19-kb contig (Minghetti et al., 1986). Figure 12 shows that the average consistency of "excellent" predictions is 57%. However, in this case only 8 out of the 14 characterized exons are successfully identified by any Grail version including all categories of prediction.

C. Database "Look Up": An Alternative Method for Finding Exons

As our theoretical ability to recognize vertebrate exons slowly improves, the problem might be solved more readily by simply looking them up in the databases. Because of the increasing pace by which new sequences (in particular ESTs) are

GRAIL loc.	Strand	GRAIL Version 1	1a	2	Validation
11 - 39	+	---	---	marg	
1776 - 1854	+	---	---	---	missed
2564 - 2621	+	---	---	marg	real [2564-2621]
4076 - 4208	+	marg	good	Exc	real [4076-4208]
6041 - 6256	+	good	---	Exc	real [6041-6252]
6802 - 6934	+	---	---	---	missed
7759 - 7856	+	---	---	---	missed
8631 - 8651	+	good	---	---	
9444 - 9573	+	---	---	---	missed
10867- 11081	+	Exc	Exc	Exc	real [10867-11081]
12481- 12613	+	---	Exc	good	real [12481-12613]
13451- 13461	+	marg	---	---	
13702- 13803	+	---	---	Exc	real [13702-13799]
14977- 15115	+	---	---	---	missed
15691- 15701	+	marg	---	---	real [15534-15757]
16950- 17080	+	Exc	Exc	Exc	real [16941-17073]
17688- 17732	+	---	---	---	missed
18251- 18271	+	marg	---	---	
9300 - 9440	-	Exc	Exc	---	

Figure 12. Grail prediction from a 19-kb contig encompassing the human Serum Albumin gene [A: 31%, T: 34%, C: 17%, G: 18%].

now accumulating, a nearly complete catalog of all of them might be available in a few years. Already, because of the ancient evolutionary relationship between vertebrates and well-represented species from other phyla (*S. cerevisiae, C. elegans, A. thaliana, E. coli*), 50% of all of the human protein-encoding genes appear to be related to a homologous "prototype" in the current databases (Green et al., 1993; Claverie, 1993). A simple local similarity search using BLASTP (against the protein sequence database) or TBLASTX (again the EST databases), i.e., taking advantage of the amino acid conservation, is thus a very reasonable and straightforward method to search for putative exons in anonymous human genomic sequences. As previously discussed, we expect that the use of all of the masking techniques will be required for the database "look up" method to be of practical use. This point is illustrated in Figure 13.

Here, we have submitted the 67-kb sequence from the human Xp22.3 region (already used in Figure 9) to an exon search by local similarity with the exhaustive nonredundant protein database NR. This interval is known to contain two short coding exons (222 and 140 coding nucleotides) and the 3′-end terminal exon (with 57 coding nucleotides) from the ADML-X/KALIG-1 gene (Legouis et al., 1991; Franco et al., 1991), the mutation of which is responsible for Kallmann syndrome.

This region is a typical human genomic region with a usual density of Alu, Line-1, simple sequence and other repeated elements.

For both strands of this sequence separately, all putative internal exons (defined as ORF ≥ 90 nucleotides flanked by AG/GT splice sites) were identified, translated, and collected into a set of amino acid sequence queries. Those queries (109 for the plus strand, 119 for the minus strand) were then used to scan the database with the BLASTP program at a highly significant score threshold (S ≥ 70 , p ≤ 5.10^{-3}). Without any query masking, those searches resulted in 7-Megabyte (plus strand) and 4-Megabyte (minus strand) outputs, indicating more than 200 "candidate" exons matching more than a total of 3500 different entries in the NR database. Such an output is clearly useless and not biologically significant.

The next step was to mask out the "low-entropy" segment of those queries (using the filter "XNU+SEG" option of BLAST) prior to the same search. Although the resulting output was notably reduced (1.6 Megabyte and 1.4 Megabyte for the plus and minus strands, respectively), it was still much too large to contain only biologically significant results. In parallel, a masking of the Alu-like sequences (using XBLAST and our Alu database, see above) reduced the outputs by half

	No masking	low entropy[1] masking	Alu repeat[2] masking	low & Alu masking	& Line-1[3] masking
			plus strand		
N match	104	77	36	13	5[4]
matched	2286	511	1407	142	43
<N>	22	6.6	39	10.9	8.6
output (kb)	7550	1690	4650	470	140
			minus strand		
N match	115	90	36	11	5
matched	1313	484	836	57	12
<N>	11.4	5.3	23	5.2	2.4
output (kb)	4330	1450	2760	190	40

Notes: [1]option: -filter "XNU+SEG"; [2]blastp alu query S=70 S2=70 I xblast + query; [3]blastp line_1 query S=70 S2=70 I xblast + query; [4]includes 3 ADML-X (Kallmann syndrome gene) proven exons. Scoring matrix: blosum62, p(S≥70) ≤ 5 10-3, there are 109 putative internal (AG-GT rule) exons of length ≥90 on the plus strand and 119 putative internal (AG-GT rule) exons of length ≥90 on the minus strand.

Figure 13. Masking low-complexity sequences and common repeats: Analysis of an anonymous human genomic sequence (67-kb contig from human chromosome X (Xp22.3)).

(Figure 13). However, combining the masking of low-entropy segments and Alu-like sequences reduced the output to 13 (plus strand) and 11 (minus strand) candidates, a manageable number. An additional masking of Line-1, the next most frequent repeat in the human genome, resulted in a handful of candidates for each strand, and a total output of 30 pages. For the plus strand, the five candidates included the three proven exons, and two short pieces of retrovirus-related sequences. The five candidates from the minus strand might include real exons from a gene unrelated to ADML-X/KALIG-1 and are being investigated. The results obtained with the plus strand are biased by the fact that the ADML-X/KALIG-1 gene product is now included in NR. If we discard the matches with the human ADML-X/KALIG-1 protein and its known homologues in other vertebrate species, the two main coding exons in the interval (222 and 140 coding nucleotides) remain identified by their local similarity with various remotely related cellular adhesion molecules.

The "look up" approach we just illustrated, using the NR database, will become increasingly effective using the TBLASTX (translation-dependent DNA sequence comparison) or TBLASTN (translation versus DNA) version of the BLAST program, for comparing the anonymous queries against the rapidly growing EST data bank. However, given the various problems associated with EST data, additional steps of query masking (e.g., including vector-, organelle- and RNA-derived sequences) are required to obtain a noise-free output.

D. The Grail/Look Up Composite Approach

A natural way to try to improve our ability to detect coding exons in anonymous mammalian DNA sequences is to use Grail and a database look up method in combination. The highest sensitivity of exon detection will be reached by considering all Grail predictions AND/OR all significant matches with the protein database as potential candidates. This approach is expected to generate many false positive identifications. In contrast, the highest specificity is achieved if we require that acceptable candidates be predicted by Grail AND correspond to a significant match in the database. In any case, it is always useful to test all Grail predictions against the databases, as a way to rank them further before experimental follow-up. Again, the proper masking of the query is necessary to make the database search truly informative.

It is worth noting that, in the context of large-scale sequencing using the shotgun strategy, the identification of exons can proceed very early in the project, much before the complete assembly of the sequence (Claverie, 1994). The method simply consists of applying the Grail/Database Look Up Composite Approach on the individual 350-bp runs as they are determined. Statistically, a twofold coverage (i.e., the ratio of the total number of sequenced nucleotides to the size of the target sequence) is sufficient to find 85% of all detectable exons in a given interval (while a sixfold coverage is theoretically necessary to assemble a 25-kb contig). This "shallow shotgun" approach to exon detection has been described in detail elsewhere (Claverie, 1994). It is applicable irrespective of the size of the target sequence. Bypassing the complete assembly can lead to a huge gain in time and effort for large target sequences (>100kb). Furthermore, the shotgun approach can be fully automated, an additional advantage over directed strategies and exon trapping.

From our own estimate, we can grant Grail a minimal success rate of 50% (e.g., for the prediction ranked "excellent" and "good"). On the other hand, 50% of all newly determined genes appear to have a significant match in the current protein database (Green et al., 1993; Claverie, 1993; Dujon et al., 1994). We could expect the most sensitive Grail/look up composite method to be able to detect more coding exons in mammalian DNA. How much more will depend on how correlated are the Grail and look up methods. If totally uncorrelated, half of the exons NOT detected by Grail will be detected by database homology, thus an overall success rate of 75%. In fact, a larger fraction of all exons might still be invisible to both methods. Because most ancient conserved protein motifs have already been identified (Green et al., 1993; Claverie, 1993), it is reasonable to postulate that the genes with no relatives in the current database must be evolving very fast, thus erasing any detectable similarity. The detection of exons by Grail is primarily based on the recognition of a bias in the usage of 6-nucleotide words in coding versus noncoding regions. This bias is due to multiple causes, such as the preferential usage of synonymous codons, the CpG depletion of noncoding regions, or other not yet identified constraints. However, fast evolving genes (those with no relatives in the database) might mutate too rapidly for those constraints to have a measurable effect. Such genes will thus be as transparent to the Grail neural network as they are to similarity searches. We have preliminary evidence that Grail performs poorly with some of the genes known to be poorly conserved in mammals (cytokines, serum albumin (Figure 12)), a property which might also be linked to A+T rich regions. A detailed cross-validation of the computer-based methods with experimental exon trapping or transcript analysis is necessary to assess our actual ability to detect exons in anonymous sequences using all current theoretical techniques.

IV. STATISTICAL SIGNIFICANCE IN SEQUENCE MOTIF SEARCHING

A. The Problem

In section I, we insisted that a program must associate a quantitative measure of biological significance to the results it produces to be useful in the context of large-scale sequence analysis. If we can build a suitable mathematical model of the property we try to identify and if the proper masking techniques are used to remove the influence of known anomalies and artifacts, a measure of statistical significance can usually be derived to approximate very well what biologists consider worthwhile. For instance, we have seen that, once using the proper query masking technique, the statistically significant scores as computed by BLAST coincide well with the score-characterizing alignments between truly related sequences. The BLAST algorithm is a simplification of the one used by FASTA (see Pearson, 1990), itself an approximation of the rigorous local alignment algorithm (Smith and Waterman, 1981). Because of its simplicity, the BLAST algorithm is both fast and mathematically tractable. The a priori distribution of random scores is computable, hence the statistical significance of the alignments (Karlin and Altschul, 1990). More than sensitivity or speed, this is what most distinguishes BLAST from other alignment programs and is the real reason behind its overwhelming (and deserved) success.

Beyond pairwise sequence comparisons, position-weight matrices have been used for quite some time to capture the information contained in the local multiple alignment of distant but (functionally and/or structurally) related molecular sequences. Position-weight matrices have been used to define and locate nucleotide sequence patterns, such as promoter, splice sites, etc. (see Stormo, 1990, for review). Various types of position-weight matrices have been introduced in the study of protein motifs (see reviews by Staden, 1990; Gribskov et al., 1990; Henikoff and Henikoff, 1993; Henikoff and Henikoff, 1994). There is a common belief that position-weight matrices contain more information than the sum of their parts (the individual sequences) and that a search with a motif should be more informative than a series of searches with individual queries, although this has never been formally proven. Also, the application of motif searches to large-scale sequence analysis has been limited by the lack of an objective threshold to determine which scores truly indicate the presence of a pattern. Those limitations are now ended, because the random distribution of best matching scores for patterns spanning 10 positions has recently been established (Claverie, 1994). From this distribution, it is now straightforward to compute the statistical significance of any matching score for any position-weight matrix and use this value as an objective threshold for large-scale searches. The computed statistical significance appears to coincide well with the expected "biological significance" and realizes a good partition between false and true positive matches.

B. Mathematical Representation of Sequence Motifs

The information contained in a block alignment (Figure 14) can be turned into a set of weights characterizing the propensity of a given symbol to be found at a given position using various methods. For nucleotide sequences, it is usual to attribute to each symbol (e.g. A, T, G or C) observed at each position p a score S_{ip} such as

$$s_{ip} = \log \frac{q_{ip}}{f_i}; \quad p = 1 \text{ to } w \quad (1)$$

where q_{ip} denotes the observed frequency of symbol i at position p in a block alignment of N sequences (Figure 14), and f_i its a priori frequency in a random sequence. Once the s_{ip} matrix has been computed, it can be used to scan other sequences (or a complete database) for the pattern it represents. The scanning

Block-alignment

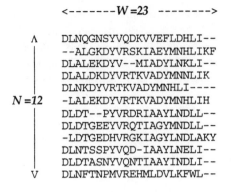

Figure 14. A block alignment (top) and the corresponding position-observed matrix representation (bottom).

algorithm simply consists of sliding the matrix along the test sequences to compute a local goodness of fit as a sum of the cognate symbol weights over a span of w positions. Eq. 1 suggests that the aggregate score S can be interpreted as a lod score between two competing hypotheses: the current positions in the test sequence i) obey the pattern encoded in the weight matrix, or ii) correspond to a random succession of symbols.

Eq. 1, however, is usually not applicable to protein sequences. Proteins consist of 20 different amino acids, and a very large number of sequences would be required in the alignment to reliably estimate each of the q_{ip}s. In practice, there are often less than 20 related sequences available for study, forcing some of the q_{ip}s to be equal to zero.

C. C, P and M Modes: Three Solutions to the Missing Data Problem

The study of protein sequences requires a specially adapted encoding scheme. This scheme must be capable of encoding the information from a block alignment of a small number ($N \geq 1$) of sequences, but should converge toward Eq. 1 when N increases. From a block of N sequences, for which we observe ob_{ip} occurrences of residue i at position p, we are now going to compute the weight matrix as

$$s_{ip} = \log \frac{Q_{ip}}{f_i} \qquad (2)$$

where Q_{ip} must have the following properties:

$$Q_{ip} \to q_{ip} = \frac{ob_{ip}}{N} \quad \text{for large N, and } Q_{ip} > 0 \text{ for } ob_{ip} = 0 \qquad (3)$$

Eq. 3 simply states that s_{ip} should retain a finite value for any sample size $N \geq 1$. We will require two additional properties to preserve a direct biological interpretation of the s_{ip} weights in terms of preference ($s_{ip} > 0$) or avoidance ($s_{ip} < 0$) of a given residue. On one hand, we will not allow a residue *not yet observed* at position p to give rise to a positive weight:

$$s_{ip}(ob_{ip} = 0) \leq 0 \Rightarrow Q_{ip}(ob_{ip} = 0) \leq f_i. \qquad (4)$$

On the other hand, a residue observed *more often* than its expected background frequency f_i should correspond to a nonnegative score:

$$s_{ip}(ob_{ip} \geq Nf_i + 1) \geq 0 \Rightarrow Q_{ip}(ob_{ip} \geq Nf_i + 1) \geq f_i \qquad (5)$$

For each independent position p, a suitable expression for Q_{ip} is

$$Q_{ip} = \frac{ob_i + \varepsilon_i}{N + \varepsilon} \qquad (6)$$

where $\varepsilon_i > 0$ denotes a "pseudocount" for amino acid i

$$\text{and } \varepsilon = \sum_{i=1}^{20} \varepsilon_i$$

We are now ready to review three different ways to compute ε_i.

1) ε_i independent of i: C ("constant") mode.
With ε_i identical for each amino acid, Eq. 6 becomes

$$Q_i = \frac{ob_i + \varepsilon/20}{N + \varepsilon} \qquad (7)$$

After incorporating the actual min $(f_i) \cong 0.01$ (for the Tryptophan residue) and max $(f_i) \cong 0.1$ (for the Leucine residue) in Eqs. (4) and (5), we found that

$$\varepsilon \leq N/4 \text{ and } \varepsilon \leq 20, \tag{8}$$

and the weight attributed to a residue not yet observed is

$$s_i = \log \frac{(\varepsilon/20)/(N+\varepsilon)}{f_i} \tag{9}$$

2) ε_i as a function of the a priori frequency of residue i: P mode.

In the absence of any other information, a natural way of distributing the pseudocounts among the various residues is according to their background probabilities f_i (Lawrence et al., 1993). Eq. (6) then becomes

$$Q_{ip} = \frac{ob_i + \varepsilon f_i}{N + \varepsilon}. \tag{10}$$

The weight attributed to a nonobserved amino acid is

$$s_i = \log \frac{\varepsilon}{N + \varepsilon} \tag{11}$$

and Eqs. (4) and (5) are now satisfied for any $\varepsilon > 0$.

3) ε_i as a function of a matrix of amino acid similarities: M mode.

Patterns are often captured (and weight matrixes built) through an iterative "bootstrap" procedure. A first set of weight s_{ip} is derived from a minimal block alignment (N small), and then used as a "seed" to detect additional related sequences. The new sequences are incorporated in an expanded block alignment and used to compute the next set of s_{ip}s. This set of weight is in turn used to scan the target sequences (e.g., the database) and the whole process is repeated until final convergence to a stable set. In such a context, we may want the computation of the pseudocounts to be updated by the information currently available about the pattern (i.e., the q_is) rather than using the random frequencies f_i. A simple way to implement such a "seeding" is to distribute each observed count ob_i among the other residues, according to their a priori similarities (e.g., polarity, size, etc.). The residue relatedness is best expressed by a transition matrix T_{ij}, such as the popular PAM transition matrices (review by Altschul, 1991). Eq. 6 can then be written as

$$Q_i = \frac{ob_i + \varepsilon \sum_{i=1}^{20} q_j T_{ij}}{N + \varepsilon} \tag{12}$$

The pseudocounts for the nonobserved amino acids ($ob_i = 0$) are now proportional to the off-diagonal T_{ij} elements they share with those already observed. It is important to note that Eq. 12 differs from

$$Q_i = \frac{\sum_{i=1}^{20} q_j T_{ij}}{N} \quad (12')$$

where the *observed* counts (instead of the pseudocounts) are distributed among related amino acids according to the similarity matrix T_{ij}. While Eq. 12' takes care of the infinity of the lod score for $ob_i = 0$ (provided all $T_{ij} > 0$), it *does not* converge towards the actual q_is for large N, unless T simultaneously converges towards the identity matrix $T_{ij} = \delta_{ij}$. Eq. (12') adds a constant noise to the information gained from the block alignment, and, in particular, fails to represent cases accurately where specific residues are forbidden at a given position.

Returning to Eq. 12, Eq. (3) and Eq. (4) imposes

$$\min(T_{ij}) > 0, \forall i \text{ and } j \neq i \quad (13)$$

$$\max(T_{ij}) \leq f_i \frac{N+\varepsilon}{\varepsilon}, \forall i \text{ and } j \neq i \quad (14)$$

thus constraining the value of the off-diagonal elements of the T_{ij} similarity matrix. On the other hand, Eq. (5) imposes that

$$\frac{1}{\varepsilon} \geq f_i(1 - T_{ii}) - (1 - f_i)\min T_{ji}, \forall i \text{ and } j \neq i \quad (15)$$

All transition matrices in the PAM1 to PAM50 range obey the simultaneous constraints in Eq.s (13) to (15) provided $\varepsilon \leq 20$ and $\varepsilon < N$. The necessity for Q_i to converge towards the observed ratio $q_i = ob_i/N$ imposes that $\varepsilon/N \to 0$ when N increases. The actual ε/N value directly determines the relative weight of the observed (ob_i) vs. estimated (pseudocounts) data and strongly influences the s_{ip} scores computed from small samples (i.e., $N > 20$). A first scheme consists in using

$$\varepsilon = N \text{ for } N \leq 20 \text{ and } \varepsilon = \text{constant} = 20 \text{ for } N > 20. \quad (16)$$

Let us analyze the case where a single residue i has been observed ($q_i = 1$) at a given position. The weight attributed for matching this residue in a sequence is

$$s_i = \log \frac{(1+T_{ii})/2}{f_i} \quad (17)$$

and for matching a residue not yet seen at this position

$$s_j = \log \frac{T_{ij}/2}{f_j} \qquad (18)$$

These weights are the same as those attributed for matching two sequences using a transition matrix computed as an average between the identity matrix and the original similarity matrix T_{ij}. Thus, using Eq. (16) for ε causes the observed and the estimated data to contribute equally until N becomes larger than 20. For $N > 20$, the influence of the estimated data then decreases linearly (1/3 for $N = 40$, ≅ 1/10 for $N = 400$). A better scheme is

$$\varepsilon = \sqrt{N} \text{ for } N \leq 20 \text{ and } \varepsilon = \text{constant} = 20 \text{ for } N > 400 \qquad (19)$$

where the influence of the estimated data is more rapidly decreasing with the sample size. Eq. (18) now becomes

$$s_j = \frac{T_{ij}}{\sqrt{N}+1} \qquad (20)$$

Here the estimated data contribute ≅ 1/2 for $N = 2$, but only ≅ 1/4 for $N = 10$, and 1/5 for $N = 20$). This latter scheme will be preferred as it corresponds to the intuitive idea that an alignment of $N \geq 20$ sequences should be sufficient to gather most of the information in a motif exhibiting a strong consensus.

4) Explicit computation of the probability of the scores

The previous section presented three ways of encoding a block alignment, as defined in Figure 14, into a matrix of $20 \times w$ weights, indicating the residue propensity at each position. Given the composition of a random sequence, one can compute the expected distribution of random scores for each individual position and, iteratively, the expected distribution of the total aggregate score S. Computer algorithms have been described that simulate the probability generating function (McLachlan, 1983; Staden, 1989).

Briefly, one takes advantage of the fact that the range of the various s_{ip} lod scores is small (e.g., [−3, +6]). Using a satisfactory approximation (e.g., 10^{-3}), the various s_{ip} (and the aggregate score S) can be scaled to positive integers and their values used to index an array where the associated probabilities are stored. As anticipated, the density distribution $p(S)$ is close to a Gaussian distribution. From $p(S)$, we can compute the cumulative distribution $c(S)$, i.e., the probability for an individual match to score $\leq S$. However this distribution is *not* the proper one to use for assessing the statistical significance of a given match in an *entire* sequence. For this we need to compute the cumulative distribution of the expected *best score* for a random sequence of length L. Given a weight matrix of length w, the probability (P) for the best score to be $>S$ in a sequence of length N is given by

$$P(S) = 1 - c(S)^{L-w}. \qquad (21)$$

The properties of $P(S)$ and of its numerically computed derivative $\mu(S) = \Delta P/\Delta S$ are discussed below.

D. Statistical Significance of Motif Scores

1) The expected best matching score per sequence follows the extreme value distribution

The values of the various s_{ip} lod scores (and thus aggregate score S) depend on the block alignment being studied (N sequences, w) and the weight computation schemes (C, P or M). To compare the distributions of scores obtained in various conditions, all scores S were expressed as reduced score $Z = (S - S_{mean})/\sigma$ for an average sequence length (L = 341 residues) in Figures (15)–(17).

First, I examined the influence of the size w on the density $\mu(Z)$ computed from Eq. (21). In Figure 15, $\mu(Z)$ was computed from "slices" of various sizes ($w = 5$ to 40) cut into a 22-sequence block alignment encompassing the glucanohydrolase

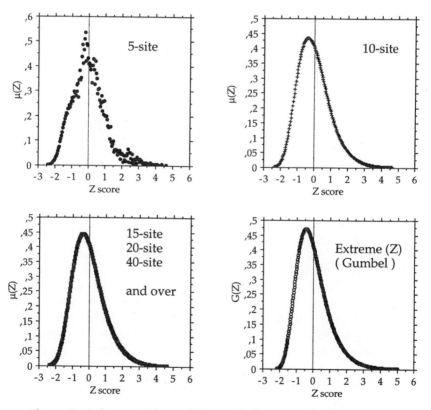

Figure 15. Influence of the motif size on the best score distribution $\mu(Z)$.

motifs (J.M. C, unpublished). For $w \geq 10$ residues, $\mu(Z)$ converges rapidly to a smooth distribution independent of w. This distribution is a close approximation of the extreme value distribution (also called Fisher–Tippet type 1, doubly exponential or Gumbel distribution (Gumbel, 1958)):

$$G(Z) = \frac{1}{\beta} e^{-(s-\alpha)/\beta} \exp[-e^{-(s-\alpha)/\beta}] \tag{22}$$

with $\alpha \cong -0.45$ and $\beta \cong 0.78$ corresponding to a $Z_{mean} = 0$, and $\sigma = 1$. The residual discrepancies (mostly around $Z = 0$) are, in part, due to the approximation of the scores as integers. In practice, these differences are negligible because the significance assessment will be made from the tail region ($Z > 2$). Thus, all for practical purposes, the random best scores can be expected to occur according to Eq. (22) for any pattern defined over a span of 10 positions or more. This result does not depend on the number of sequences used to compute the s_{ip} and thus applies to the simplest case of single sequence matching over a window of 10 residues (or more).

The extreme value distribution (more specifically its asymptotic form for high scores) was previously encountered in the context of the alignment of two sequences by the maximal paired segment algorithm (Karlin & Altschul, 1990). This algorithm (for which the window length w is not a priori fixed) is used in BLAST. Now, we have shown that the extreme value distribution also characterizes the scores of the best random matching segments obtained from a fixed-window scanning algorithms.

2) The details of the s_{ip} lod score computations (C, P or M) do not influence the shape of the expected best score distribution

The C, P or M computation schemes result in different (although similar) weight matrices (and thus different aggregate scores S) when applied to the same block alignment. A threshold on the value of the s_{ip} lod scores can also be applied, e.g., only to retain the residues and positions with the highest information content. Figure 16 shows $\mu(Z)$ for a variety of related weight matrices computed using the C, P and M (with PAM matrix 23 or 50). All of them are superimposable on $G(Z)$ [Eq. (22)]. Moreover, applying a threshold to the weights ($|s_{ip}| > 2$) or only retaining positive ones ($s_{ip} > 0$) (e.g., the "consensus" residues) has no measurable effect.

3) The range of "biologically significant" matches is well estimated from the distribution of statistical significance

Figure 17 shows the difference between $p(S)$, the density distribution of individual matching scores (bell-shaped curve), and $\mu(S)$, the random *best score* distribution for the same weight matrix (again computed from a slice of a glucanohydrolase block alignment). The top ($w = 10$) and bottom ($w = 40$) plots illustrate the influence of the pattern size w. Looking at $p(S)$, may (wrongly) lead us to conclude that a match scoring $S = -10$ (top) or $S = -40$ (bottom) is highly "significant" ($p(S) < 0.01$). However, $\mu(S)$ shows that the random occurrence of equal or higher *best scores* is not so improbable. For a random sequence of average length (341

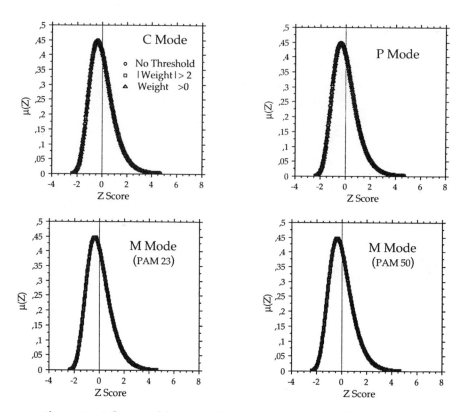

Figure 16. Influence of the various lod score computation schemes on $\mu(Z)$.

residues) scanned by the same matrices, the average expected best score is, in fact, close to 0 for $w = 10$ and -20 for $w = 40$.

The distance between the maxima of $p(S)$ and $\mu(S)$ appears to be remarkably constant and approximately equal to 4 σ (of the $p(S)$ distribution) for any window size $w > 5$. This fortuitous property is at the origin of the "6 σ" popular rule of thumb (Lipman and Pearson, 1985; Doolittle, 1990) used to interpret the results of sequence similarity searches: scores greater than $S_{6\sigma} = S_{mean} + 6\,\sigma$ (as computed from the misleading $p(S)$ distribution) are reputed "significant." Indeed, although $p(S)$ is the wrong reference to use, $S_{6\sigma}$ will always fall in the high tail (significant) of the correct best score statistics $\mu(S)$. We now see that the proper $\mu(S)$ allows an accurate delineation of the range of significant scores, as illustrated in Figure 18. Here, two weight matrices were computed (using the C and P mode) from a 26-sequence block alignment of a 72-residue well-conserved region of proteins with glucanohydrolase activity. The DBSITE program (Claverie, unpublished) was then used to scan the entire Swiss-Prot (Bairoch & Boeckmann, 1994) protein

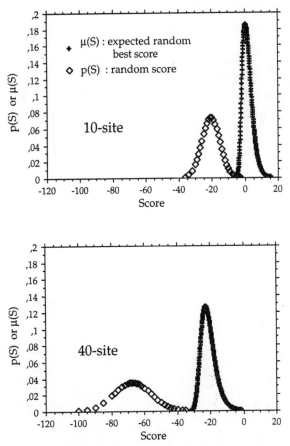

Figure 17. Difference between $p(S)$, the density distribution of individual matching score (bell-shaped curve), and $\mu(S)$, the random best score distribution for the same weight matrix.

sequence database for this motif. The complete (experimental) distribution of the observed best scores was recorded. The distributions exhibit two distinct domains: a plateau, where a slowly increasing number of bona fide glucanohydrolase-related proteins are progressively revealed, and a steep region corresponding to a high number of low scoring matches with unrelated sequences ("false positive") and ultimately including the whole database. This two-slope feature is characteristic of all weight matrices derived from an actual, biologically significant motif. After computing $\mu(S)$ [Eq. (21)], we can determine the scores corresponding to $p < 0.01$ and $p < 0.001$. These usual significance threshold values (indicated in Figure 18) very accurately correspond to the sharp slope transition between the "true positive"

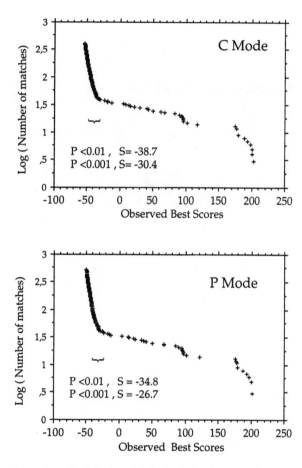

Figure 18. Statistical vs. biological significance of scores.

and "false positive" domains. Similarly good results are obtained with a wide variety of protein motifs for all log score computation schemes (C, P or M).

E. Motif Vs. Blast Search: The Whole Vs. the Sum of Its Parts

We are now capable of computing the statistical significance of the score of any best alignment between a motif and a sequence. Thus, we can finally compare BLAST and a motif search in terms of specificity and sensitivity and answer the question: are we gaining anything by building position-weight matrices from sequence alignments? The protocol we used for this comparison is illustrated in Figure 19.

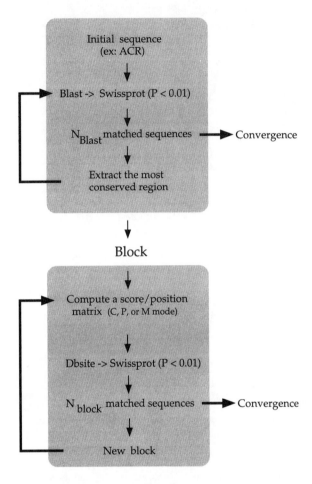

Figure 19. Protocol to compare the performance of BLAST vs. Motif iterative searches.

We initiated the whole process with a single sequence query known to contain a well-conserved functional motif, capable of identifying a large family of related proteins. In the following example, I chose a member of the ACR set, the α-amylase from *Drosophila*. This sequence was used as a query to scan the whole Swiss-Prot database using the BLASTP program. The BLASTP search was set to report matches with statistical significance level up to 0.01. A total of 25 different Swiss-Prot sequences were matched with this first pass. The protocol then followed two alternative pathways:

Large-Scale Sequence Analysis

```
ISNYNDANEVRNCELVGLRDLNQGNSYVQDKVVEFLDHLIDLGVAGFRVDAAKHMWPADLAVIYGRLKNLNT
IENYNDATQVRDCRLSGLLDLALGKDYVRSKIAEYMNHLIDIGVAGFRIDASKHMWPGDIKAILDKLHNLNS
IESYNDPYQVRDCQLVGLLDLALEKDYVRSMIADYLNKLIDIGVAGFRLDASKHMWPGDIKAVLDKLHNLNT
INNYNDANQVRNCRLSGLLDLALDKDYVRTKVADYMNNLIDIGVAGFRLDAAKHMWPGDIKAVLDKLHNLNT
IENYQDAAQVRDCRLSGLLDLALEKDYVRTKVADYMNHLIDIGVAGFRLDASKHMWPGDIKAILDKLHNLNT
IDNYNDAYQVRNCRLTGLLDLALEKDYVRTKVADYMNHLIDIGVAGFRLDAAKHMWPRDIKAVLDKLHNLNT
INDYGNRANVQNCELVGLADLDTGEPYVRDRIAAYLNDLLLLGVDGFRIDAAKHMPAADLTAIKAKVGNGST
ITDYQDRWNVQHCELVGLADLDTGEEYVRQTIAGYMNDLLSLGVDGFRIDAATHIPAEDLANIKSRLSNPN.
ISNYQDRANVQNCELVGLQLPDLDTGEDHVRGKIAGYLNDLASLGVDGFRIDAAKHMPAADLANIKSRLTNPN.
ITNWNDKEVQHCELVGLADLKTSSPYVQDRIAAYLNELIDLGVAGFRIDAAKHIPEGDLQAILSRLKNVH.
....NDRYRVQNCELVGLADLDTASNYVQNTIAAYINDLQAIGVKGFRFDASKHVAASDIQSLMAKV.....
IKNWSDRWDVTQNSLLGLYDWNTQNTQVQSYLKRFLDRALNDGADGFRFDAAKHI................
VTNWNDFFQVKNHNLFNLSDLNQSNTDVYQYLLDGSKFWIDAGVDAIRIDAIKHM................
...YTDRLELTYYSMGGLPDVDTENTGFQQYFYEFLKDCVYLGADGFRIDTAKHI................
..............LYDLADLNHNNSTIDTYFKNAIRLWLDMGIDGIRVDAVKHM................
..............LYDLADLNHNNSSVDVYLKDAIKMWLDLGVDGIRVDAVKHM................
..............LYDLADLNHNNSTVDTYLKDAIKMWLDLGIDGIRMDAVKHM................
..............LFDLADLNQQNSTIDSYLKSAIKVWLDMGIDGIRLDAVKHM................
..............LYDLADLNHNNSTIDTYFKDAIKLWLDMGIDGIRVDAVKHM................
..............LYDLADINQNNNTIDSYLKESIQLWLNLGVDGIRFDAVKHM................
..............DLNYRNPAVVEEMKNVLRYWLDRGVSGFRIDAVPYLFESDI...............
..............LFDLADLNHQNPVIDRYLKDAVKMWIDMGIDGIRVDAVKHM................
..............LYDLADFNHHNNATIDKYFKDAIKLWLDMGVDGIRVDAVKHM................
..............DLNFDNPKVREEVKKIAKFWIEKGVDGFRLDAAKHIYDDD.............
..............DLNFTNPMVREHMLDVLKFWLDRGVDGFRIDAVPHIY................


# pattern source= amylase.ali.155.227
# mode= 1, threshold= 0, from 25 to 7 seqs, width= 73
# range of perfect matches from 240 to 92.3
# Average Score= -126.735 , Std. Deviation= 15.2446
#-----------------------------------------------------------
#           For an average (341 aa) Sequence
#-----------------------------------------------------------
# Median     of best      score= -61.4
# 1/100      significant  score= -43.7
# 1/1000     significant  score= -35.2
# 1/10000    significant  score= -27.3
# 1/100000   significant  score= -19.8
# ---------------------- Verbose Mode ----------------------
# Average  of best      score= -60.5501
# Std dev. of best      score=   5.68633
1:A -2.9 C -1 D -2.4 E -2.7 F -2 G -2.8 H -1.2 I  3.8 ...   X -1.9
2:A -2.9 C -1 D  0.2 E  2 F -2 G -2.8 H -1.2 I -2.5 K ......X -1.2
.............................................................
```

Figure 20. Definition of a glucanohydrolase motif.

i) On one hand, we collected the homologous (matched) regions of those 25 sequences and, with each of them separately, ran a BLASTP search (still at $p \leq 0.01$), again collected the matched sequences (after removing the redundancy), and did it again until convergence (i.e., no more new target sequences matched). This iterative BLASTP search allowed the identification of 43 different Swiss-Prot sequences, including 1 false positive (i.e., a protein unlikely to be related to the general glucanohydrolase family).

ii) On the other hand, we used the result of the first BLASTP search to build (using the MKSITE program) a position-weight matrix (centered on the multialignment of the longest homologous region; see Figure 20.). Using the threshold score computed (By MKSITE) to correspond to $p \leq 0.01$ [Eq. (21)], we then scanned Swiss-Prot with the DBSITE program and collected all the matched sequences. The target sequences not initially contained in the alignment were then incorporated in a refined version of the weight matrix, a new score threshold (at $p \leq 0.01$) computed, and the DBSITE search run again. The whole process was performed iteratively until convergence, using the C, M and P log score computation schemes.

In all cases, the iterative weight-matrix searches allowed the identification of more glucanohydrolase family members than the iterative BLASTP searches at the same $p \leq 0.01$ significance level. However, there were marked differences in the performances of the C, P, and M modes. The C mode allowed the correct identification of 80 glucanohydrolase family members plus 14 false positives. Using the M mode, 51 family members were identified plus 5 false positives. Finally, the P mode allowed the identification of 48 family members plus 14 false positives. Thus, in all cases, the weight matrix search appears more sensitive than a multiple BLASTP search. However, the simplest log score computation scheme C (first proposed by the 17th century French mathematician Blaise Pascal) performs much better than the M and P mode. In all cases, the weight matrix searches resulted in a number of false positives larger than BLASTP. Although some of the false positive identification may represent the discovery of a distant evolutionary relationship, it is reasonable to conclude from this study (and from other comparable computer experiments with various protein families) that "the whole is indeed better than the sum of the parts." Defining weight matrices is more than a convenient way to summarize the functional motif within a multialignment. Associated with the computation of the proper (i.e., statistically significant) score threshold, it leads to a more sensitive (although less specific) detection of distant relationships between protein sequences.

ACKNOWLEDGMENTS

I thank my former colleagues at the National Center for Biotechnology Information (NIH, Bethesda, MD) for stimulating discussions and Drs. Chantal Abergel and Daniel Gautheret for their help in improving earlier versions of this manuscript.

REFERENCES

Adams, M. D., Kelley, J. M., Gocayne, J. D., Dubnick, M., Polymeropoulos, M. H., Xiao, H., Merril, C. R., Wu, A., Olde, B., Moreno, R. F. et al. (1991). Complementary DNA sequencing: expressed sequence tags and human genome project. Science 252, 1651–1656.

Adams, M. D., Dubnick, M., Kerlavage, A. R., Moreno, R. F., Kelley, J. M., Utterback, T. R., Nagle, J. W., Fields, C. A., & Venter, J. C. (1992). Sequence Identification of 2,375 human brain genes. Nature 355, 632–634.

Adams, M. D., Kerlavage, A. R., Fields, C., & Venter, J. C. (1993a). 3,400 new expressed sequence tags identify diversity of transcripts in human brain. Nature Genet. 4, 256–267.

Adams, M. D., Soares, M. B., Kerlavage, A. R., Fields, C., & Venter, J. C. (1993b). Rapid cDNA sequencing (expressed sequence tags) from a directionally cloned human infant brain cDNA library. Nature Genet. 4, 373–380.

Albrecht, J. C., Nicholas, J., Biller, D., Cameron, K. R., Biesinger, B., Newman, C., Wittmann, S., Craxton, M. A., Coleman, H., Fleckenstein, B., & Honess, R. W. (1992). Primary structure of the *herpesvirus saimiri* genome. J. Virol. 66, 5047–5058.

Altschul, S. F. (1991). Amino acid substitution matrices from an information theoric perspective. J. Mol. Biol. 219, 555–565.

Altschul, S. F., Gish, W., Miller, W., Myers, E. W., & Lipman, D. J. (1990). Basic local alignment search tool. J. Mol. Biol. 215, 403–410.

Ayres, M. D., Howard, S. C., Kuzio, J., Lopez-Ferber, M., & Possee, R. D. (1994). The complete DNA sequence of *Autographa californica* nuclear polyhedrosis virus. Virology 202, 586–605.

Baer, R. J., Bankier, A. T., Biggin, M. D., Deininger, P. L., Farrell, P. J., Gibson, T. J., Hatfull, G. F., Hudson, G. S., Satchwell, S. C., Seguin, C., Tuffnell, P. S., & Barrell, B. G. (1984). DNA sequence and expression of the B95-8 Epstein-Barr virus genome. Nature 310, 207–211.

Bairoch, A. & Boeckmann, B. (1994). The SWISS-PROT protein sequence database: current status. Nucleic Acids Res. 22, 3578–3580.

Bairoch, A. & Bucher, P. (1994). PROSITE, recent developments. Nucleic Acids Res. 22, 3583–3589.

Barker, W. C., George, D. G., Mewes, H.-W., Pfeiffer, F., & Tsugita, A. (1993). The PIR-International database. Nucleic Acids Res. 21, 3089–3092.

Benson, D. A., Boguski, M., Lipman, D. J., & Ostell, J. (1994). GenBank. Nucleic Acids Res. 22, 3441–3444.

Bernardi, G. (1989). The isochore organization of the human genome. Annu. Rev. Genet. 23, 637–661.

Blattner, F. R., Burland, V., Plunkett, G., III, Sofia, H. J., & Daniels, D. L. (1993). Analysis of the *Escherichia coli* genome. IV. DNA sequence of the region from 89.2 to 92.8 minutes. Nucleic Acids Res. 21, 5408–5417.

Boguski, M. S., Lowe, T. M., & Tolstoshev, C. M. (1993). dbEST—database for "expressed sequence tags." Nature Genet. 4, 332–333.

Boguski, M. S., Tolstoshev, C. M., & Bassett, D. E. Jr. (1994). Gene discovery in dbEST. Science 265, 1993–1994.

Burglin, T. R. & Barnes, T. M. (1992). Introns in sequence tags. Nature 357, 367–367.

Burland, V., Plunkett, G., III, & Blattner, F. R. (1994). Analysis of the *Escherichia coli* genome. VI. DNA sequence of the region from 92.8 through 100 minutes (338534 nucleotides) (direct submission to GenBank, Aug. 22, 1994).

Cawthon, R. M., Weiss, R. B., Xu, G., Viskochil, D., Culver, M., Stevens, J., Robertson, M., Dunn, D., Gesteland, R., O'Connell, P., & White, R. (1990). A major segment of the neurofibromatosis type 1 gene: cDNA sequence, genomic structure and point mutations. Cell 62, 193–201.

Chakrabarti, D., Reddy, G. R., Dame, J. B., Almira, E. C., Laipis, P. J., Ferl, R. J., Yang, T. P., Rowe, T. C., & Schuster, S. M. (1994). Analysis of Expressed Sequence Tags from *Plasmodium falciparum*. Mol. Biochem. Parasitol. 66, 97–104.

Chee, M. S., Bankier, A. T., Beck, S., Bohni, R., Brown, C. M., Cerny, R., Horsnell, T., Hutchison III, C. A., Kouzarides, T., Martignetti, J. A., Preddie, E., Satchwell, S. C., Tomlinson, P., Weston, K. M., & Barrell, B. G. (1990). Analysis of the protein-coding content of the sequence of human cytomegalovirus strain AD169. Curr. Top. Microbiol. Immunol. 154, 125–169.

Cherry, J. M., Cartinhour, S. W., & Goodman, H. M. (1992). AAtDB, an *Arabidopsis thaliana* database. Plant Mol. Biol. Rep. 10, 308–309.

Chissoe, S. L., Bodenteich, A., Wang, Y.-F., Wang, Y.-P., Freeman, A., Burian, D., Clifton, S., Groffen, J., Heisterkamp, N., & Roe, B. A. (1994). Human breakpoint cluster region (BCR) gene region (*152141 nucleotides*) (direct submission to GenBank Feb. 22, 1994).

Claverie, J.-M. (1986). Correct translation of protein coding regions in GenBank. Trends Biochem. Sci. 11, 381–382.
Claverie, J.-M. (1993). Detecting frame shifts by amino acid sequence comparison. J. Mol. Biol. 234, 1140–1157.
Claverie, J.-M. (1993). Database of ancient sequences. Nature 364, 19–20.
Claverie, J.-M. (1994). Large scale sequence analysis. In: Automated DNA Sequencing and Analysis techniques (Adams, M. D., Fields, C., & Venter, J. C., Eds.), Chapter 36, pp. 267–279, Academic, New York.
Claverie, J.-M. (1994). A streamlined random sequencing strategy for finding coding exons. Genomics 23, 575–581.
Claverie, J.-M. (1994). Some useful statistical properties of position-weight matrices. Comput. Chem. 18, 287–294.
Claverie, J.-M. & Bougueleret, L. (1986). Heuristic Informational Analysis of Sequences. Nucl. Acids Res. 14, 179–196.
Claverie, J. M. & Makalowski, W. (1994). Alu alert. Nature 371, 752–752.
Claverie, J.-M., Sauvaget, I., & Bougueleret, L. (1990). k-tuple frequency analysis: from intron/exon discrimination to T-cell epitope mapping. Meth. Enzym. 183, 237–252.
Claverie, J. M. & States, D. (1993). Information enhancement methods for large-scale sequence analysis. Comput. Chem. 17, 191–201.
Davies, R. W., Roberts, A. B., Morris, A. J., Griffith, G., Jerecic, J., Ghandi, S., Kaiser, K., & Savioz, A. (1994). Enhanced access to rare brain cDNAs by prescreening libraries: 206 new mouse brain ESTs (direct submission to Genbank, Aug. 1994).
Desprez, T., Amselem, J., Chiapello, H., Caboche, M., & Hofte, H. (1993). The *Arabidopsis thaliana* transcribed genome: the GDR cDNA program (direct submission to GenBank, Aug. 1993).
Doolittle, R. F. (1990). Searching through sequence databases. Meth. Enzym. 183, 99–110.
Dujon, B., Alexandraki, D., Andre, B., Ansorge, W., Baladron, V., Ballesta, J.P., Banrevi, A., Bolle, P.A., Bolotin-Fukuhara, M., Bossier, P. et al. (1994). Complete DNA sequence of yeast chromosome XI. Nature 369, 371–378.
Durbin, R. & Thierry-Mieg, J. (1991). A *C. elegans* Database. Documentation, code and data available from anonymous FTP servers at lirmm.lirmm.fr, cele.mrc-lmb.cam.ac.uk and ncbi.nlm.nih.gov.
Duret, L., Mouchiroud, D., & Gouy, M. (1994). HOVERGEN: a database of homologous vertebrate genes. Nucleic Acids Res. 22, 2360–2365.
Fickett, J. W. (1982). Recognition of protein coding regions in DNA sequences. Nucl. Acids Res. 10, 5303–5018.
Fickett, J. W. & Tung, C.-S. (1992). Assessment of protein coding measures. Nucl. Acids Res. 20, 6441–6450.
Franco, B., Guioli, S., Pragliola, A., Incerti, B., Bardoni, B., Tonlorenzi, R., Carrozzo, R., Maestrini, E., Pieretti, M., Taillon-Miller, P., Brown, C. J., Willard, F. H., Lawrence, C. B., Persico, G. M., Camerino, G., & Ballabio, A. (1991). A gene deleted in Kallmann's syndrome shares homology with neural cell adhesion and axonal path-finding molecules. Nature 353, 529–536.
Fujita, N., Mori, H., Yura, T., & Ishihama, A. (1994). Systematic sequencing of the *Escherichia coli* genome: analysis of the 2.4–4.1 min (110,917–193,643 bp) region. Nucleic Acids Res. 22, 1637–1639.
Gish, W. & States, D. J. (1993). Identification of protein coding regions by database similarity search. Nature Genet. 3, 266–272.
Ghosh, D. (1993). Status of the transcription factors database (TFD). Nucleic Acids Res. 21, 3117–3118.
Goebel, S. J., Johnson, G. P., Perkus, M. E., Davis, S. W., Winslow, J. P. & Paoletti, E. (1990). The complete DNA sequence of Vaccinia virus. Virology 179, 247–266.
Grausz, J. D. & Auffray, C. (1993). Strategies in cDNA programs. Genomics 17, 530–532.
Green, P., Lipman, D., Hillier, L., Waterston, R., States, D., & Claverie, J.-M. (1993). Ancient conserved regions in new gene sequences and the protein databases. Science 259, 1711–1716.

Gribskov, M., Lüthy, R., & Eisenberg, D. (1990). Profile analysis. Meth. Enzym. 183, 146–159.
Gumbel, E. J. (1958). Statistics of Extremes. Columbia Univ., New York.
Gutell, R. R. (1993). Collection of small subunit (16S- and 16S-like) ribosomal RNA structures. Nucleic Acids Res. 21, 3051–3054.
Gutell, R. R., Gray, M. W., & Schnare, M. N. (1993). Collection of large subunit (23S- and 23S-like) ribosomal RNA structures: 1993. Nucleic Acids Res. 21, 3055–3074.
Henikoff, S. & Henikoff, J. G. (1993). Performance evaluation of amino acid substitution matrices. Proteins 17, 49–61.
Henikoff, S. & Henikoff, J. G. (1994). Protein family classification based on searching a database of blocks. Genomics 19, 97–107.
Hofmann, K. & Stoffel, W. (1993). TMBASE—A database of membrane spanning protein segments. Biol. Chem. Hoppe-Seyler 374, 166–166.
Honore, N., Bergh, S., Chanteau, S., Doucet-Populaire, F., Eiglmeier, K., Garnier, T., Georges, C., Launois, P., Limpaiboon, T., Newton, S., et al. (1993). Nucleotide sequence of the first cosmid from the *Mycobacterium leprae* genome project: structure and function of the Rif-Str regions. Mol. Microbiol. 7, 207–214.
Iris, F. J. M., Bougueleret, L., Prieur, S., Caterina, D., Primas, G., Perrot, V., Jurka, J., Rodrigez-Tome, P., Claverie, J.-M., Cohen, D., & Dausset, J. (1993). Dense Alu clustering and a potential new member of the NF-kappa B family within a 90 kb HLA class III segment. Nature Genet. 3, 137–145.
Jurka, J., Walichiewicz, J., & Milosavljevic, A. (1992). Prototypic sequences for human repetitive DNA. J. Mol. Evol. 35, 286–291.
Karlin, S. & Altschul, S. F. (1990). Methods for assessing the statistical significance of molecular sequence features by using general scoring schemes. Proc. Natl. Acad. Sci. USA 87, 2264–2268.
Karlin, S., Dembo, A., & Kawabata, T. (1990). Statistical composition of high-scoring segments from molecular sequences. Ann. Stat. 18, 571–581.
Khan, A. S., Wilcox, A. S., Polymeropoulos, M. H., Hopkins, J. A., Stevens, T. J., Robinson, M., Orpana, A. K., & Sikela, J. M. (1992). Single pass sequencing and physical and genetic mapping of human cDNAs. Nature Genet. 2, 180–185.
Kolakowski, L. F. (1994). GCRDb: A G-protein--coupled receptor database, Receptors and Channels. The International Journal of Receptors, Channels and Transporters (in press).
Koop, B. F., Wilson, R. K., Wang, K., Vernooij, B., Zallwer, D., Kuo, C. L., Seto, D., Toda, M., & Hood, L. (1992). Organization, structure, and function of 95 kb of DNA spanning the murine T-cell receptor C alpha/C delta region. Genomics 13, 1209–1230.
Koop, B. F., Rowen, L., Wang, K., Kuo, C. L., Seto, D., Lenstra, J. A., Howard, S., Shan, W., Deshpande, P., & Hood, L. (1994). The human T-cell receptor TCRAC/TCRDC (C alpha/C delta) region: organization, sequence, and evolution of 97.6 kb of DNA. Genomics 19, 478–493.
Kristensen, T., Lopez, R. S., & Prydz, H. (1992). An estimate of the sequencing error frequency in the DNA sequence databases. DNA Seq. 2, 343–346.
Kunst, F. & Devine, K. (1991). The project of sequencing the entire *Bacillus subtilis* genome. Res. Microbiol. 142, 905–912.
Lamperti, E. D., Kittelberger, J. M., Smith, T. F., & Villa-Komaroff, L. (1992). Corruption of genomic databases with anomalous sequences. Nucleic Acids Res. 20, 2741–2747.
Lawrence, C. E., Altschul, S. F., Boguski, M. S., Liu, J. S., Neuwald, A. F., & Wootton, J. C. (1993). Detecting subtle sequence signals: a Gibbs sampling strategy for multiple alignment. Science 262, 208–214.
Legouis, R., Hardelin, J-P., Levilliers, J., Claverie, J.-M., Compain, S., Wunderle, V., Millasseau, P., Le Paslier, D., Cohen, D., Caterina, D., Bougueleret, L., Lutfalla, G., Weissenbach, J., & Petit, C. (1991). The candidate gene for the X-linked Kallmann syndrome encodes a protein related to adhesion molecules. Cell 67, 423–435.

Linder, P., Doelz, R., Mosse, M.-O., Lazowska, J., & Slonimski, P. P. (1993). LISTA, a comprehensive compilation of nucleotide sequences encoding proteins from the yeast *Saccharomyces*. Nucleic Acids Res. 21, 3001–3002.

Lipman, D. J. & Pearson, W. R. (1985). Rapid and sensitive protein similarity searches. Science 227, 1435–1441.

Lopez, R., Kristensen, T., & Prydz, H. (1992). Database contamination. Nature 355, 211–211.

Lopez, R., Larsen, F., & Prydz, H. (1994). Evaluation of the exon prediction of the Grail software. Genomics 24, 133–136.

Martin-Gallardo, A., McCombie, W. R., Gocayne, J. D., FitzGerald, M. G., Wallace, S., Lee, B. M., Lamerdin, J., Trapp, S., Kelley, J. M., Liu, L.-I., Dubnick, M., Johnston-Dow, L. A., Kerlavage, A. R., de Jong, P. J., Carrano, A., Fields, C. A., & Venter, J. C. (1992). Automated DNA sequencing and analysis of 106 kilobases from human chromosome 19q13.3. Nature Genet. 1, 34–39.

Massung, R. F., Esposito, J. J., Liu, L.-I., Qi, J., Utterback, T. R., Knight, J. C., Aubin, L., Yuran, T. E., Parsons, J. M., Loparev, V. N., Selivanov, N. A., Cavallaro, K. F., Kerlavage, A. R., Mahy, B. W. J. & Venter, J. C. (1993). Potential virulence determinants in terminal regions of variola smallpox virus genome. Nature 366, 748–751.

McCombie, W. R., Adams, M. D., Kelley, J. M., FitzGerald, M. G., Utterback, T. R., Khan, M., Dubnick, M., Kerlavage, A. R., Venter, J., & Fields, C. (1992). *Caenorhabditis elegans* expressed sequence tags reveal gene families and potential disease gene homologues. Nature Genet. 1, 124–131.

McGeoch, D. J., Dalrymple, M. A., Davison, A. J., Dolan, A., Frame, M. C., McNab, D., Perry, L. J., Scott, J. E., & Taylor, P. (1988). The complete DNA sequence of the long unique region in the genome of herpes simplex virus type 1. J. Gen. Virol. 69, 1531–1574.

McLachlan, A. D. (1983). Analysis of gene duplication repeats in the myosin rod. J. Mol. Biol. 169, 15–30.

Minghetti, P. P., Ruffner, D. E., Kuang, W. J., Dennison, O. E., Hawkins, J. W., Beattie, W. G., & Dugaiczyk, A. (1986). Molecular structure of the human albumin gene is revealed by nucleotide sequence within q11-22 of chromosome 4. J. Biol. Chem. 261, 6747–6757.

Monaco, A. P. (1994). Isolation of genes from cloned DNA. Curr. Opinion Genet. Dev. 4, 360–365.

Mount, D. W. & Schatz, B. R. (1994). A genomic database of *Escherichia coli*: total information on a given organism. In: Biocomputing. Informatics and genome projects (Smith, D.W., Ed.), pp. 249–268, Academic, New York.

Neefs, J.-M., Van de Peer, Y., De Rijk, P., Chapelle, S., & De Wachter, R. (1993). Compilation of small ribosomal subunit RNA structures. Nucleic Acids Res. 21, 3025–3049.

Newman, T. (1993) (direct submission to GenBank).

Okubo, K., Hori, N., Matoba, R., Niiyama, T., Fukushima, A., Kojima, Y., & Matsubara, K. (1992). Large scale cDNA sequencing for analysis of quantitative and qualitative aspects of gene expression. Nature Genet. 2, 173–179.

Oliver, S.G., van der Aart, Q. J., Agostoni-Carbone, M. L., Aigle, M., Alberghina, L., Alexandraki, D., Antoine, G., Anwar, R., Ballesta, J. P., Benit, P. et al. (1992). The complete DNA sequence of yeast chromosome III. Nature 357, 38–46.

Pearson, W. R. (1990). Rapid and sensitive sequence comparison with FASTP and FASTA. Meth. Enzym. 183, 63–98.

Posfai, J. & Roberts, R. J. (1992). Finding errors in DNA sequences. Proc. Natl. Acad. Sci. USA 89, 4698–4702.

Prediger, E. A., Hoffman, S., Edelman, G. M., & Cunningham, B. A. (1988). Four exons encode a 93-base-pair insert in three neural cell adhesion molecule mRNAs specific for chicken heart and skeletal muscle. Proc. Natl. Acad. Sci. USA 85, 9616–9620.

Reddy, G. R., Chakrabarti, D., Schuster, S. M., Ferl, R. J., Almira, E. C., & Dame, J. B. (1993). Gene sequence tags from *Plasmodium falciparum* genomic DNA fragments prepared by the genease activity of mung bean nuclease. Proc. Natl. Acad. Sci. USA 90, 9867–9871.

Rice, C. M. & Cameron, G. N. (1994). Submission of nucleotide sequence data to EMBL/GenBank/DDBJ. Methods Mol. Biol. 24, 355–366.

Richards, S., Eichler, E. E., Lu, F., King, J., Pizzuti, A., Nelson, D. L., & Gibbs, R. A. (1994). Complete Sequence of the Human FMR-1 locus (61613 nucleotides) (direct submission to GenBank, Nov. 8, 1994).

Romano, P., Aresu, O., Parodi, B., Manniello, A., Campi, G., Angelini, G., Romani, M., Iannotta, B., Rondanina, G., Ruzzon, T., & Santi, L. (1993). Molecular Probe Data Base: a database on synthetic oligonucleotides. Nucleic Acids Res. 21, 3007–3009.

Rowen, L., Koop, B. F., & Hood, L. (1994). Sequence of the human T cell receptor beta locus (684973 nucleotides) (direct submission to GenBank, Oct. 29, 1994).

Rudd, K. E. (1993). Maps, genes, sequences, and computers: an *Escherichia coli* case study. ASM News 59, 335–341.

Savakis, C., & Doelz, R. (1993). Contamination of cDNA sequences in databases. Science 259, 1677–1678.

Senapathy, P., Shapiro, M. B., & Harris, N. L. (1990). Splice junctions, Branch point sites, and exons: sequence statistics, identification, and applications to genome project. Methods Enzymol. 183, 252–278.

Shchelkunov, S. N., Blinov, V. M., & Sandakhchiev, L. S. (1993). Genes of variola and vaccinia viruses necessary to overcome the host protective mechanisms (185575 nucleotides). FEBS Lett. 319, 80–83.

Shumyatsky, G. & Reddy, R. (1993). Compilation of small RNA sequences. Nucleic Acids Res. 21, 3017–3017.

Sikela, J. M. & Auffray, C. (1993). Finding new genes faster than ever. Nature Genet. 3, 189–191.

Slightom, J. L., Siemieniak, D. R., Sieu, L. C., Koop, B. F., & Hood, L. (1994). Nucleotide sequence analysis of 77.7 kb of the human V beta T-cell receptor gene locus: direct primer-walking using cosmid template DNAs. Genomics 20, 149–168.

Smith, D. R. (1994). (unpublished, Genome Therapeutics Corp., Waltham MA).

Smith, T. F. & Waterman, M. S. (1981). Identification of common molecular subsequences. J. Mol. Biol. 147, 195–197.

Sofia, H. J., Burland, V., Daniels, D. L., Plunkett, G. III, & Blattner, F. R. (1994). Analysis of the *Escherichia coli* genome. V. DNA sequence of the region from 76.0 to 81.5 minutes (225419 nucleotides) (direct submission to GenBank, Mar. 25, 1994).

Staden, R. (1989). Methods for calculating the probabilities of finding patterns in sequences. Comput. Appl. Biosci. 5, 89–96.

Staden, R. (1990). Searching for patterns in protein and nucleic acid sequences. Methods Enzym. 183, 193–211.

Stormo, G. D. (1990). Consensus patterns in DNA. Methods Enzym. 183, 211–221.

Sulston, J., Du, Z., Thomas, K., Wilson, R., Hillier, L., Staden, R., Halloran, N., Green, P., Thierry-Mieg, J., Qiu, L., et al. (1992). The *C. elegans* genome sequencing project: a beginning. Nature 356, 37–41.

Telford, E. A. R., Watson, M. S., McBride, K., & Davison, A. J. (1992). The DNA sequence of equine herpesvirus 1 (150223 nucleotides) (direct submission to GenBank, June 11, 1992).

Toguchida, J., McGee, T. L., Paterson, J. C., Eagle, J. R., Tucker, S., Yandell, D. W., & Dryja, T. P. (1993). Complete genomic sequence of the human retinoblastoma susceptibility gene. Genomics 17, 535–543.

Tugendreich, S., Boguski, M. S., Seldin, M. S., & Hieter, P. (1993). Linking yeast genetics to mammalian genomes: identification and mapping of the human homolog of CDC27 via the expressed sequence tag (EST) database. Proc. Natl. Acad. Sci. USA 90, 10031–10005.

Uberbacher, E. C. & Mural, R. J. (1991). Locating protein-coding regions in DNA sequences by a multiple sensor-neural approach. Proc. Natl. Acad. Sci. USA 88, 11261–11265.

Verkerk, A. J. M. H., Pieretti, M., Sutcliffe, J. S., Fu, Y.-H., Kuhl, D. P. A., Pizzuti, A., Reiner, O., Richards, S., Victoria, M. F., Zhang, F., Eussen, B., van Ommen, G.-J. B., Blonden, L. A. J., Riggins, G. J., Chastain, J. L., Kunst, C. B., Galjaard, H., Caskey, C. T., Nelson, D. L., Oostra, B. A., & Warren, S. T. (1991). Identification of a gene (FMR-1) containing a CGG repeat coincident with a breakpoint cluster region exhibiting length variation in fragile X syndrome. Cell 65, 905–914.

Waterston, R., Martin, C., Craxton, M., Huynh, C., Coulson, A., Hillier, L., Durbin, R. K., Green, P., Shownkeen, R., Halloran, N., Hawkins, T., Wilson, R., Berks, M., Du, Z., Thomas, K., Thierry-Mieg, J., & Sulston, J. (1992). A survey of expressed genes in *Caenorhabditis elegans*. Nature Genet. 1, 114–123.

Wilson, R., Ainscough, R., Anderson, K., Baynes, C., Berks, M., Bonfield, J., Burton, J., Connell, M., Copsey, T., Cooper, J., et al. (1994). 2.2 Mb of contiguous nucleotide sequence from chromosome III of *C. elegans*. Nature 368, 32–38.

Wootton, J. C. & Federhen, S. (1993). Statistics of local complexity in amino acid sequences and sequence databases. Comput. Chem. 17, 149–163.

Xu, Y., Einstein, J. R., Mural, R. J., Shah, M. B., & Uberbacher, E. C. (1994). Recognizing exons in genomic sequence using Grail II. In: Genetic engineering: principles and Methods, (Setlow, J., Ed.), Plenum, New York.

Yuzo, M. & Takuji, S. (1993). Rice cDNA from root and callus (direct submission to GenBank, Nov. 1993).

Zinn, A. R., Alagappan, R. K., Brown, L. G., Wool, I., & Page, D. C. (1994). Structure and function of ribosomal protein S4 genes on the human and mouse sex chromosomes. Mol. Cell Biol. 14, 2485–2492.

COMPUTATIONAL GENE IDENTIFICATION:
UNDER THE HOOD

James W. Fickett

	Abstract	209
I.	Introduction	209
II.	Scope of the Problem	211
III.	Database Similarity Searches	211
IV.	Statistical Regularities of Coding Regions	213
	A. Principles	213
	B. Assessment	215
	C. Related Issues and Future Developments	216
V.	Pattern Recognition of Functional Sites	217
	A. Principles	217
	B. Basal Gene Biochemistry	220
	C. Regulation of Gene Expression	221
VI.	Integration of Fragmentary Results	222
	A. The Goal	223
	B. Kinds of Integration	223
	C. Syntactical Integration	224
	D. Logical Integration	225
	E. Efficient Computation	226
	F. Summary	227

Advances in Computational Biology
Volume 2, pages 209–237
Copyright © 1996 by JAI Press Inc.
All rights of reproduction in any form reserved.
ISBN: 1-55938-979-6

VII.	Accuracy and Availability	227
	A. Standards	227
	B. Evaluation	228
	C. Interfaces	229
VII.	Summary	230
	References	231

ABSTRACT

This paper is a review, for the scientific layman, of computational techniques for identifying genes in DNA sequences. Our purpose is to describe the working principles, the capabilities, and the limitations of gene identification software. Some attention is also given to likely future developments. The emphasis is on eukaryotes, as this is the application domain in which the problem is of the most interest and difficulty.

I. INTRODUCTION

Two types of computational analysis are normally performed on essentially every newly determined DNA sequence. The first is a database search to compare the new sequence with existing collections (nucleotide sequence, amino acid sequence, or motif). The second, the topic of this study, is a search for protein coding regions or genes.

In recent years there has been a tremendous proliferation of computational tools and services, and increasing complexity in the kinds of analysis offered. This abundance can be confusing, but behind the complexity of computational gene identification tools, there is a relatively simple collection of basic techniques. Our purpose in this review is to provide an overview of these techniques for the person who would like to understand, at a high level, how computational gene identification is done. (A related review, intended for the algorithm developer, may be found in Fickett (1995a).

We will also discuss the reliability of current methods. There have been few thorough and objective benchmarking studies. However, important qualitative information is well known to practitioners in the field. We will describe known benchmark results, as well as some rules of thumb on how well the main techniques work in various contexts. An important emerging problem is the degree to which current algorithms generalize to genes unlike those in the current databases. There are hints that many current algorithms work well on highly conserved or highly expressed genes, but to some extent break down on many other genes.

This review is primarily a guide to current techniques, rather than to current tools. It is restricted primarily to published work, though unpublished developments may be mentioned briefly. In most sections, coverage is limited to techniques that are either widely used or seem to us to be particularly important for future developments. Although the number of papers cited is large, there are many others that could have been included. A basic background in molecular biology is assumed. Much of the mathematical background is omitted as well, but an attempt has been made to make the main flow and conclusions of the text comprehensible even if the more mathematical sections must be skipped. Experimental approaches to gene identification are assuming an increasing importance, but are not covered here. For a recent overview see Church et al. (1994).

A number of related reviews exist. A few of the more recent works on sequence analysis in general are Doolittle (1990), Gribskov & Devereux (1991), Adams, Fields & Venter (1994), Griffin & Griffin (1994), Konopka (1994a), Gelfand (1995), and Waterman (1995). Staden (1990) and Gelfand (1990b) give overviews of the gene identification problem. Doolittle (1986) and Gish and States (1993) discuss the interpretation of similarity searches in the context of gene identification.

We will begin by describing the scope of the problem. Next we will treat in turn the three primary means of gathering clues about the existence, location, and function of genes, namely, database similarity search, statistical regularities of coding regions, and pattern recognition of functional sites. In particular circumstances, each of these means can give a relatively accurate picture of a gene, yet each often fails, as well. The next topic discussed, the integration of fragmentary results, is, therefore, one of the most active in current computational research. In the last two sections, some higher level issues are considered.

II. SCOPE OF THE PROBLEM

The fundamental purpose of gene identification programs is to provide a tentative report on the structure of genes, the main goal being to deduce the derived amino acid sequence via conceptual translation. Most older methods provide mainly a map of protein coding regions of exons, while newer programs generally attempt to group these into overall gene structures.

Developers of computational methods for gene identification have long hoped that such methods would provide information not just on the products of translation, but on the expression of genes, for example, by accurately identifying sites on the DNA involved in transcription, splicing, and translation, and perhaps even by helping to elucidate the conditions under which expression occurs. The fulfillment of this hope remains largely for the future, though some progress has been made and will be discussed below.

Of course, the goal driving both sequencing and sequence analysis is often to understand the function of a fragment of a genome in the biochemistry of the cell (or, of course, the likely function of a synthetic piece of DNA). Although it is quite possible in the future that gene identification algorithms will provide clues to function directly from the sequence, most such information now comes indirectly from homology between the deduced protein product and the contents of protein and motif databases.

III. DATABASE SIMILARITY SEARCHES

There is an emerging issue, possibly of fundamental importance, in the development of techniques for gene identification, which might best be expressed as the tension between template methods and lookup methods (termed "intrinsic" and "extrinsic" approaches in Borodovsky, Koonin, & Rudd (1994) and Borodovsky, Rudd, & Koonin (1994). Template methods attempt to compose more or less concise and elegant descriptions of prototype objects, and then identify genes by matching to such prototypes. A good example is the use of consensus sequences in identifying promoter elements or splice sites. Template methods often make important contributions to our understanding, but usually leave out important exceptions and ambiguities, most likely because genomes are not elegantly designed from scratch, but are a collection of contraptions honed by experience.

On the other hand, lookup methods simply rely on what is, without attempting to summarize it neatly, and attempt to identify a gene or gene component by finding a similar known object in available databases. The basis for lookup methods is, of course, sequence conservation due to functional constraint. The most basic lookup method is to search for regions of similarity between the sequence under study (or its conceptual translation) and the sequences of known genes (or their protein products). As molecular biological data stores have increased, lookup methods have gained in importance. A recent, large scale example of applying similarity search to gene finding, clearly illustrating both its power and its difficulties, may be found in Robison, Gilbert, & Church (1994).

A clear advantage to searching for genes by similarity is that, if a significant similarity is found, it may yield clues as to the function, as well as the existence, of the new gene. In addition, if the search is carried out at the amino acid, rather than the nucleotide level, the additional advantage may be had of lowered sensitivity to the "noise" of neutral mutations. The obvious disadvantage of this method is that when no homologues to the new gene are found in the databases, similarity search will yield little or no useful information.

The question naturally arises, then, of the likelihood that the databases will contain a homologue of a gene awaiting discovery. In an early attempt to answer this question, Seely et al. (1990) took one half of GenBank release 56 as a test set, introduced "mutations", "introns", and "intergenic DNA" to make the test set resemble new genomic data, and searched for genes in this test set by comparing it to the remaining half of GenBank as a reference set. In this experiment, they found that approximately three-quarters of the genes could be clearly identified. Thus one might hope that the majority of new genes could be found by means of simple similarity searches in the database.

When the complete sequence of yeast chromosome III (Oliver et al., 1992) was first reported, 26% of the putative protein products (conceptual translations of all open reading frames over 300 bp in length) were found to have significant similarity with some other known sequence. Similarly, in reporting analysis of three cosmid sequences from *C. elegans*, Sulston et al. (1992) state that roughly a third of the putative genes show clear homology to sequences already in the databases. Both of these estimates have rather large error bounds, as the list of tentative genes depends primarily on computational, not experimental, evidence. Yet these studies do seem to suggest that the conclusions of the Seely et al. study are perhaps too optimistic. Probably the disparity between the simulation study and the results of actual genomic sequencing is due to the biased nature of the databases. For example, both of the halves of GenBank used in the experiment of Seely et al. are much richer in highly expressed genes than is a eukaryotic genome in toto.

One overall lesson from a long line of work studying amino acid sequence motifs and blocks from related sets of proteins (cf. Gribskov, McLachlan, & Eisenberg (1987), Posfai, Bhagwat, Posfai, & Roberts (1989), Smith, Annau, & Chandresegaran (1990), Smith & Smith (1990), Henikoff & Henikoff (1991), Bairoch (1992), and Ogiwara et al. (1992)) is that database searches seem to be much more sensitive if carried out with meaningful patterns such as motifs or profile matrices. When Bork et al. (1992a,b) studied the yeast chromosome III sequence using more permissive cut-off scores, multiple alignment methods, and motif searches, 42% of the putative genes were found to be similar to a known sequence or motif. Later, Koonin et al. (1994) revised the list of putative genes, again using the most recent and sensitive known algorithms, and found that 61% of the putative proteins exhibited significant similarities to known proteins or motifs. This increase is due in part to revisions in the list of putative proteins, in part to more complete databases, and in part to improvements in computational methods.

In another vein, current efforts to sequence (at least fragments of) all transcribed sequences from a number of genomes (e.g., Adams et al., 1991; Williamson, Elliston & Sturchio, 1995) concentrate much of the genomic information necessary for gene identification. Boguski, Tolstoshev, & Bassett (1994) collected the 32 human disease gene sequences that have been positionally cloned to date and found that 85% of them showed homology to an entry in the dbEST collection (Boguski, Lowe, & Tolstoshev, 1993) of expressed sequence tags. This is a small sample, but the indication still seems strong that cDNA sequence collections will be an important resource for gene identification. Note, however, that for most of the sequences in dbEST, the only information available is that they are transcribed; mapping and functional data will surely come, but are presently accumulating much more slowly than the sequences themselves.

How fast will the fraction of genes identifiable by similarity search increase? Green et al. (1993) and Claverie (1993a) compare recently determined sequences both to each other and to older sequences in the databases and conclude that (1) most *ancient conserved regions* (or ACRs, roughly defined as regions of protein sequences showing highly significant homologies across phyla) of the protein universe are already known and may be found in current databases, (2) roughly 20–50% of newly found genes contain an ancient conserved region that is represented in the databases (cf. also Borodovsky, Koonin, & Rudd (1994)), and (3) rarely expressed genes are less likely to contain an ancient conserved region than moderately or highly expressed ones.

Taken together, these results seem to suggest that on the order of one-half of all new genes may be discovered, perhaps some functional information determined on the basis of similarity to known sequences or motifs, and that this fraction will continue to rise. Due to the larger variety in non-ACR-containing proteins, however, the rise will likely be rather slow.

Gish and States (1993) discuss the effects of sequencing errors on gene identification by similarity searching and the interpretation of BLASTX search results. Claverie (1992) also discusses practical aspects of such searching, in particular providing a means to eliminate the most common source of high scoring similarities *not* due to gene function, namely repeats.

IV. STATISTICAL REGULARITIES OF CODING REGIONS

A. Principles

At the core of most gene recognition algorithms are one or more *coding measures*. Coding measures are numerical or tabular correlates of protein coding function. Common examples of coding measures include codon usage tables, base composition, and measures of imperfect periodicity (for example, some type of Fourier transform of the sequence). Because codons are used nonrandomly in true coding regions, the typical values of coding measures are different on coding and noncoding regions. Separation is of course imperfect (e.g., it is possible to find regions of DNA that are almost surely noncoding, but that have triplet counts very similar to the codon counts from some coding regions), but the average values of coding measures give template properties for exons in general.

Where a coding measure yields multiple numbers, for example, a table of codon usage, these numbers are usually combined in a weighted sum (which we will call a discriminant), with the more important numbers receiving a higher weight. There are standard statistical methods for choosing the weights (e.g., linear discriminant analysis (Manly, 1986) or logistic regression (Hosmer & Lemeshow, 1989)); artificial intelligence methods (e.g., neural nets (Lippman, 1987)) are also used. Coding measure discriminants are often used with a "sliding window" approach, where successive subsequences of fixed length are examined and the discriminant calculated for each. Something on the order of a hundred bases is required to gain significant information from a coding measure discriminant, so that this method gives a rather low resolution picture of gene boundaries. In integrated gene recognition algorithms, candidate endpoints of coding regions are chosen by other means, and coding measure discriminants are applied to the tentative coding regions.

Coding measures, which have a long and rich history, have been reviewed and benchmarked in Fickett and Tung (1992) (cf. also Gelfand (1990b). Very brief descriptions of the measures tested there are given below (for details and citations see the review).

- Codon frequencies and hexamer frequencies (the latter in-phase or in arbitrary position)
- Amino acid and diamino acid frequencies of conceptual translations
- Prototypical codon patterns (e.g., predominance of purines in the first codon position)
- Base composition at each codon position
- Bias in the frequency distribution of all dinucleotides in the window
- Occurrence of open reading frames
- Expected change in hydrophobicity of the conceptually translated protein under all possible mutations
- Asymmetry of base composition among the three (hypothetical) codon positions
- Bias in the dinucleotide distributions at each successive pair of codon positions (as compared to the sequence overall)
- Randomness (entropy in the sense of information theory) of the base composition in each of the three codon positions
- Periodicity as measured by an autocorrelation function, as measured by a Fourier transform, or as measured by the bias in the frequency distribution of successive dinucleotides versus the bias in the frequency distribution of successive trinucleotides
- Number of runs of different lengths of different nucleotides or classes (e.g., purine) of nucleotides
- Counts of just those hexamers that are typical of common repeats

B. Assessment

The benchmark used in Fickett & Tung (1992) may be summarized as follows. Homogeneous (fully coding or fully noncoding) windows of fixed size were taken from the international nucleotide sequence collection. The data corpus was split in half, and the first part was used as a training set. Discriminant analysis was used to define a weighted sum discriminant. A threshold was then set to equalize the error rates on the coding and noncoding training sets. The discriminant so defined was evaluated on the other half of the data as test set. The average accuracy on the coding and noncoding parts of the test set was taken as the overall accuracy of the measure. The whole process was carried out both for a region-specific definition of coding and for a phase-specific definition.

There is a great deal of redundancy in the suite of measures proposed to date. In some cases two measures are sensing very similar things (e.g., autocorrelation and Fourier). In many cases one measure is derivable from, or a specialization of, another (e.g., position-specific base composition can be derived from codon frequency counts). Taking into account both derivability and accuracy, the review found that essentially all of the information in the collection of coding measures used to date may be had by using only six of the measures: the Fourier transform, the "run" measure, open reading frame occurrence, and frequencies of in-phase hexamer counts.

A surprising result of the study was that a measure which seems to embody little biological understanding—counts of in-phase hexanucleotides—is in fact the most effective one. In-phase word count measures have a long history. The first use we know of the codon usage measure in a published algorithm is in Staden & McLachlan (1982). Separate word counts of different lengths for each phase were considered by Borodovsky et al. (1986a, 1986b, and 1986c). These papers considered words of length 1, 2 and 3 (limited data were available at that time). More recently the same author (Borodovsky & McIninch, 1993) has extended his work to include words of length 6. Claverie, Sauvaget, & Bougeleret (1990) was the first published use of in-phase hexamer count measures.

Combining several measures does improve accuracy. The highest score of any measure in the region-specific prediction of coding function on 108 base human windows was 76.6%. But E. Uberbacher kindly applied the Coding Recognition Module of GRAIL (Uberbacher & Mural, 1991) to the 108 base human test set (using only the first 100 bases of each window), and, when a threshold was set to equalize sensitivity and specificity, the resulting accuracy was 79%. For phase-specific discrimination, we combined the six measures just discussed, again using classical linear discriminant analysis, and obtained 87.8% accuracy on human 108 base windows (compared to 84.9% for the most accurate individual measure). This last combination was also applied to human 54 base windows, giving 82.4% accuracy (compared to 80.7% accuracy for the highest individual measure).

C. Related Issues and Future Developments

Since the above survey, other measures have been proposed. Snyder and Stormo (1993) used the average complexity of octamers (measured by entropy in the sense of information theory; cf. Konopka & Owens (1990), which takes a somewhat different approach towards entropy than does (Almagor, 1985), reviewed in Fickett & Tung (1992)). Extending the in-phase hexamer approach in a direction that takes on some characteristics of similarity search, Solovyev and Lawrence (1993) report that in-phase octamers and nonamers give even higher accuracy. Ossadnik et al. (1994) suggest a measure based on fluctuations in purine/pyrimidine window content (in a rather large window; >800 bp suggested by the authors). Often, when new coding measures are introduced, it is difficult to tell whether the measures are, in themselves, better or worse than existing ones, or whether, on the other hand, the context in which they are applied gives better performance. It would be interesting to apply the above benchmark to these new measures.

In a related vein, experimentalists often use the length of an open reading frame as primary evidence for the existence of a gene, particularly in organisms like yeast, where splicing is rare. In (Fickett, 1994) and (Fickett, 1995b) means are introduced for quantitatively evaluating the strength of such evidence.

We will likely continue to see incremental improvements in coding measures, particularly in two areas: First, Guigó and Fickett (1994) have shown that dependence of most measures on C+G content is high and that mere base compositional differences can cause fluctuations in the values of coding measures larger than the differences between coding and noncoding regions. So tailoring the measures to differing base compositions may well improve accuracy. In this regard, (Xu et al., 1994) adopted the strategy (not separately evaluated) of measuring in-phase hexamer counts for "high" and "low" CG content reference sets, and then using linear interpolation to make a set of counts intended to be appropriate for the CG content of the test sequence. Second, one wonders whether the many variables of some of the above coding measures (for example, the 4096 variables of each hexamer measure) are all making important and independent contributions to discrimination. It might be, for example, that the signal-to-noise ratio of the measure could be improved by pruning out the less informative variables.

The means by which a discriminant is derived from a coding measure show great variety. In the case of in-phase hexamers, for example, Claverie, Sauvaget, and Bougeleret (1990) weighted the observed count of each hexamer by the ratio of its frequency in coding regions to that in all DNA. Farber, Lapedes, and Sirotkin (1992) used a neural net with 4096 inputs to derive a discriminant. Borodovsky and McIninch (1993) derived two nonhomogeneous (frame-dependent) five-step Markov models, one for the coding regions of each strand and a homogeneous model for noncoding regions to calculate the probability of observing a window under each of the seven corresponding hypotheses, and then used Bayes' theorem to derive the posterior probability of each hypothesis, given the window. (It is worth noting that, in most algorithms, the method is applied separately to the two strands, and the results are combined in a post-processing step. In the work of Borodovsky and McIninch, on the other hand, the seven relevant hypothesis—coding in each of six possible frames, or noncoding—are directly compared in one step.) Thomas and Skolnick (1994) considered seven classes of nucleotides: those in the three codon positions, those in intergenic regions, and those in introns breaking the coding sequence at each of the three possible codon positions. Assuming a one-step Markov model for the state variable, and that the probability distribution of the bases at each position of the sequence depends only on the bases and states in the immediate vicinity, they use Bayes' theorem to make a maximum likelihood estimate of the state at each base of a given sequence. There is very limited information on which of these methods (or the many others that have been used with these measures, other measures, or combinations of measures) is best. The general feeling among developers is that the differences are usually small, but comparative objective testing would be very valuable.

V. PATTERN RECOGNITION OF FUNCTIONAL SITES

A. Principles

The coding measures considered above are all closely related to patterns of codon usage. In what has now become common usage, Staden (1990) termed the use of such measures "gene search by content". Of course codon usage is merely a side effect of the biochemistry of organisms. It will be more enlightening when we are able to recognize the locations in a genome where the gene expression machinery interacts with the nucleic acid, and so recognize the genes in a way parallel to the action of the cell. This approach Staden termed "gene search by signal".

Any portion of the DNA whose binding by another biochemical plays a key role in transcription is variously called a signal, a binding site, a recognition site, or a sequence element. Regions on a genome that correspond directly to regions on an mRNA or pre-mRNA with analogous function in splicing or translation are also referred to by the same terms.

The collection of all specific instances of some particular kind of signal, for instance, the set of all intron donor sites in human genes, will normally tend to be recognizably similar. In the early days of sequence analysis, it was hoped that this similarity could be captured adequately by a *consensus sequence*. To form the consensus that is, one aligns all the specific sequences, and then takes the most commonly occurring base at each aligned position. Then, it was hoped, the actual sites would be differentiated from spurious sites simply by distance (e.g., number of bases different) from the consensus. This approach turned out to be too simple, though the consensus sequences at various sites are still useful for their mnemonic value.

It is now most common to summarize the commonalities in (that is, form a template for) a particular signal by recording the frequencies of each nucleotide at each aligned position, rather than simply recording the most frequent one. That is, the individual sequences are aligned, and the frequency of base b at position i is tabulated as $f(b,i)$. Then a *position weight matrix m* is derived from f, most often by $m(b,i) = \log[f(b,i)/p(b)]$, where $p(b)$ is the genomic frequency of base b (reviewed in Stormo, 1990). Any sequence to be tested for signal function is represented analogously, with $s(b,i) = 1$ if the i-th base of the sequence is b, and 0 otherwise. Then the test value of a sequence is the dot product of these two matrices, $\Sigma_{b,i}[m(b,i)^*s(b,i)]$. (Because of the form of representing the information, this approach is sometimes called, among computational biologists, the "matrix method").

This approach is justified by several theoretical studies of protein-DNA binding (e.g., Berg & von Hippel (1988), von Hippel (1994), and references therein), and by a number of experiments in which a DNA signal sequence is systematically varied and the activity of the variants measured (e.g., Mulligan et al. (1984), Takeda, Sarai, & Rivera (1989), and Barrick et al. (1994)).

Overall, we may summarize the results of these studies as follows. The activity of a signal sequence is determined by the proportion of the time that the sequence is bound, which in turn depends on the abundance of the binding molecule (typically protein or RNA) and its binding specificity, that is, the degree to which the binding molecule "prefers" the signal sequence to pseudosites. In comparing the activity of different signal sequences for the same binding molecule, or in attempting to distinguish the signal sequences from pseudosites, we may take as constant all factors affecting the availability of the binding molecule (overall abundance, the frequency of pseudosites, and the average affinity of the pseudosites) and deal simply with the binding energy of the binding molecule to the site at hand. The first major result from experiment is that this binding energy is often closely approximated by simply summing the contributions of the individual base positions, as if they were independent. This of course means that activity can be predicted reasonably well by some matrix calculation as described above, though it does not determine the form of the matrix.

If we assume that $f(b,i)/p(b)$, as defined above, represents the ratio of bound to free reaction concentrations for base b in its interaction with a specific site on the binding molecule, then the logarithms in the position-weight matrix are proportional to the free energies of binding for each base. This is one way of justifying the particular form of the position-weight matrix. Alternatively, one may note that the sum in the dot product above, from a statistical point of view, is just the log likelihood ratio of the test sequence being found given (1) the hypothesis that the sequence comes from a set in which the bases at position i have probability distribution $f(b,i)$ and (2) the hypothesis that the sequence comes from a set in which the bases occur with frequencies $p(b)$.

In many cases, the dot product of the position-weight matrix with the sequence seems to be a relatively good predictor of signal sequence activity. In Barrick et al. (1994), for example, 185 clones with randomized ribosome binding sites were selected, and for each the activity was measured and the binding site sequenced. A matrix was first determined by multiple linear regression. The regression matrix predicted actual activity with a correlation coefficient of 0.89 (when cases with alternate start codons were eliminated, this rose to 0.92). Further, when a position-weight matrix was calculated from natural sites, the correlation coefficient between the two matrices was 0.88.

However, position-weight matrices do not always work well, and it must be recognized that a number of simplifying assumptions underlie their use. The use of position-weight matrices ignores the availability of the DNA or RNA (the effects of chromosome packaging and secondary structure), nonindependence between bases (important, for example, in conformational changes due to base stacking), different versions or conformations of the binding molecule, and interactions between multiple binding molecules.

Nonindependence between bases may be taken into account by a relatively simple extension of the position-weight matrix, namely, using a larger matrix where columns correspond to the various possible oligomers at various positions, rather than to individual bases. An example will be seen below in the work of Solovyev, Salamov, and Lawrence. Of course, the longer the oligomers, the more data is needed to calculate the matrix reliably.

To date the use of position-weight matrices in recognizing key elements of eukaryotic genes, namely, splice sites and promoter sequences, has led to relatively limited success. All of the above limitations of the method probably play a role here. However we would hazard the guess that the main factor is the cooperativity among multiple binding molecules. It is rare in eukaryotes, for example, for large numbers of genes to have precisely the same complement of proteins involved in initiating transcription. We will return to this point below.

Where applicable, the consensus and position-weight matrix methods have the advantage of being relatively simple and well understood. Assessing the significance of search results has been treated in Waterman (1989) for approximate matches to a consensus pattern, and in Claverie (1994a) and Goldstein & Waterman (1994) for searches using position weight matrices.

Many other methods, difficult to summarize in a limited space, have been proposed to recognize signal sequences in genomes. Most of these are not widely used and the reader is referred to Gelfand (1995) for more details. One method which has seen fairly extensive use is that of neural networks (Lippmann, 1987). When the network has only one layer, it produces a linear discriminant function usually fairly close to the position-weight matrix derived by the methods described above. However, when the network has multiple layers with hidden units, the function encoded is more complex. The use of neural networks in analyzing nucleotide (and amino acid) sequences was reviewed in Hirst & Sternberg (1992). The neural network algorithms reviewed showed better performance than more statistical approaches in a number of cases. However it is not altogether clear whether the improvement was due to integration of several kinds of evidence (discussed below) or to the neural network means of integration.

One difficulty with neural nets, and in fact, with machine learning methods in general, is the distance between the understanding in the machine and the understanding in the human expert. Most such algorithms are designed to begin from a randomized state, that is, without the benefit of any knowledge already gained by experiment or other methods. And, when the algorithm has finished the training stage, it is typically rather difficult to retrieve the "understanding" that has been captured. In this regard, Towell and Shavlik (1994) have made interesting progress by developing neural net methods that can start from an intelligible base of rules and, after training, can return a refined set of rules.

Many methods of sequence signal recognition require a set of sequences with functional sequence elements already precisely located and aligned. However, it is often the case that experimental work has only approximately located the sequence element and that the best alignment is unclear. Thus several groups have developed methods to optimize the localization of the sequence elements, the alignment, and a weight matrix or other discriminant, simultaneously; see, for example, Cardon & Stormo (1992), Lawrence et al. (1993), Borodovsky & Peresetsky (1994), and Krogh et al. (1994). To date these methods have been applied primarily to other problems, but show significant promise for identifying eukaryotic signal sequences.

B. Basal Gene Biochemistry

Gene signal recognition work to date has dealt with the problem of recognizing the signals common to essentially all genes. For example Bucher (1990) has defined weight matrices to characterize partially four elements common to most eukaryotic pol II promoters: the TATA-box, cap-signal, CCAAT-, and GC-box. These were derived from the Eukaryotic Promoter Database (Bucher, 1988). In Cavener & Ray (1991) sequences flanking translational initiation and termination sites have been compiled and statistically analyzed for various eukaryotic taxonomic groups. The polyadenylation reaction is relatively well-understood now (Wahle and Keller, 1992), and information on translation termination sites has been collected in the Translational Termination Signal Database (Brown et al., 1993). All of this information is useful in helping to recognize the beginnings and ends of genes, however, computational methods for such recognition are in their infancy and are mostly of uncharacterized reliability.

Consensus sequences for splice junctions have been recognized for many years (Breathnach & Chambon, 1981). A comprehensive collection of splice junctions and weight matrices, commonly referred to, may be found in Senapathy, Shapiro, and Harris (1990). Consensus sequences alone give rather unsatisfactory results. The best successes to date in predicting splice junctions come from integrating several kinds of evidence. Shapiro and Senapathy (1987) combined base frequency information at the splice site with a check for an open reading frame on the correct side and an evaluation of a potential polypyrimidine tract near the acceptor. Including a requirement for related patterns (e.g., a branch point within a specified distance upstream of the acceptor and no AG dinucleotide between these two sites (Ohshima & Gotoh, 1988), (Gelfand, 1989)) seems to improve accuracy. At true splice sites, coding measures should give values characteristic of coding regions on one side of the splice and values characteristic of noncoding regions on the other. Thus in Nakata, Kanehisa and DeLisi (1985) and Brunak, Engelbrecht and Knudsen (1991) information concerning splice sites per se, for example, positional frequencies and binding energies, are combined with the values of coding measures on either side of each potential splice site to give improved splice site prediction. Solovyev, Salamov, and Lawrence (1994b) give an excellent overview of the literature and a careful synthesis of existing techniques. They report what appears to be the most accurate algorithm for human sequences to date, using triplet counts (due to Mural, Mann, & Uberbacher, 1990) at significant positions near the branch point and splice junctions, octamer counts on either side of the junction, counts of

G, GG, and GGG downstream of potential donor sites, and counts of T and C upstream of potential acceptor sites, all combined using linear discriminant analysis. Taking the sets of GT and AG dinucleotides as the set of all potential splice sites, Solovyev, Salamov, and Lawrence reported 96% sensitivity and 97% specificity for donors, and 96% sensitivity and 96% specificity for acceptors. (These methods are combined, using linear discriminant analysis, with oligonucleotide-based recognition methods for coding regions and the beginnings and ends of genes to produce an exon recognition algorithm FEX.)

In as many as 90% of the vertebrate mRNAs, the first AUG codon is the unique initiation site, and in the exceptional cases a number of factors have been elucidated that govern the probability of translation initiation at a particular ATG. These include neighboring nucleotides, leader length, distance to other ATGs, ORF length, and secondary structure; cf. Kozak (1991).

C. Regulation of Gene Expression

The complexity of gene regulation naturally increases greatly with the number of tissue and cell types in an organism. Thus, although some universal commonalities have been identified in the known genes of some prokaryotes, it would now appear unlikely that any simple characterization will be found for the gene promoters of *Homo sapiens* (or, probably, of any other differentiated metazoan species). Thus, although the regulation of eukaryotic gene expression has attracted relatively little attention to date from developers of gene identification algorithms, in the future, such algorithms will almost take into account the complex signals for transcription initiation of specific classes of genes.

Utilizing this sort of information will bring an added advantage, in that specific transcription elements provide important clues to gene function. This is an opportune time to begin making use of information on gene regulation, for a remarkable amount of information is now appearing in new papers daily on gene-specific, tissue-specific, stage-specific, and stimulus-specific transcription signals.

Several collections of sequence elements for transcription factors have appeared, including the Transcription Factor Database (Ghosh, 1990), the collections in Locker and Buzard (1990) and Faisst & Meyer (1992), TRRD (Kel et al., 1995), and TRANSFAC (Knueppel et al., 1994; Wingender, 1994). As far as we know, the last two are the only collections being actively maintained. In addition to incorporating the sequences of individual signal instances, these collections also include consensus sequences or weight matrices.

It is not clear at this point to what extent consensus sequences or weight matrices can differentiate true from false transcription elements. This remains a research area, as does the problem of how best to use the transcription element information in gene identification algorithms. One clue may be had from the practice of experimental biologists, who place considerable weight on co-occurrence of related elements (cf. Fickett (1996) and Prestridge (1995)).

VI. INTEGRATION OF FRAGMENTARY RESULTS

As a whole the field is making a transition from studying primarily components of genes to studying genes and genomes in their entirety. Thus the issue of choosing an appropriate language in which to express and integrate the knowledge gained from less global calculations is one of the most active areas in computational gene identification.

It is well known that gene expression *in vivo* involves considerable interaction and interdependence among various components of the transcription and translation machinery. Examples include coordinate binding of multiple transcription factors and mutations in a 5' splice site resulting in the skipping of an upstream 3' site. Thus it is not surprising that programs incorporating some overall model of gene structure give increased accuracy even for the recognition of individual gene components. In the case of intron splice sites, the integrated methods discussed above improve results by roughly a factor of 10 over recognition by consensus or matrix methods. Another example is seen in Einstein et al. (1992), where it is shown that 60% of exons under 50 bp missed by the original GRAIL e-mail server may be detected by a logical analysis of splicing and frame.

A number of programs have appeared in the last few years that are integrated in the sense of taking gene structure into account to predict exons (SORFIND: Hutchinson and Hayden, 1992, 1993; FEX: Solovyev, Salamov, and Lawrence, 1994a,b) or genes (GM: Fields and Soderlund, 1990; Soderlund et al., 1992; the Gelfand program: Gelfand, 1990a; Gelfand & Roytberg, 1993; GeneID: Guigó et al., 1992; Knudsen, Guigó & Smith, 1993; GenViewer: Milanesi et al., 1993; GeneParser: Snyder & Stormo, 1993; GRAIL II: Uberbacher et al., 1993; Xu et al., 1994; GenLang: Dong & Searls, 1994 (cf. also Searls, 1992); and the program of Krogh, Mian, Haussler, 1994). (There are other gene prediction algorithms not yet published. In one prominent case, the analysis of the *C. elegans* genomic sequencing group (cf. Wilson et al., 1994) makes use of an algorithm GeneFinder developed by P. Green.)

A. The Goal

The goal of a gene identification algorithm has usually been taken as obvious: to assemble all of the components of a gene and report an integral gene to the user. In the long run, this must be extended to meet the practical need of analyzing incomplete sequences. Current algorithms typically expect to find all components of each gene and, sometimes, of only one gene. In practice, however, a sequence presented for analysis may have no genes, partial genes, multiple genes, or genes with multiple expression patterns. Thus it will be necessary to develop algorithms that can produce a feature table of relevant gene features in whatever combinations they happen to occur.

In addition, it is now widely recognized (but only beginning to be addressed) that an important part of the goal is to recognize when a small change in sequence will result in a large change in function. This is important for recognizing nonfunctional alleles of "disease genes", pseudogenes, and genes in first pass (and error prone) sequence data (cf. Claverie, 1993b; Krogh, Mian and Haussler, 1994; Fields, 1994; Xu et al., 1995).

B. Kinds of Integration

Gene identification algorithms typically begin by attempting to evaluate possible component objects or aspects of genes, proceed to integrate these into exons, and finally integrate the exons into genes. At both the exon level and the gene level there are two very different kinds of integration involved. The first is primarily biological, taking into account the syntax of genes, for example, typical spacing of components and the partitioning of the primary transcript into alternate exons and introns. The second is primarily logical and statistical, taking into account the relative importance of different kinds of evidence and the combining of scores into overall measures of optimality in gene models. We will take these up in turn.

C. Syntactical Integration

All integrated gene identification programs make use of the high level syntax of genes resulting from our basic understanding of transcription, splicing, and translation. Taking "exon" in the coding sense, rules similar to the following are normally used:

- The first coding exon begins with the start codon and ends with a donor site (or the stop codon, if there are no internal exons).
- Any internal exons begin with acceptor sites and end with donor sites.
- The last exon begins with an acceptor site (or the start codon) and ends with the stop codon.
- The primary transcript consists of the transcription initiation site, a 5'UTR, alternating exons and introns, the 3'UTR, and the transcription termination site.
- When the introns are excised and the combined exons read in frame, no internal stop codons are found.

In addition to this syntax of order, there is also some information on distance, as, for example, appears in known size distributions for exons and introns (cf. Naora & Deacon, 1982: Hawkins, 1988; Smith, 1988).

Although this basic syntax is clear enough, biology is of course far more complex and less well-understood than these simple rules would imply. Such facets of gene syntax as alternative splicing, overlapping genes, and promoter structure remain beyond the reach of the current generation of algorithms.

In many of the algorithms available today, the rules of gene syntax are implicit in the structure of the algorithm, but no "gene grammar" is explicitly listed. Two groups have, however, taken a more linguistic approach, making an explicit grammar the foundation of the algorithm.

D. Searls suggested, some years ago, that a linguistic approach to the analysis of features in DNA sequences could be beneficial (for an overview, see Searls, 1992). This approach is first applied to the identification of protein coding genes in Dong and Searls (1994), where a formal, definite clause grammar of genes is described. Partial scores are passed up the parse tree and combined by rules stored as part of the grammar. A training procedure is used to alter the score combination rules to optimize accuracy. Standard parse tools are used to find correct and high scoring parses of a sequence.

Krogh, Mian, and Haussler (1994) use a Hidden Markov Model (HMM) to integrate gene components into overall gene models for *E. coli* sequences. In essence, this means that they construct a probabilistic finite automaton that assigns a probability to every possible parse of a sequence into promoter, start, coding, stop, and intergenic regions. The Expectation Maximization algorithm is used to estimate the parameters of the HMM. Then the Viterbi algorithm is used to find the most probable parse of the sequence.

D. Logical Integration

A variety of evidence is typically employed in computer searches for protein coding genes. Gelfand (1990a) was the first to discuss explicitly the question of providing a natural framework for integrating coding measures, matrix scores of signals, and overall syntactical requirements. The approach chosen was basically statistical. To avoid dependence of score on the length of the gene, raw scores are taken as the average donor score, the average acceptor score, and the average TESTCODE window score (Fickett, 1982) over the exons. Then all scores are put on the same scale by expressing them in standard deviation units about the means of their observed distributions. The sum of these normalized scores is the score for the gene.

Several other authors have also taken a fundamentally probabilistic/statistical approach. The discriminant analysis approach of Solovyev, Salamov, and Lawrence (1994a,b) is of course statistical. Stormo and Haussler (1994) suggest a general probabilistic framework where one is partitioning a sequence into two classes of intervals (e.g., exons and introns), has a number of scores for each possible classification of each possible interval, and is combining these scores as a linear weighted sum. They suggest interpreting the scores and the sum as log probabilities. They then give efficient algorithms for scaling the scores so that the probabilities will sum to one, for calculating the probabilities, for choosing the weights in order to maximize the probability of given ("training set") sequence parses, and for finding the top ranked optimal and suboptimal parses. (Compare also States and

Gish (1994), where codon bias is integrated into BLAST searches using a likelihood approach.)

A particular advantage of the HMM approach of Krogh, Mian, and Haussler (1994) is that it naturally provides a joint probability distribution over sequences and parses of those sequences. The HMM thus provides a very natural vehicle for considering the possibility of introducing a sequence correction to get a more probable parse.

The salient advantage of taking a probabilistic point of view is that it may be possible to assign a natural meaning to the scores. It would seem to be very desirable to apply the probabilistic point of view consistently to the sequence interpretation problem, in a way that allowed one to provide answers for questions such as: "How likely is it that at least one exon of this predicted gene is completely correct?," "How likely is it that the correct gene and this predicted gene have at least 90% of the translated protein in common?", or "How likely is it that this is, in fact, the most commonly used translation initiation site?"

Applying probabilistic notions consistently is, however, very difficult because of our limited knowledge. Most authors, therefore, have taken what might be termed a machine learning approach, in which scores of various aspects of putative genes are meaningless numbers, and the rules for combining these numbers may therefore be manipulated at will to improve the accuracy of prediction. The advantage of this point of view, successfully exploited by a number of investigators, is that purely empirical machine learning techniques may be used to improve the algorithms by which scores are assigned. Thus, for example, both Guigó et al. (1992) and Snyder and Stormo (1994) used a neural net to revise the weights by which different atomic measures are combined, and Dong and Searls (1994) used an *ad hoc* training procedure to revise the score-combining rules associated with each node of the parse tree. In these cases it is reported that such training significantly improves performance.

Orthogonal to the choice of a probabilistic or a machine learning approach to the interpretation of scores, there is also the issue of organizing one's evidence. Most gene identification algorithms recursively construct gene models from partial subassemblies. For instance, atomic components may be scored first, then exons constructed and scored, and finally genes assembled from exons and a final score assigned. Further, most evidence gathered by gene identification algorithms fits neatly into this recursive hierarchy. Thus Dong and Searls (1994) elegantly summarized the basic approach of most investigators by attaching the scoring rules directly to nodes of the gene parse tree.

Unfortunately, however, not all of the evidence that one needs to take into account is directly related to a subassembly of the gene. For example, if the translated protein from a candidate gene contains a region similar to a known protein motif and this region corresponds to parts of each of two exons, it is not obvious how this should affect either the scores of the exons or the score of the gene overall. Dong and Searls solved this problem by specifying a grammar in which not all components of a parse are components of the gene; for example, one parse component is the average exon quality. Another common approach is to append postprocessor rules to the main algorithm. Thus GRAIL (Uberbacher et al., 1993; Xu et al., 1994) incorporates a number of heuristic rules for finding the boundaries of exons, and Krogh, Mian, and Haussler (1994) complete independent analyses of the complementary DNA strands and then combine them by means of a small set of rules.

E. Efficient Computation

The number of possible genes to construct, score, and rank, even in a sequence of a few kilobases, is quite large. Snyder and Stormo (1992) and, independently, Gelfand and Roytberg (1993) introduced dynamic programming algorithms to solve this problem. Guigó et al. (1992) introduced the idea of exon equivalence—using one exon to represent a class of roughly equivalent exons—as an alternative (and possibly coordinate) approach.

Despite significant advances in sequencing technology, it still takes longer to produce a sequence than it does to submit it to the analysis of even the slowest gene identification algorithms. What may be an even more serious bottleneck is the human attention required to interpret and integrate the output from the several kinds of important computational analyses. Thus in addition to efficient computation, significant attention should be devoted to the problem of building algorithms to truly integrate all the evidence for gene location and function and to give accurate answers to biologically meaningful questions.

F. Summary

As will be clear from even this short overview, the area of whole-gene recognition is moving rapidly, with advances being made on several fronts. Divergent, sometimes even conflicting, innovations are being made by different groups. Particular techniques are rarely evaluated in isolation, and each pair of programs usually differs in many aspects. It is likely that there will continue to be many exciting advances, and that these advances will continue to be somewhat difficult to follow.

VII. ACCURACY AND AVAILABILITY

The biologist who uses computational means to help identify genes needs, in addition to a good algorithm, packaging in intuitive software and a means to evaluate the reliability of the predictions that the algorithm makes.

A. Standards

Gene prediction algorithms are evaluated by comparing their predictions to true genes. But how is "true" to be defined? Even for well-studied sequences, revisions in our picture of the encoded genes may come at any time. In particular. it is very dangerous to assume, on the basis of a lack of positive evidence to date, that there is no gene in a given region (compare, for example, Robison, Gilbert, and Church, 1994). Even if we take community consensus to be the gold standard, it is beyond the scope of the nucleotide sequence databases to maintain a reflection of current biological understanding in the features recorded on all known sequences. Thus the databases, the most common source of reference data, are often incomplete and sometimes incorrect.

The most common solution to this difficulty is to take a set of a few tens of sequences, verify the annotation in detail for this set, and then use it for algorithm development and evaluation. The advantage of this approach is, of course, that one can be personally assured that the data are as good as one can reasonably make them. A disadvantage is that the variety in such a set is rather limited. Thus the evaluation given may not apply to new data. Another solution is to accept the databases as they are, perhaps removing some large classes of entries likely to confuse one's study (for example, entries with no annotation, or duplicates) and take the incompleteness of annotation into account in interpreting results. A compromise between these two approaches (not used as often as it should be) is to take advantage of one of a number of specialized, curated databases of intermediate size. One such, of particular relevance to the development of gene identification algorithms, is the collection of Functionally Equivalent Sequence sets (including, for example, a number of specialized collections of exons and introns) described in Konopka (1994a).

Genes whose sequences are currently known are, generally speaking, either highly expressed or well conserved. At least in this sense, then, they are not typical of genomes as a whole. There are hints that this presents a serious limitation to both the development and evaluation of gene identification algorithms and that accuracies now reported may not extend to many new sequences. For example both Lopez, Larsen, and Prydz (1994) and Burset and Guigó (1996) report that the accuracy of current algorithms on recently determined sequences is significantly lower than on the original test sets. It is quite possible that similar results will be found for other tools.

It is unlikely that there will soon be a clear solution to these problems. In the meantime, anyone who is concerned with algorithm accuracy needs to examine very closely the way in which the data for development and evaluation were chosen.

B. Evaluation

It is also important to compare carefully the definition of "accuracy" given by a developer with that appropriate to any particular use of the algorithm. Meanings assigned by different investigators to this term include (1) the fraction of coding "windows" correctly predicted (giving sensitivity but not specificity), (2) the fraction of genes that are clearly detected (strongly dependent on the definition of the degree of match required between the predicted and known genes), (3) the fraction of bases correctly classified (in humans, this largely ignores the performance of the algorithm on the small portion of known sequences that code for protein), (4) the fraction of exons for which the overlap between the predicted and known extent is good, by some defined criterion (the definition of overlap must be appropriate to the planned use of the algorithm), and many others.

Guigó et al. (1992) made an important advance by suggesting that the accuracy of integrated algorithms be evaluated on a nucleotide-by-nucleotide basis. This suggestion has been followed by most developers of integrated gene identification algorithms and provides at least a rudimentary basis for comparison among the many current algorithms. On this basis, the reported accuracy of most current algorithms is fairly similar and may be summarized so: When a new (not seen before by the algorithm) sequence is chosen that contains all of one gene and its flanking regions (and no other genes or partial genes) and this sequence is presented to the algorithm, the predicted gene will typically largely overlap the known gene, so that about 85–90% of the predicted coding bases are in the known gene, and about 85–90% of the known coding bases will be in the predicted gene, that is, the predicted gene will look very much like the known one, but there will usually be significant differences as well. (A recent, comprehensive, third-party benchmarking study by Burset & Guigó (1996) finds higher variation than this in performance.)

The base-by-base accuracy numbers may be combined into a single performance measure (again a suggestion of Guigó et al. (1992)) by using the set-theoretic correlation coefficient (Cramer, 1946; Matthews, 1975) between the set of true coding nucleotides and the set of predicted coding nucleotides. This number is useful, but has the disadvantage of emphasizing the performance of the algorithm on whichever class of nucleotides—coding or noncoding—is more common in the test set.

Performance of algorithms is, of course, in part dependent on the quality and contiguity of the sequences presented. Claverie (1994c) evaluates the performance of GRAIL when raw, single sequencing runs are analyzed, and suggests that it is unlikely for the use of first pass, fragmented data, in itself, to lead to failed detection of genes (cf. also Kamb et al., 1995 and Xu et al., 1995).

Users need to know not only how good an algorithm is on average, but how to interpret a particular score. In the case of SORFIND, which predicts internal exons, Hutchinson and Hayden (1992, 1993) divided the range of the output score into four ranges and for each reported the actual frequency with which the algorithm correctly reports exons in that score range. The situation becomes more complicated when the output consists of genes (or feature tables) rather than exons. By considering many suboptimal solutions, Snyder and Stormo (1992) attempted to give the user a feel for which parts of a predicted gene are most likely to be correct.

Both Singh and Krawetz (1994) and Burset and Guigó (1996) compared the performance of multiple algorithms on a common test set. This sort of objective, third-party, comparative performance measurement is very valuable and unfortunately rare.

C. Interfaces

It is a remarkable fact about the field of gene identification today that many, perhaps most, of the best algorithms are not widely available. This is first of all simply because many developers have not had the time to develop an intuitive interface for those whose primary business is experimental biology. Indeed, one of the most important factors in the widespread use of GM and GRAIL is the effort that its developers have put into interface development and community education.

A second limitation on availability is less obvious but no less real, that most algorithms today are organism specific, in implementation even if not in concept. To overcome this problem, research on the degree of generality of various techniques is needed. For example, are in-phase hexamer counts, the single most useful coding measure, fairly stable only within species? Or can discriminant vectors for this measure be meaningfully calibrated for all mammals or even for some wider group, in one step? If most techniques are highly specific to relatively small parts of the taxonomic tree (similar remarks apply to classes of genes), then a way needs to be found to allow the typical computational support person in larger biological laboratories to tailor existing algorithms to a particular context.

VII. SUMMARY

There has been a great deal of progress in gene identification methods in the last few years. At least in the case of sequence data from mammals, *C. elegans*, and *E. coli*, the older coding region identification methods have mostly given way to methods that can suggest the overall structure of genes. And for all organisms, computational methods are sufficiently accurate that they give practical help in many projects of biological and medical import.

Yet there is still room for significant improvement. Many of the better algorithms are not widely available. Investigators studying organisms other than those mentioned above may find that only the older algorithms are available to them. For the more advanced algorithms, it is still the case that predicted genes, while largely overlapping expressed natural genes, are typically incorrect in a number of details. Further, it is not clear that current algorithms, developed on the very atypical gene sample available in current databases, will perform as well on genes more typical of the biological universe as a whole. Essentially all current algorithms depend heavily on codon usage bias, but it has been shown that this bias is less informative in genes with low-level expression (McLachlan, Staden, & Boswell, 1984; Sharp et al., 1988; States and Gish, 1994).

Perhaps the single greatest opportunity in the development of gene identification algorithms is the inclusion of more detailed biological knowledge, relying less on techniques that attempt to provide a single elegant description valid for all cases. The description of (say) human genes inherent in any of the current gene recognition programs could be written in a few pages. Given the extent to which evolution is opportunistic and haphazard and given the prevalence of exceptions to essentially all general principles in molecular biology and biochemistry, it seems most unlikely that essential aspects of any genome will be described in such simple terms. Greater emphasis should probably be placed, then, on lookup methods over template methods; more richness is needed in the modeling of eukaryotic gene regulation. In general, a trend may be expected toward gene identification algorithms becoming interfaces, with a general model of gene syntax, to a large number of databases of specific facts. First steps in this direction may be found in Borodovsky, Rudd, and Koonin (1994), Claverie (1994b), and States and Gish (1994).

The single most important area where specific aspects of genes are important, even to discover the coding regions, is control of gene expression. Further, control of gene expression is very closely connected to product function. Thus, in addition to providing greater accuracy, bringing gene identification algorithms closer to models of underlying biological mechanisms will also bring them closer to answering what is, in the end, the more important questions: not just "Where are the genes in this sequence?", but "How do they determine the biochemistry of the cell?"

REFERENCES

Adams, M. D., Kelley, J. M., Gocayne, J. D., Dubnick, M., Polymeropoulos, M. H., Xiao, H., Merril, C. R., Wu, A., Olde, B., Moreno, R. F., Kerlavage, A. R., McCombie, W. R., & Venter, J. C. (1991). Complementary DNA sequencing: expressed sequence tags and human genome project. Science 252, 1651–1656.

Adams, M. D., Fields, C., & Venter, J. C., Eds. (1994). Automated DNA sequencing and analysis. Academic, San Diego, CA.

Altman, R., Brutlag, D., Karp, P., Lathrop, R., & Searls, D., Eds. (1994). Proc. 2nd int. conf. intelligent systems for molecular biology. AAAI, Menlo Park, CA.

Altschul, S. F., Gish, W., Miller, W., Myers, E. W., & Lipman, D. J. (1990). Basic local alignment search tool. J. Mol. Biol. 215, 403–410.

Altschul, S. F., Boguski, M. S., Gish, W., & Wootton, J. C. (1994). Issues in searching molecular sequence databases. Nature Genetics 6, 119–129.

Bairoch, A. (1992). PROSITE: A dictionary of sites and patterns in proteins. Nucl. Acids Res. 20, 2013–2018.

Barrick, D., Villanueba, K., Childs, J., Kalil, R., Schneider, T. D., Lawrence, C. E., Gold, L., & Stormo, G. D. (1994). Quantitative analysis of ribosome binding sites in $E.\ coli$. Nucl. Acids Res. 22, 1287–1295.

Berg, O. G. & von Hippel, P. H. (1988). Selection of DNA binding sites by regulatory proteins. Trends Biochem. Sci. 13, 207–211.

Boguski, M. S., Lowe, T. M. J., & Tolstoshev, C. M. (1993). dbEST - database for "expressed sequence tags". Nature Genetics 4, 332–333.

Boguski, M. S., Tolstoshev, C. M., & Bassett, D. E., Jr. (1994). Gene discovery in dbEST. Science 265, 1993 (see also http://www.ncbi.nlm.nih.gov/dbEST/dbEST_genes/ for supplementary data).

Bork, P., Ouzounis, C., Sander, C., Scharf, M., Schneider, R., & Sonnhammer, E. (1992a). What's in a genome? Nature 358, 287.

Bork, P., Ouzounis, C., Sander, C., Scharf, M., Schneider, R., & Sonnhammer, E. (1992b). Comprehensive sequence analysis of the 182 predicted open reading frames of yeast chromosome III. Prot. Sci. 1, 1677–1690.

Borodovsky, M. Y., Sprizhitskii, Y. A., Golovanov, E. I., & Aleksandrov, A. A. (1986a). Statistical patterns in primary structures of the functional regions of the genome in *Escherichia coli*. I. Frequency characteristics. Molekulyarnaya Biologiya 20, 1014–1023.

Borodovsky, M. Y., Sprizhitskii, Y. A., Golovanov, E. I., & Aleksandrov, A. A. (1986b). Statistical patterns in primary structures of the functional regions of the genome in *Escherichia coli*. II. Nonuniform Markov models. Molekulyarnaya Biologiya 20, 1024–1033.

Borodovsky, M. Y., Sprizhitskii, Y. A., Golovanov, E. I., & Aleksandrov, A. A. (1986c). Statistical patterns in primary structures of the functional regions of the genome in *Escherichia coli*. III. Computer recognition of coding regions. Molekulyarnaya Biologiya 20, 1390–1398.

Borodovsky, M. & McIninch, J. (1993). GENMARK: parallel gene recognition for both DNA strands. Proc. 2nd int. conf. open problems in computational biology, A. Konopka, Ed. Published as Comput. Chem. 17(2), 123–134.

Borodovsky, M. & Peresetsky, A. (1994). Deriving nonhomogeneous DNA Markov-chain models by cluster analysis algorithm minimizing multiple alignment entropy. In Konopka, A. K. (1994b). Proc. 3rd intl. conf. on open problems in computational biology. Published as Comput. Chem. 18(3).

Borodovsky, M., Koonin, E. V., & Rudd, K. E. (1994). New genes in old sequence: A strategy for finding genes in the bacterial genome. Trends Biochem. Sci. 19, 309–313.

Borodovsky, M., Rudd, K. E., & Koonin, E. V. (1994). Intrinsic and extrinsic approaches for detecting genes in a bacterial genome. Nucl. Acids Res. 22, 4756–4767.

Breathnach, R. & Chambon, P. (1981). Organization and expression of eukaryotic split genes for coding proteins. Annu. Rev. Biochem. 50, 349–393.

Brown, C. M., Dalphin, M. E., Stockwell, P. A., & Tate, W. P. (1993). The translational termination signal database. Nucl. Acids Res. 21, 3119–3123.

Brunak, S., Engelbrecht, J., & Knudsen, S. (1991). Prediction of human mRNA donor and acceptor sites from the DNA sequence. J. Mol. Biol. 220, 49–65.

Bucher, P. (1988). The eukaryotic promoter database of the Weizmann Institute of Science. EMBL Nucleotide Sequence Data Library 17, Postfach 10.2209, D-6900 Heidelberg.

Bucher, P. (1990). Weight matrix descriptions of four eukaryotic RNA polymerase II promoter elements derived from 502 unrelated promoter sequences. J. Mol. Biol. 212, 563–578.

Burset, M. & Guigó, R. (1996). Evaluation of gene structure prediction programs. Genomics, to appear.

Cardon, L. R. & Stormo, G. D. (1992). Expectation maximization algorithm for identifying protein-binding sites with variable lengths from unaligned DNA fragments. J. Mol. Biol. 223, 159–170.

Cavener, D. R. & Ray, S. C. (1991). Eukaryotic start and stop translation signals. Nucl. Acids Res. 19, 3185–3192.

Church, D. M., Stotler, C. J., Rutter, J. L., Murrell, J. R., Trofatter, J. A., & Buckler, A. J. (1994). Isolation of genes from complex sources of mammalian genomic DNA using exon amplification. Nature Genet. 6, 98–105.

Claverie, J.-M., Sauvaget, I., & Bougueleret, L. (1990). k-tuple frequency analysis: from intron/exon discrimination to T-cell epitope mapping. In: Doolittle, R. F., Ed., Molecular Evolution: Computer Analysis of Protein and Nucleic Acid Sequences. Methods Enzymol. 183.

Claverie, J.-M. (1992). Identifying coding exons by similarity search: Alu-derived and other potentially misleading protein sequences. Genomics 12, 838–841.

Claverie, J.-M. (1993a). Database of ancient sequences. Nature 364, 19–20.

Claverie, J.-M. (1993b). Detecting frame shifts by amino acid sequence comparison. J. Mol. Biol. 234, 1140–1157.

Claverie, J.-M. (1994a). Some useful statistical properties of position-weight matrices. In: Konopka, A. K. (1994b). Proc. 3rd intl. conf. on open problems in computational biology. Published as Comput. Chem. 287–294.

Claverie, J.-M. (1994b). Large-scale sequence analysis. Chapter 36 in Adams, Fields, and Venter (1994). 267–279.

Claverie, J.-M. (1994c). A streamlined random sequencing strategy for finding coding exons. Genomics 23, 575–581.
Cramer, H. (1946). Mathematical methods of statistics. Princeton University. Princeton, NJ.
Dong, S. & Searls, D. B. (1994). Gene structure prediction by linguistic methods. Genomics 23, 540–551.
Doolittle, R. F. (1986). Of URFs and ORFs. University Science. Mill Valley, CA.
Doolittle, R. F., Ed. (1990). Molecular Evolution: Computer Analysis of Protein and Nucleic Acid Sequences. Methods Enzymol. 183 (special issue).
Einstein, J. R., Mural, R. J., Guan, X., & Uberbacher, E. C. (1992). Computer-based construction of gene models using the GRAIL gene assembly program. Oak Ridge National Laboratory report TM-12174.
Faisst, S. & Meyer, S. (1991). Compilation of vertebrate-encoded transcription factors. Nucl. Acids Res. 20, 3–26.
Farber, R. B., Lapedes, A. S., & Sirotkin, K. M. (1992). Determination of eukaryotic protein coding regions using neural networks and information theory. J. Mol. Biol. 226, 471–479.
Fickett, J. W. (1982). Recognition of protein coding regions in DNA sequences. Nucl. Acids Res. 10, 5303–5318.
Fickett, J. W., Torney, D. C., & Wolf, D. R. (1992). Base compositional structure of genomes. Genomics 13, 1056–1064.
Fickett, J. W. & Tung, C.-S. (1992). Assessment of protein coding measures. Nucl. Acids Res. 20, 6441–6450.
Fickett, J. W. (1994). Inferring genes from open reading frames. In: Konopka, A. K. (1994b). Proc. 3rd intl. conf. on open problems in computational biology. Published as Comput. Chem. 203–205.
Fickett, J. W. (1995a). The gene identification problem: an overview for developers. Proc. 4th Int. Workshop on Open Problems in Computational Molecular Biology, A. Konopka, Ed. (to appear).
Fickett, J. W. (1995b). ORFs and genes: How strong a connection? J. Comp. Biol. 2, 117–123.
Fickett, J. W. (1996). Coordinate Positioning of MEF2 and Myogenin Binding Sites. Gene-COMBIS (at http://www.elsevier.nl.locate/genecombis) and Gene, to appear.
Fields, C. A. (1994). Integrating computational and experimental methods for gene discovery. In: Adams, Fields, and Venter (1994). 321–325.
Fields, C. A. & Soderlund, C. A. (1990). GM: A practical tool for automating DNA sequence analysis. Comput. Appl. Biosci. 6, 263–270.
Gelfand, M. S. (1989). Statistical analysis of mammalian pre-mRNA splicing sites. Nucl. Acids Res. 17, 6369–6382.
Gelfand, M. S. (1990a). Computer prediction of exon-intron structure of mammalian pre-mRNAs. Nucl. Acids Res. 18, 5865–5869.
Gelfand, M. S. (1990b). Global methods for the computer prediction of protein-coding regions in nucleotide sequences. Biotechnol. Software 7, 3–11.
Gelfand, M. S. (1992). Prediction of protein-coding regions in DNA of higher eukaryotes. In: Gindikin (1992). 87–98.
Gelfand, M. S. (1995). Prediction of function in DNA sequence analysis. J. Comp. Biol. 2, 87–115.
Gelfand, M. S. & Roytberg, M. A. (1993). A dynamic programming approach for prediction of the exon-intron structure. Bio Systems 30, 173–182 (special issue on computer genetics).
Ghosh, D. (1990). A relational database of transcription factors. Nucl. Acids Res. 18, 1749–1756.
Gish, W. & States, D. J. (1993). Identification of protein coding regions by database similarity search. Nature Genet. 3, 266–272.
Goldstein, L. & Waterman, M. S. (1994). Approximations to profile score distributions. J. Comput. Biol. 1, 93–104.
Green, P., Lipman, D., Hillier, L., Waterston, R., States, D., & Claverie, J.-M. (1993). Ancient conserved regions in new gene sequences and the protein databases. Science 259, 1711–1716.
Gribskov, M. & Devereux, J. (1991). Sequence analysis primer. Stockton. New York.

Gribskov, M., McLachlan, A. D., & Eisenberg, D. (1987). Profile analysis: Detection of distantly related proteins. Proc. Natl. Acad. Sci. USA 84, 4355–4358.

Griffin, A. & Griffin, H. G. (1994). Computer analysis of sequence data (2 vol.). Humana, Totowa, NJ.

Guigó, R., Knudsen, S., Drake, N., & Smith, T. (1992). Prediction of gene structure. J. Mol. Biol. 226, 141–157.

Guigó, R. & Fickett, J. W. (1994). Distinctive sequence features in protein coding, non-coding, and intergenic human DNA. J. Mol. Biol. 253, 51–60.

Hawkins, J. D. (1988). A survey on intron and exon lengths. Nucl. Acids Res. 16, 9893–9908.

Henikoff, S. & Henikoff, J. G. (1991). Automated assembly of protein blocks for database searching. Nucl. Acids Res. 19, 6565–6572.

Hirst, J. D. & Sternberg, M. J. E. (1992). Prediction of structural and functional features of protein and nucleic acid sequences by artificial neural networks. Biochemistry 31, 7211–7219.

Hosmer, D. W. & Lemeshow, S. (1989). Applied logistic regression. Wiley, New York.

Hutchinson, G. B. & Hayden, M. R. (1992). The prediction of exons through an analysis of spliceable open reading frames. Nucl. Acids Res. 20, 3453–3462.

Hutchinson, G. B. & Hayden, M. R. (1993). SORFIND: A computer program that predicts exons in vertebrate genomic DNA. In: Lim et al. (1993), 513–520.

Jurka, J., Walichiewicz, J., & Milosavljevic, A. (1992). Prototypic sequences for human repetitive DNA. J. Mol. Evol. 35, 286–291.

Kamb, A., Wang, C., Thomas, A., DeHoff, B. S., Norris, F. H., Richardson, K., Rine, J., Skolnick, M., & Rosteck, P. R., Jr. (1995). Software trapping: a strategy for finding genes in large genomic regions. Computers and Biomedical Research 28, 140–153.

Kel, O. V., Romachenko, A. G., Kel, A. E., Naumochkin, A., & Kolchanov, N. A. (1995). Structure of data representation in TRRD - database of transcription regulatory regions on eukaryotic genomes. In Proceedings of the 28th Annual Hawaii International Conference on System Sciences (HICSS) Wailea, Hawaii, January 4–7, 1995.

Knudsen, S., Guigó, R., & Smith, T. (1993). GeneID—A computer server for prediction of genes in DNA sequences. In: Lim et al. (1993), 545–553.

Knueppel, R., Dietze, P., Lehnberg, W., Frech, K., & Wingender, E. (1994). TRANSFAC retrieval program: A network model database of eukaryotic transcription regulating sequences and proteins. J. Comput. Biol. 1, 191–198.

Konopka, A. K. & Owens, J. (1990). Complexity charts can be used to map functional domains in DNA. Gene Anal. Technol. Appl. 7, 35–38.

Konopka, A. K. (1994a). Sequences and codes: Fundamentals of biomolecular cryptography. In: Biocomputing: Informatics and Genome projects. Academic Press, New York.

Konopka, A. K., Ed. (1994b). Proc. 3rd Int. Conf. open problems in Computational Biology. Published as Comput. Chem. 18(3).

Koonin, E. V., Bork, P., & Sander, C. (1994). Yeast chromosome III: New gene functions. EMBO J. 13, 493–503.

Kozak, M. (1991). An analysis of vertebrate mRNA sequences: Intimations of translational control. J. Cell Biol. 115, 887–903.

Krogh, A., Mian, I. S., & Haussler, D. (1994). A hidden Markov model that finds genes in *E. coli* DNA. Nucl. Acids Res. 22, 4768–4778, and (in a slightly expanded version) UCSC Report CRL-93-33.

Krogh, A., Brown, M., Mian, I. S., Sjoelander, K., & Haussler, D. (1994). Hidden Markov models in computational biology. Applications to protein modeling. J. Mol. Biol. 235, 1501–1531.

Lawrence, C. E., Altschul, S. F., Boguski, M. S., Liu, J. S., Neuwald, A., & Wootton, J. C. (1993). Detecting subtle sequence signals: A Gibbs sampling strategy for multiple alignment. Science 262, 208–214.

Lim, H. A., Fickett, J. W., Cantor, C. R., & Robbins, R. J. (1993). Proc. 2nd Int. Conf. Bioinformatics, Supercomputing, and Complex Genome Analysis. World Scientific, Singapore.

Lippmann, R. P. (1987). An introduction to computing with neural nets. IEEE ASSP Magazine, April, 4–22.
Locker, J. & Buzard, G. (1990). A dictionary of transcription control sequences. J. DNA Sequencing and Mapping 1, 3–11.
Lopez, R., Larsen, F., & Prydz, H. (1994). Evaluation of the exon predictions of the GRAIL software. Genomics 24, 133–136.
Manly, B. F. J. (1986). Multivariate statistical methods. Chapman and Hall, London.
Matthews, B. W. (1975). Comparison of the predicted and observed secondary structure of T4 phage lysozyme. Biochem. Biophys. Acta 405, 442–451.
McKeown, M. (1992). Alternative mRNA splicing. Annu. Rev. Cell. Biol. 8, 133–155.
McLachlan, A. D., Staden, R., & Boswell, D. R. (1984). A method for measuring the nonrandom bias of a codon usage table. Nucl. Acids Res. 12, 9567–9575.
Milanesi, L., Kolchanov, N. A., Rogozin, I. B., Ischenko, I. V., Kel, A. E., Orlov, Y. L., Ponomarenko, M. P., & Vezzoni, P. (1993). GenViewer: A computing tool for protein-coding regions prediction in nucleotide sequences. In: Lim et al. (1993), 573–587.
Mitchell, P. J. & Tjian, R. (1989) Transcriptional regulation in mammalian cells by sequence-specific DNA binding proteins. Science 245, 371–378.
Mulligan, M. E., Hawley, D. K., Entriken, R., & McClure, W. R. (1984). *Escherichia coli* promoter sequences predict in vitro RNA polymerase selectivity. Nucl. Acids Res. 12, 789–800.
Mural, R. J., Mann, R. C., & Uberbacher, E. C. (1990). Pattern recognition in DNA sequences: The intron-exon junction problem. In: The First International Conference on Electrophoresis, Supercomputing and the Human Genome (Cantor, C. R. & Lim, H. A., Eds.), pp. 164–172, World Scientific, London.
Nakata, K., Kanehisa, M., & DeLisi, C. (1985). Prediction of splice junctions in mRNA sequences. Nucl. Acids Res. 13, 5327–5340.
Naora, H. & Deacon, N. J. (1982). Relationship between the total size of exons and introns in protein-coding genes of higher eukaryotes. Proc. Natl. Acad. Sci. USA 78, 6196–6200.
Ogiwara, A., Uchiyama, I., Seto, Y., & Kanehisa, M. (1992). Construction of a dictionary of sequence motifs that characterize groups of related proteins. Protein Eng. 5, 479–488.
Ohshima, Y. & Gotoh, Y. (1988). Signals for the selection of a splice site in pre-mRNA. J. Mol. Biol. 195, 247–259.
Oliver et al. (1992). The complete DNA sequence of yeast chromosome III. Nature 357, 38–46.
Ossadnik, S. M., Buldyrev, S. V., Goldberger, A. L., Havlin, S., Mantegna, R. N., Peng, C.-K., Simons, M., & Stanley, H. E. (1994). Correlation approach to identify coding regions in DNA sequences. Biophys. J. 67, 64–70.
Posfai, J., Bhagwat, A. S., Posfai, G., & Roberts, R. J. (1989). Predictive motifs derived from cytosine methyltransferases. Nucl. Acids Res. 17, 2421–2435.
Prestridge, D. S. (1995). Predicting Pol II promoter sequence using transcription factor binding sites. J. Mol. Biol. 249, 923–932.
Robison, K., Gilbert, W., & Church, G. M. (1994). Large scale bacterial gene discovery by similarity search. Nature Genet. 7, 205–214.
Searls, D. B. (1992). The linguistics of DNA. Am. Scientist 80, 579–591.
Seely, O., Jr., Feng, D.-F., Smith, D. W., Sulzbach, D., & Doolittle, R. F. (1990). Construction of a facsimile data set for large genome sequence analysis. Genomics 8, 71–82.
Senapathy, P., Shapiro, M. B., & Harris, N. L. (1990). Splice junctions, branch point sites, and exons: sequence statistics, identification, and applications to genome project. In: Doolittle (1990), 252–278.
Shapiro, M. B. & Senapathy, P. (1987). RNA splice junctions of different classes of eukaryotes: Sequence statistics and functional implications in gene expression. Nucl. Acids Res. 15, 7155–7174.
Sharp, P. M., Cowe, E., Desmond, G. H., Shields, D. C., Wolfe, K. H., & Wright, F. (1988). Codon usage patterns in *Escherichia coli, Bacillus subtilis, Saccharomyces cerevisiae, Schizosaccharomyces*

pombe, Drosophila melanogaster, and *Homo sapiens*; a review of the considerable within-species diversity. Nucl. Acids Res. 16, 8207–8211.

Singh, G. B. & Krawetz, S. A. (1994). Computer-based exon detection: An evaluation metric for comparison. Int. J. Genome Res. 1, 321–338.

Smith, M. W. (1988). Structure of vertebrate genes: A statistical analysis implicating selection. J. Mol. Evol. 27, 45–55.

Smith, H. O., Annau, T. M., & Chandresegaran, S. (1990). Finding sequence motifs in groups of functionally related proteins. Proc. Natl. Acad. Sci. USA 87, 826–830.

Smith, R. F. & Smith, T. F. (1990). Automatic generation of primary sequence patterns from sets of related protein sequences. Proc. Nat. Acad. Sci. USA 87, 118–122.

Snyder, E. E. & Stormo, G. D. (1993). Identification of coding regions in genomic DNA sequences: an application of dynamic programming and neural networks. Nucl. Acids Res. 21, 607–613.

Soderlund, C., Shanmugam, P., White, O., & Fields, C. (1992). In: Proc. 25th Hawaii Int. Conf. System Sciences, V. (Milutinovic, V. & Shriver, B., Eds.), pp. 653–662, IEEE Computer Society, Los Alamitos, CA.

Solovyev, V. V. & Lawrence, C. B. (1993). Identification of human gene functional regions based on oligonucleotide composition. In: The First Int. Conf. Intelligent Systems for Molecular Biology (Hunter, L., Searls, D., & Shavlik, J., Eds.), AAAI, Menlo Park, CA.

Solovyev, V. V., Salamov, A. A., & Lawrence, C. B. (1994a). The prediction of human exons by oligonucleotide composition and discriminant analysis of spliceable open reading frames. In: Altman et al. (1994), pp. 354–362.

Solovyev, V. V., Salamov, A. A., & Lawrence, C. B. (1994b). Predicting internal exons by oligonucleotide composition and discriminant analysis of spliceable open reading frames. Nucl. Acids Res. 22, 5156–5163.

Staden, R. & McLachlan, A. D. (1982). Codon preference and its use in identifying protein coding regions in long DNA sequences. Nucl. Acids Res. 10, 141–156.

Staden, R. (1990). Finding protein coding regions in genomic sequences. In: Doolittle (1990), pp. 163–179.

States, D. J. & Gish, W. (1994). Combined use of sequence similarity and codon bias for coding region identification. J. Comp. Biol. 1, 39–50.

Stormo, G. D. (1990). Consensus patterns in DNA. In: Doolittle (1990), 211–220.

Stormo, G. D. & Haussler, D. (1994). Optimally parsing a sequence into different classes based on multiple types of evidence. In: Altman et al. (1994), pp. 369–375.

Sulston, J., Du, Z., Thomas, K., Wilson, R., Hillier, L., Staden, R., Halloran, N., Green, P., Thierry-Mieg, J., Qiu, L., Dear, S., Coulson, A., Craxton, M., Durbin, R., Berks, M., Metzstein, M., Hawkins, T., Ainscough, R., & Waterston, R. (1992). The C. elegans genome sequencing project: A beginning. Nature, 356, 37–41.

Takeda, Y., Sarai, A., & Rivera, V. M. (1989). Analysis of the sequence-specific interactions between Cro repressor and operator DNA by systematic base substitution experiments. Proc. Natl. Acad. Sci. USA 86, 439–443.

Thomas, A. & Skolnick, M. H. (1994). A probabilistic model for detecting coding regions in DNA sequences. IMA J. of Math. Applied in Med. & Biol. 11, 149–160.

Towell, G. G. & Shavlik, J. W. (1994). Knowledge-based neural networks. Artificial Intelligence 70, 119–165.

Uberbacher, E. & Mural, R. J. (1991). Locating protein-coding regions in human DNA sequences by a multiple sensor-neural network approach. Proc. Natl. Acad. Sci. USA 88, 11261–11265.

Uberbacher, E. C., Einstein, J. R., Guan, X., & Mural, R. J. (1993). Gene recognition and assembly in the GRAIL system: Progress and challenges. In: Lim et al. (1993), pp. 465–476.

von Hippel, P. H. (1994). Protein-DNA recognition: New perspectives and underlying themes. Science 263, 769–770.

Wahle, E. & Keller, W. (1992). The biochemistry of 3' end cleavage and polyadenylation of messenger RNA precursors. Annu. Rev. Biochem. 61, 419–40.

Waterman, M. (1989). Consensus patterns in sequences. In: Mathematical methods for DNA sequences, (Waterman, M., Ed.), pp. 93–116, CRC, Boca Raton, FL.

Waterman, M. (1995). Introduction to computational biology: Maps, sequences and genomes. Chapman Hall, New York.

Williamson, A. R., Elliston, K. O., & Sturchio, J. L. (1995). The Merck gene index is a public resource for genomics research. J. of NIH Res. 7, 61–63.

Wilson, R., Ainscough, R., Anderson, K., Baynes, C., Berks, M., Burton, J., Connell, M., Boonfield, J., & Copsey, T. (1994). 2.2 Mb of contiguous nucleotide sequence from chromosome III of *C. elegans*. Nature 368, 32–38.

Wingender, E. (1994). TRANSFAC database. 31 October posting to human-genome-program @net.bio.net.

Xu, Y., Einstein, J. R., Mural, R. J., Shah, M., & Uberbacher, E. C. (1994). An improved system for exon recognition and gene modeling in human DNA sequences. In: Altman et al. (1994), pp. 376–383.

Xu, Y., Mural, R. J., & Uberbacher, E. C. (1995). Correcting sequencing errors in DNA coding regions using a dynamic programming approach. Comp. Appl. Biosci. 11, 117–124.

THE ARCHITECTURE OF LOOPS IN PROTEINS

Anna Tramontano

	Abstract	239
I.	Introduction	240
II.	The Protein Folding Problem	240
III.	Prediction of Loop Structure	242
IV.	The Architecture of Short Loops and Their Sequence Patterns	242
V.	Medium-Sized Loops and Their Stabilization	246
VI.	Database Searching for the Prediction of Loop Conformations	247
VII.	*Ab Initio* Calculation of Protein Fragment Conformations	248
VIII.	Loop Design and Transplant: A New Route to Therapy	249
IX.	Conclusions	251
	Acknowledgments	252
	References	252

ABSTRACT

The difficulty in correctly predicting the conformation of loops represents a serious stumbling block for modeling the structure of unknown proteins. This chapter will review our current understanding of the structural characteristics of loops and of the requirements for their stability and will discuss critically the different methodologies currently used to produce three-dimensional models of these nonrepetitive segments of protein structures, with particular emphasis on the structural reasons behind their successes and failures. An improved understanding of loop structure can open the road to a number of potentially very interesting practical applications directed toward the development of new drugs and vaccines, some examples of which will be described briefly.

I. INTRODUCTION

The study of protein architecture is a fascinating and challenging task: structural molecular biology has unquestionably been, and will remain for some time, amongst the most rapidly evolving fields of biology and one of those attracting most interest from computational biologists. Both the complexity of protein structure and the rapidly increasing size of the collection available has revealed a number of computational problems related to the storage, analysis and classification of the data.

The ultimate aim of structural molecular biology is surely to be able to manipulate protein structures at will, modifying their amino acid sequences or designing novel ones to achieve the desired structural and functional characteristics.

To be able to properly address the problem of the relationship between structure and function of a protein, and consequently to be able to change the former and obtain predictable modifications of the latter, we need to understand the rules that govern the architecture of proteins.

A great deal of progress has been achieved toward this goal, but, as I will illustrate in this chapter, there is one particular objective which is still eluding us, at least in general terms, and that is a detailed comprehension of those parts of a protein structure which are less regularly organized and are more subject to changes during evolution, namely, loops.

Although these have been extensively investigated, since the early days of protein structural analysis, it is now clear that we have still not gained an understanding of their architecture detailed enough to be able to derive rules to predict their structure.

This chapter is intended to give an overview of the many interesting results in the field of protein loop analysis, prediction, and design, and of their applications, with particular emphasis on some of the prediction techniques and on the assumptions they are based upon in order to assess critically their adequacy for treating the problem at hand.

II. THE PROTEIN FOLDING PROBLEM

The folding problem has been considered a quite impenetrable enigma for a long time. Nevertheless, interesting achievements are being continuously reported. The most successful approaches to this end are homology modeling, threading, and *ab initio* prediction.

In homology modeling, the structure of a related protein sharing a significant sequence similarity with the protein to be predicted is used as a template. It is possible to estimate the reliability of the model in the well packed interior "core" of the structure from the percent of residue identity between the template and the modeled protein (Chothia & Lesk, 1986). The prediction of those regions which are not part of the core of the protein is certainly less reliable, and although a number of reasonably successfully cases have been reported in the literature (Benner, 1992; Cohen & Cohen, 1994), much work has still to be done to make models obtained by homology reliable in all cases, not only in special instances (Chothia et al., 1987; Chothia et al., 1989), and to improve the accuracy of the details of the models produced.

Threading (Bowie et al., 1991; Jones et al., 1992; Godzik et al., 1992; Sippl & Weitckus, 1992; Bowie & Eisenberg, 1993; Skolnick et al., 1993; Ouzonis et al., 1993) is an interesting example of a case where posing the question in a different way can often be substantially helpful in solving a problem. In this approach the question "Which structure will my protein have?" is reformulated as "Which of these enumerated folds, if any, is my protein more likely to assume?" The result of a threading procedure is the identification of the fold of the protein and once again, the major elements of the core structure of the protein can be identified but the details of the structure will have to be built, if possible, using different alternative strategies.

The wording "*ab initio* prediction" is used to describe attempts to use computationally intensive simulations of the folding pathway of proteins and statistical methods to predict the secondary structure of the unknown protein. The first approach has gained power as computational tools have became faster and more affordable (Hagler & Honig, 1978; Levitt, 1983; Moult & James, 1986), the second by the intuition that much information is in fact stored in protein evolution, so that better accuracy is obtained when sequence changes among families of homologous proteins are taken into account (Rost et al., 1993; Rost & Sander, 1993). Neither technique relies on the existence of an homologous protein of known structure nor on the existence of the fold in the database of solved protein structures, but they share the drawback of being restricted to defining partial regions of the protein molecule. Now only small elements of proteins can be properly simulated and the secondary structure elements that can be predicted are the regular ones, helices and strands. In both cases the predicted elements still need to be properly positioned in space and the remaining parts modeled to give the final predicted structure of the protein.

It is very likely that combining all of these approaches, taking advantage of the ever increasing number of known protein structures and sequences and the availability of more and more powerful computer tools will provide us with a reasonable way to construct reliable and accurate models of the protein core structure in the near future.

As anticipated above however, a major obstacle to the correct prediction of a complete protein structure is the modeling of loops, regions of the polypeptide chain that have a lower degree of regularity and where insertions and deletions frequently occur during evolution. On the other hand, loops are very important parts of protein structures which have been shown in some cases to be the nucleation sites for protein folding and very often are involved in the catalytic mechanisms of enzymes or in interaction surfaces between different proteins.

Loops are intrinsically irregular structures compared to the regular arrangement of hydrogen bonds and dihedral angles in α-helices and β-strands, and they frequently accommodate insertions and deletions. From a statistical survey of the insertions and deletions occurring in a data bank of multiple sequence alignments, it has been calculated (Pascarella & Argos, 1992) that nearly every loop in an ancestral structure is a possible target for insertions and deletions during evolution.

A better understanding of loops in proteins and an improved ability to predict their conformation is certainly needed now. Unfortunately, most of the widely used techniques for loop modeling fall short of providing a general answer to the problem, as will be illustrated by a number of examples.

III. PREDICTION OF LOOP STRUCTURE

The routes to predict loop structures explored so far are essentially the identification of sequence patterns in the loop itself, database searching techniques, and *ab initio* calculations of their conformations.

Sequence patterns have been used mainly in the case of short loops, especially those connecting adjacent strands of antiparallel β sheets and in the special case of immunoglobulin hypervariable loops. Both will be described briefly below.

Database searching is the default technique used in many model building experiments and is based on the observation that segments of similar conformation occur in both related and unrelated proteins. It has been calculated that all eight-residue regions of main chain in a given conformation are similar in conformation to one of a set of about 60 structures (Lesk, 1994). However, the exploitation of such a property of protein structure for the prediction of loops, requires that some information about the structure of the loop is contained in the regions surrounding it, so that the latter can be used to identify the former. So far, only the information contained in the regions adjacent to the loop in the primary structure of the protein have been taken into account in this approach.

Other methods for loop structure prediction include molecular simulations, combinatorial searches, and subsequent evaluation of loop candidates using conformational energy estimation and a combination of this latter method with database searching techniques (Rose et al., 1985; Moult & James, 1986; Bruccoleri et al., 1988; Martin et al., 1989).

IV. THE ARCHITECTURE OF SHORT LOOPS AND THEIR SEQUENCE PATTERNS

An exhaustive classification of loops is very useful to help in defining the sequence patterns specific for each type of loop and in deducing predictive rules from them. Consequently, since the early work of Venkatachalam (1968) much effort has been devoted to this task.

A turn is generally defined as a loop that allows the polypeptide chain to change its direction by 180°. In his analysis of three-residue stretches in proteins, Venkatachalam defined a turn as being characterized by the formation of a hydrogen bond between the main-chain carboxylic oxygen of the first residue and the amide proton of the third. He identified three conformations (called I, II and III) according to the main-chain dihedral angles of residues in the turn. The mirror images of these turns (denoted I', II', III') are also possible but disfavored. β-turns connecting adjacent strands of an antiparallel β-sheet constitute a subclass of β-turns, the β-hairpin, carefully analysed by Sibanda and Thornton (1985). Two-residue β-hairpins are often found in type I' and II' conformations, rather than I and II as happens for most nonhairpin turns (Figure 1). It is interesting to mention that these conformations are probably preferred for β-hairpins because they allow the correct twisting of the two adjacent β-strands.

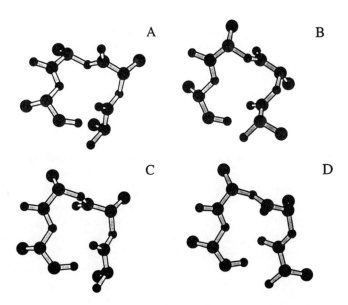

Figure 1. Ideal type I (A), type II (B), type I' (C) and type II' (D) β-turn conformation.

By adopting a less stringent definition of turns (not based on the existence of the hydrogen bond, but on the distance of the C_α between the first and last residue) the number of examples of turns increased (Lewis et al., 1973) leading to the classification of 10 different turn types, reduced to seven in 1981 by a more rational definition of turn types by Richardson (1981).

γ-turns are defined by the presence of a hydrogen bond between the carbonyl group of one residue and the amino group of the amino acid two residues ahead in the sequence. They can be classified in two types, named classic and inverse, which are the mirror image of each other (Rose, 1978; Rose et al., 1985; Milner-White et al., 1988; Milner-White, 1990). It has been shown that inverse γ-turns are much more frequent than classic ones.

ω-loops have been described and classified (Leszczynski & Rose, 1986) as irregular segments of chain where sequentially distant N- and C-terminal residues are spatially close.

Many attempts have been made to link sequence patterns to the described classifications of short loops, much along the same lines as statistical methods for the prediction of α-helices and strands. In this case the task may be simplified by the fact that short loops have to introduce a sharp change in the direction of the polypeptide chain and this implies restrictions on the dihedral angles of their residues which can in turn be correlated to the presence of special amino acids, such as glycines and prolines. For example, an analysis of the sequence patterns of β-hairpins in proteins (Wilmot & Thornton, 1988) has led to the widely accepted view that type I and type II turns are often associated with the presence of a glycine as the third or fourth amino acid of the loop, respectively.

Similarly, sequence patterns have been associated with the various turn types and to ω- and γ-turns.

It is important to realize that any method to relate the sequence of a loop to its structure relies on assuming that the structure of the loop is determined by local rather than tertiary interactions. This is not necessarily true and I will use a few examples to demonstrate how tertiary interactions can be determinant for turn conformation.

Immunoglobulins are multidomain proteins consisting of two chains each including a variable domain (Alzari et al., 1988; Lesk & Tramontano, 1993). The antigen binding site is formed by six loops, clustering in space to form the antigen binding site (Figure 2). The high sequence variability of these loops allows immunoglobulins to recognize different antigens. The comparative analysis of immunoglobulin structures has revealed that different sequences in different antibody loops do not always generate different conformations in both the main chain and side chains of these regions. It has been demonstrated (Chothia & Lesk, 1987) that five of these six loops can only assume a limited number of main-chain conformations, called "canonical structures". Most sequence variations only affect the side chains of the loops, consequently modifying the antigen binding site surface, without, however, changing the backbone structure of the loop. Only some specific sequence changes

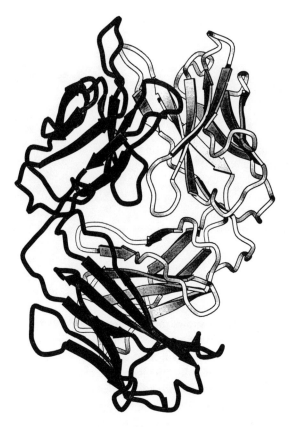

Figure 2. Molscript (Kraulis, 1991) representation of the Fab fragment of an immunoglobulin. The heavy chain is shown in lighter gray on the right.

at a limited set of positions produce a change in the main-chain conformation of the loops. This observation is interesting from a theoretical point of view and has a number of implications for the understanding of the immunoglobulin structure-function relationship.

Through a careful analysis of the available structures, Chothia and Lesk (1987) were also able to describe specifically the relationship between sequence changes and canonical structures, that is, they were able to identify the residues that, through their packing, hydrogen bonding or ability to assume unusual values of their main-chain torsional angles, are responsible for the occurrence of each canonical structure.

This implies that it is possible, for five of the six immunoglobulin hypervariable loops, to define precisely the sequence-structure relationship. This model has been repeatedly validated through blind tests, in which immunoglobulin structures were

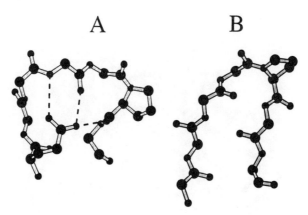

Figure 3. Example of two different canonical structures of the L3 loop of immunoglobulins.

correctly predicted prior to their experimental determination (Chothia et al., 1987; Chothia et al., 1989).

As an example, Figure 3 shows the canonical structures for the six-residue L3 loop of immunoglobulins. Important determinants for the occurrence of one conformation or the other are the position of a proline residue and the nature of the interaction of the residue preceding the loop with atoms of the loop itself (Chothia & Lesk, 1987). In the canonical structure indicated by A in the figure, there is a proline in position 96 (Kabat numbering scheme, Kabat et al., 1991) with a *cis* peptide bond and the side chain of the residue immediately preceding the loop forms hydrogen bonds with the main-chain atoms of the loop. In the other canonical structure (Figure 3B) the proline is instead in position 94, and its peptide bond is in a *trans* conformation.

So far, six-residue L3 loops in immunoglobulin of known structure have always been found to belong to canonical structure A or B, and in each case the respective sequence pattern was observed.

Analogous situations are observed in all other loops, except for H3, for which such a clear sequence-structure correlation has not been found. These loops certainly represent a special case and their structural features are quite unique to this class of molecules, but can nevertheless be exploited for a great number of practical applications as will be discussed later.

The L3 illustration above is a case where the determinants of the conformation of the loop are in the loop itself or in residues adjacent to the loop. Analysis of immunoglobulin loops has, however, also been of great utility for understanding how tertiary interactions can in general overrule sequence patterns, as illustrated by the H2 loop.

Loops in Proteins

Table 1. Amino Acid Sequences of Four-Residue H2 Loops of Immunoglobulins

	52a	53	54	55
HyHEL-5	Pro	Gly	Ser	Gly
J539	Pro	Asp	Ser	Gly
KOL	Asp	Asp	Gly	Ser

The second loop of the heavy chain (H2) is a β-hairpin, often short (three or four residues). The expectation, from the discussion about hairpin loops, is that for these short loops a correlation exists between its sequence pattern and its conformation.

An interesting result was obtained by comparing H2 loops of different immunoglobulins of known structure, for example, those of HyHEL-5, KOL and J539 (Tramontano et al., 1990). They all form a four-residue hairpin turn, and their sequence is shown in Table 1. If one observes the position of glycines and remembers what was discussed above, the conclusion would be that the conformation of the loop of J539 and that of HyHEL-5 should be very similar, in that they both have a glycine in the fourth position, and differing from that of KOL, where the glycine is in third position. What is instead observed is that the conformation of the J539 loop is much more similar to that of the corresponding KOL loop. A careful analysis of the interactions of this loop with the rest of the immunoglobulin structure has shown that the determinants of the conformation of this loop involve tertiary interactions, in particular, they are dependent on the size of residue 71, a residue far away in the sequence from the loop, and part of the conserved immunoglobulin β-framework (Tramontano & Lesk, 1992). When position 71 contains a small- or medium-sized residue, the conformation of four-residue H2 loops is similar to that illustrated in Figure 4A; when residue 71 is an arginine, a different packing of side chains arises in the loop region and the main chain of the loop has the conformation illustrated in Figure 4B.

The implications of this observation are many and relevant to our understanding of loop architecture and for practical purposes. First, as stated before, a tertiary interaction can be responsible for the loop conformation, so that our ability to predict the structure of short hairpins may very well be impaired by our limited understanding of the overall stability requirements of proteins. Second, the ability to transplant loop regions from one protein to another, for example, from antibodies of nonhuman origin into human frameworks (Reichmann et al., 1988), relies on assuming that the conformation of the loops is independent of the rest of the structure, which is obviously not generally true.

Since this early observation, many other cases where loop transplant in antibodies had to be combined with mutation of the framework residues, so as to be effective, have been reported (Foote & Winter, 1992), showing the relevance of tertiary interactions in determining the structure of the antibody loops.

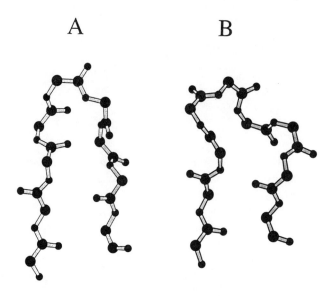

Figure 4. The difference between the structures of the H2 loops of immunoglobulins HyHEL-5 (A) and KOL (B) is determined by the size of the amino acid in position 71.

V. MEDIUM-SIZED LOOPS AND THEIR STABILIZATION

When loops are medium-sized, it is much more difficult to define them according to their main-chain dihedral angles, and consequently the classification becomes less rigorous. It is, however, possible to derive some rough classification of these loops, based on the type of interactions that stabilize them (Tramontano et al., 1989). For loops that form compact substructures, the major conformation determinant is the formation of hydrogen bonds to main-chain atoms of the loop. For loops having more extended conformations, the required stabilization is obtained by packing an inward pointing hydrophobic side chain of the loop between the secondary structure elements connected by the loop.

Immunoglobulins can again be used to exemplify both types. The interesting question that arises is of course how conserved are such interactions and whether the stabilizing elements can be detected and used to predict the structure of the loop.

In an analysis of the occurrence of loops similar in conformation to immunoglobulins, a number of interesting results were derived.

The H1 loop of immunoglobulins connects strands which are part of the two β-sheets in the heavy-chain variable domain of the molecule. The conformation of this loop, illustrated in Figure 5A, resembles a distorted helix, with a central hydrophobic side chain packed inside the protein and buried between the β-sheets. There is at least one loop very similar to this in the database of solved structures and that is formed by residues 111–117 of an insect globin (Figure 5B). The

Loops in Proteins

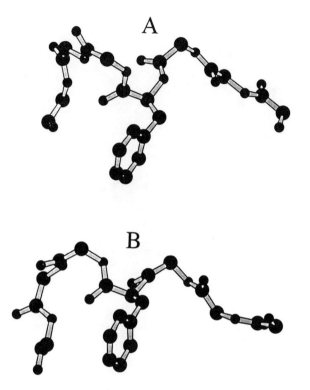

Figure 5. Structure of the H1 loop of McPC603 (A) and of a similar loop formed by residues 111–117 of *Chironomus erythrocruorin* (B).

superposition of the two loops is very good, with a root-mean-square deviation of 0.7 Å for all main-chain atoms. Also in the globin loop, the central side chain is large and hydrophobic, a phenylalanine, and packs in a cavity of the protein. What is interesting to observe is that while the H1 loop connects two strands, the globin loop connects two helices. The two helices linked by the loop form a cavity equivalent to that provided by the β-sheet framework of the immunoglobulins, but obviously unrelated from a structural point of view.

The other example of stabilization of medium-sized loops is obtained by hydrogen bonding of inward pointing main chain polar atoms, as in the case of some of the L3 loops in immunoglobulins, for example, in the case shown in Figure 3A and described above.

Loops with a structure very similar to this L3 canonical structure have been found to occur (Figure 6) in both Tomato bushy stunt virus coat protein (residues 355–360) and in cytochrome c from *desulfovibrio vulgaris* (residues 12–17). What do these loops have in common?

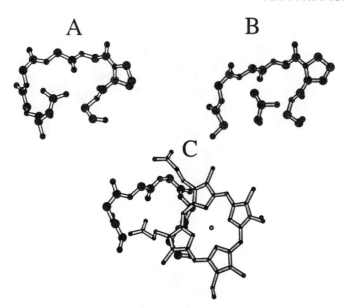

Figure 6. Structure of the L3 loop of McPC603 (A) and of two similar loops in the data base: residues 355–360 of tomato stunty bush virus coat protein (B) and residues 12–17 of cytochrome c3 from *Desulfovibrio vulgaris* (C).

First, the viral loop also has a *cis* proline in a position equivalent to that of the immunoglobulin. Second, both the viral and the cytochrome loops are stabilized by hydrogen bonds similar to those observed in immunoglobulins. The interesting observation, however, is that such hydrogen bonds are formed by the residues of the loop with completely unrelated partners. As mentioned before, in the immunoglobulin the partner for these interactions is the side chain of the residue preceding the loop. In the virus the partner is the main chain of an alanine distant in the primary structure, and in the cytochrome it is the propionyl group of a heme. In all cases the hydrogen bond partners occupy the same position in space relative to the loop (Figure 6). Furthermore, the structural context of these three loops is once again completely different. In immunoglobulins and in the cytochrome, the loop is a hairpin (joins two adjacent strands of the same antiparallel β-sheet); in the virus it connects strands from different sheets.

In these cases, as in the case of the H2 loop, tertiary interactions seem to be much more important in determining the conformation of the loop than its own sequence.

The conclusions that can be derived from the previous examples are that the conformation of the loop dictates the interactions required to stabilize it, but in different proteins a variety of different topologies can be used to provide these interactions. This implies that it is unlikely that rules relating sequence to structure can be identified in medium-sized loops.

VI. DATABASE SEARCHING FOR THE PREDICTION OF LOOP CONFORMATIONS

A widely used technique to predict loop conformation relies on the early work of Jones and Thirup (1986). The technique has been extended and applied in many instances and is generally included as a tool in a number of commercial modeling packages (Jones, 1985; Dayringer et al., 1986; Vriend, 1990; Jones et al., 1990).

The basic idea consists of searching in the database of solved protein structures for two regions closely matching the segments preceding and following the loop to be modeled ("stems") and separated by the same number of residues as those forming the loop. The assumption is that the structure of the loop is correlated to the structure of its "stem" regions. In order for this technique to be useful, three things have to demonstrated. The first is that the loop to model exists in the database, second, that similar loops have similar stems, and third that an equivalent geometric relationship exists between the stems and the loops in the modeled and template structures.

The first hypothesis is very likely to be correct, as discussed before, but the other two still need to be proven.

In an attempt to define more rigorously the applicability of such a technique for model building, we (Tramontano & Lesk, 1992) simulated a model building experiment with immunoglobulin loops. The basic steps consisted of

1. searching the database for loops similar in conformation to those selected for the experiment to show that the loop could indeed be predicted by using a knowledge-based approach,
2. searching the database for regions matching the stems of each loop separated by the appropriate number of residues, and
3. comparing the loops selected in step 2 with those selected in step 1.

It has to be mentioned that such a simulation has more chance of being successful than a real modeling experiment, in that in our test case, the stem structures are experimentally determined and not predicted, as would be the case in a model building experiment.

We used 41 hypervariable regions from eight different immunoglobulins of known structures. The result of the loop search in the database showed, with rare exceptions, that, for all the selected loops, it was possible to find structurally similar regions in the database of known structures, both among immunoglobulin structures and unrelated proteins.

When searching for regions matching the stems of the selected loops, a more complicated situation arose. In the large majority of the cases, a good fit of the stems corresponds to a good fit of the intervening regions (Figure 7A), but this is not generally true. In most cases a good fit of the stem does not imply a good fit of the loop (Figure 7B) or vice versa (Figure 7C). There are also cases where, although

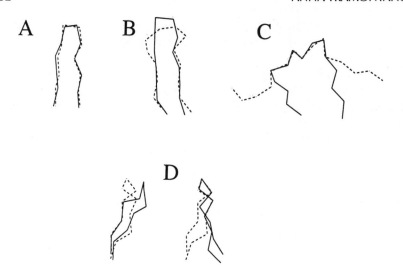

Figure 7. Superposition of loops from different proteins. (A) Residues 87–98 of the light chain of J539 (solid) and residues 13–24 of α-amylase inhibitor 1HOE (broken) have a very similar structure and the similarity extends to the four residues adjacent to the loops on each side. (B) The L3 loop of McPC603 (solid) is different from the loop 15–22 of α-amylase inhibitor (broken) but the four adjacent residues flanking the loops show a very good fit. (C) The H1 loop of NEWM (solid) is very similar to residues 42–48 of actinoxanthin (1ACX) (broken), but the flanking regions diverge outside the loop region. (D) Both the loops and the stems of the H2 loop of 2HFL (solid) and of residues 52–63 of 3CNA (broken) are similar in structure, but their geometric relationship is different, as can be seen by superimposing separately the stems (left) and the loops (right).

both the fit of the stems and the loop are good, but the geometric relationship between the two is different (Figure 7D). These results indicate that, although in most cases loops of the desired conformation exist in nonhomologous proteins, the information contained in the structure of the adjacent regions is not sufficient to identify them.

VII. *AB INITIO* CALCULATION OF PROTEIN FRAGMENT CONFORMATIONS

The folding of short segments of the chain can be simulated using semiempirical calculations. Methods based on the diffusion/collision process (Karplus & Weaver, 1976), local compactness (Lesk & Rose, 1981), contact density (Montelione & Scheraga, 1989), burial of hydrophobic area (Moult & Unger, 1991) interresidue interactions in lattice models (Skolnick & Kolinski, 1989), and systematic conformational search followed by energy evaluation (Martin et al., 1989) have been developed.

Because it has been shown that pentapeptides of identical sequences can have completely different structures in different proteins (Kabsch & Sander, 1984) and, as discussed above, segments of similar structure cannot in general be identified by their sequence, the problem with these approaches is that predicting the fold of a protein segment without taking the rest of the structure into account is bound to lead to incorrect results in many instances.

In the case of protein loops, because a model for the rest of the structure might be derived, one could use either conformational searches or molecular dynamics calculations to complete the model. The computational difficulties with this approach arise from the fact that the model of the interatomic interactions is neither complete nor exact and from the further fact that, even if it was, the part of the structure into which one is trying to build the loop is affected by an error, no matter what method has been used to predict it, which might seriously affect the result of a detailed free energy calculation.

However, the time required for a protein to fold is of the same order of magnitude as the time calculated for the folding of a short segment of about 20 amino acids (Wetlaufer, 1973). It has, therefore, been proposed that short segments could be nucleation sites for the folding process and that they assume the final conformation independently of the rest of the protein structure. It has also been shown that there is a correlation between the strongest nucleation sites experimentally identified and their area buried upon folding. Methods have been developed to try to predict nucleation sites from sequence signals.

Consequently, if one is able to identify regions of the chain which will adopt their final conformation early in the folding process, the problem of simulating the folding process can be simplified and would become a step-by-step procedure which starts from a few predicted regions to build up the complete structure (Vasquez & Scheraga, 1985).

These techniques could be a very interesting approach to the folding problem when combined with other methods to identify the general fold of the protein to be predicted.

VIII. LOOP DESIGN AND TRANSPLANT: A NEW ROUTE TO THERAPY

As shown before, there appears to be an intrinsic limitation to our ability to deduce the structure of a loop solely by knowledge of the conformation of its flanking residues. Although this is a serious limitation to our ability to predict loops, it is still possible to utilize loop transplant as a technique for practical applications, in those cases where the tertiary interactions responsible for a loop's conformation can be elucidated.

The possibility, for example, of transplanting loops from one antibody molecule to another opens the road to "humanizing" antibodies, that is raising an antibody in mice or rats against the desired antigen and then using the information in the antibody sequence to transfer its specificity to a human antibody, thus avoiding immunogenic problems.

The assumption on which this strategy relies is that the conformation of antibody hypervariable loops, that is, of the antigen binding site, is either independent of the structure of the framework (and we have seen that this is in general not true), or rather can be correlated to specific tertiary interactions. The elucidation of such interactions has, in fact, made the humanization of antibodies a successful route to therapy (Reichmann et al., 1988).

Another more challenging and interesting therapeutical strategy is to replace the sequences of loops with random sequences and use suitable technologies, such as phage display, to select among a variety of random sequences for those bearing the desired properties (Smith, 1991; Smith & Scott, 1993; Felici et al., 1991). Selected loop sequences could consequently be used as a lead toward the development of nonpeptidic molecules preserving the desired function. Once again an example of applications of both steps in this strategy comes from antibodies and will be discussed here.

Random peptide libraries can be displayed on filamentous phage and used to select sequences reacting with a given receptor. Such lead molecules can be used to develop nonpeptidic reagents based upon the amino acid sequence of the selected peptides. However, the intrinsic conformational flexibility of peptides implies that information about their active conformation can only be obtained by either complex structural studies or by testing differently constrained versions of them.

An alternative approach is to constrain the peptide structure "before" the selection procedure, so that the structural information about the active conformation is known a priori. The need arises for a substructure able to tolerate sequence changes without undergoing major structural changes, in other words, a substructure sharing the properties of immunoglobulin hypervariable loops.

For practical purposes, a large multidomain protein is not ideal for phage-derived selection of peptidic leads, so that an effort has been devoted to design a protein with a novel β-fold, whose model shown in Figure 8, is based on the structure of a variable domain of an antibody of known structure (Pessi et al., 1993). This 61-residue protein (named Minibody) includes two of the three heavy-chain hypervariable loops of the original immunoglobulin. The Minibody scaffold, containing a number of residue modifications with respect to the original immunoglobulin sequence, was obtained both by solid-phase synthesis and by expression in bacteria (Bianchi et al., 1993b; Bianchi et al., 1993a; Bianchi et al., 1994). The resulting molecule was biochemically characterized and its behavior shown to be consistent with the expected properties required by the design (Pessi et al., 1993). Further modifications of the molecule were successful in improving some important properties, for example, its solubility in aqueous media (Bianchi et al., 1994).

The Minibody was subsequently displayed on the surface of the filamentous phage f1, fused at the N-terminus of its minor coat protein. The sequences of the regions corresponding to the hypervariable loops of immunoglobulins were ran-

Figure 8. Predicted structure of the Minibody, a 61-residue designed protein. Black regions correspond to the hypervariable regions H1 and H2 of the immunoglobulin used as a template for the design.

domized obtaining a repertoire of 50 million Minibodies displayed on phage (Venturini et al., 1994). From this library it was possible to select a molecule specifically able to bind a cytokine, Interleukin 6 (IL-6). The selected Minibody was also shown to be an effective inhibitor of the biological activity of IL-6 (Martin et al., 1994). The sequence of this molecule differs from that of the original Minibody, shown not to interact with the cytokine, only in the second randomized region, namely, the residues corresponding to the H2 loop of immunoglobulins. A synthetic linear peptide with the same sequence as the selected loop, however, is unable to interact with the cytokine, showing that the structure assumed by the loop is important for its function.

This result proves that it is possible to select conformationally constrained peptides with the desired activity, and because there is no reason to believe that IL-6 represents a unique example, this type of approach can represent a novel route to drug discovery.

There are a number of problems that limit the usefulness of proteinaceous material as therapeutic agents. For example, peptides often have poor bioavailability and are subject to proteolysis.

Consequently, the next step in the development of an inhibitor based on a peptidic lead is to prove that it is indeed possible to convert such a template structure into a nonpeptidic molecule. Although it has still to be proven for the Minibody loop case, such a strategy has already been successful in the case of immunoglobulin loops.

Saragovi and co-workers (Saragovi et al., 1991) produced a small organic molecule mimicking the L2 hypervariable loop of an immunoglobulin (MAb 87.92.6) able to bind to the cell surface receptor for reovirus type 3. Functional tests showed that the nonpeptidic mimetic was able to bind to the receptor and inhibit cellular proliferation. As expected the mimetic is insensitive to proteolytic enzymes and certainly constitutes a valid example of the usefulness of the selection of structurally constrained peptide ligands, followed by the generation of their nonpeptidic mimetics for therapy and diagnosis (Saragovi et al., 1992).

IX. CONCLUSIONS

The discussion about loop structures and their prediction methods presented here cannot of course be considered complete, but it is meant to give the reader a sense of the present level of our knowledge and of the difficulties that we have faced and will face while attempting to complete our understanding of protein structures.

The important question is whether we can extrapolate from here and forecast what will happen in this field of computational biology in the future.

Of course, the number of available structures will continue to grow, and maybe we will reasonably soon see all possible folds that a protein structure can assume. Many prediction techniques will benefit enormously from this. Model building by homology certainly will benefit, because the likelihood of finding the structure of a protein similar to the one under study will increase, and also the sequence-to-fold assignment prediction techniques are likely to improve.

One prediction is certainly easy to make today, that we will both ask and need more and more details from our modeling experiments. This part of the job will not necessarily be simplified by the availability of an increased number of "solved examples" of the folding problem. It is much more likely that we will need to develop different techniques altogether to be able to consider our models reliable and accurate. What can certainly be foreseen is that the availability of a larger number of structures will allow us to test our assumptions on a larger number of cases. And, as I have shown here in the case of loop searching, this can sometimes very effectively highlight and anticipate future problems of the selected technique.

The progress of molecular biology and especially the interest in "constructing" nonnatural proteins can, however, represent the key step. Natural proteins are, as mentioned, examples of the solution of the problem achieved through an enormous number of steps of evolution and selection, and we can only observe one of the possible solutions raised by an intractably large number of independent variables. Nonnatural proteins, designed in the laboratory, are usually far less perfect than the natural ones, but can allow us to study "intermediate steps" in the pathway toward the "right" fold and might highlight properties of protein structure which still elude our comprehension.

ACKNOWLEDGMENTS

I gratefully acknowledge the advice and continuous scientific support of Profs. A.M. Lesk and R. Cortese and am especially grateful to Dr. Andrew Wallace for discussions and critical reading of the manuscript.

REFERENCES

Alzari, P. M., Lascombe, M. B., & Poljak, R. J. (1988). Three-dimensional structure of antibodies. Annu. Rev. Immunology 6, 555–580.
Benner, S. A. (1992). Predicting de novo the folded structure of proteins. Curr. Opinion Struct. Biol. 2, 402–412.
Bianchi, E., Sollazzo, M., Tramontano, A., & Pessi, A. (1993a). Chemical synthesis of a beta protein through the flow polyamide method. Int. J. Peptide Protein Res. 41, 385–393.
Bianchi, E., Sollazzo, M., Tramontano, A., & Pessi, A. (1993b). Affinity purification of a difficult sequence protein. Int. J. Peptide Protein Res. 42, 93–96.
Bianchi, E., Venturini, S., Pessi, A., Tramontano, A., & Sollazzo, M. (1994). High level expression and rational mutagenesis of a designed protein, the minibody: from an insoluble to a soluble molecule. J. Mol. Biol. 236, 649–659.
Bowie, J. U., Luethy, R., & Eisenberg, D. (1991). A method to identify protein sequences that fold into a known three-dimensional structure. Science 253, 164–170.
Bowie, J. U., & Eisenberg, D. (1993). Inverted protein structure prediction. Curr. Opinion Struct. Biol. 3, 437–444.
Bruccoleri, R. E., Haber, E., & Novotny, J. (1988). Structure of antibody hypervariable loops reproduced by a conformational search algorithm. Nature 335, 564–568.
Chothia, C. & Lesk, A. M. (1987). Canonical structures for the hypervariable regions of immunoglobulins. J. Mol. Biol. 196, 901–917.
Chothia, C., Lesk, A. M., Levitt, M., Amit, A. G., Mariuzza, R. A., Philips, S. E. V., & Poljak, R. (1987). The predicted structure of immunoglobulin D1.3 and its comparison with the crystal structure. Science 196, 901–918.
Chothia, C., Lesk, A. M., Tramontano, A., Levitt, M., Smith-Gill, S. J., Air, G., Sheriff, S., Padlan, E. A., Davies, D., Tulip, W., & Colman, P. (1989). The conformation of immunoglobulin hypervariable regions. Nature 342, 877–883.
Chothia, C. & Lesk, A. M. (1986). The relation between divergence of sequence and structure in proteins. EMBO J. 5, 823–826.
Cohen, B. I. & Cohen, F. E. (1994). Prediction of protein secondary and tertiary structure. In: Biocomputing. Informatics and Genome Projects (Smith, D.W., Ed.), pp. 202–232, Academic, New York.
Dayringer, H. E., Tramontano, A., Sprang, S. R., & Fletterick, R. J. (1986). Interactive program for visualization and modelling of proteins, nucleic acids and small molecules. J. Mol. Graphics 4, 82–87.
Felici, F., Castagnoli, L., Musacchio, A., Jappelli, R., & Cesareni, G. (1991). Selection of antibody ligands from a large library of oligopeptides expressed on a multivalent exposition vector. J. Mol. Biol. 222, 301–310.

Foote, J. & Winter, G. (1992). Antibody framework residues affecting the conformation of the hypervariable loops. J. Mol. Biol. 224, 487–499.

Godzik, A., Kolinski, A., & Skolnick, J. (1992). A topology fingerprint approach to the inverse protein folding problem. J. Mol. Biol. 227, 227–238.

Hagler, A. T. & Honig, B. (1978). On the formation of protein tertiary structure on a computer. Proc. Natl. Acad. Sci. USA 75, 554–558.

Jones, D. T., Taylor, W. R., & Thornton, J. M. (1992). A new approach to protein fold recognition. Nature 358, 86–89.

Jones, T. A. (1985). Interactive computer graphics: FRODO. Methods Enzymol. 115, 157–171.

Jones, T. A. & Thirup, S. (1986). Using known structures in protein model building and crystallography. EMBO J. 5, 819–822.

Jones, T. A., Bergdoll, M., & Kjeldgaard, M. (1990). In: Crystallographic and Modeling Methods in Molecular Design (Bugg, C. E. & Ealick, S. E., Eds.), pp. 189–199, Springer, New York.

Kabat, E. A., Wu, E. T., Perry, H. M., Gottesman, K. S., & Foeller, C. (1991). Proteins of immunological interest. U. S. Department of Health and Human Service NIH, Bethesda, MD.

Kabsch, W. & Sander, C. (1984). On the use of sequence homologies to predict protein structure: identical pentapeptides can have completely different conformations. Proc. Natl. Acad. Sci. USA 81, 1075–1078.

Karplus, M. & Weaver, D. L. (1976). Protein-folding dynamics. Nature 260, 404–406.

Kraulis, P. J. (1991). MOLSCRIPT. A program to produce both detailed and schematic plots of protein structures. J. Appl. Crystallogr. 24, 946–950.

Lesk, A. M. & Rose, G. D. (1981). Folding units in globular proteins. Proc. Natl. Acad. Sci. USA 78, 4304–4308.

Lesk, A. M. & Tramontano, A. (1993). An atlas of antibody combining sites. In: Structure of Antigens (Van Regenmortel, M. H. V., Ed.), pp. 1–29, CRC, Ann Arbor.

Lesk, A. M. (1994). Computational molecular biology. In: Encyclopedia of Computer Science and Technology (Kent, A. & Williams, J. G., Eds.), pp. 101–165, Marcel Dekker, New York.

Leszczynski, J. F. & Rose, G. D. (1986). Loops in globular proteins: a novel category of secondary structure. Science 234, 849–855.

Levitt, M. (1983). Protein folding by restrained energy minimization and molecular dynamics. J. Mol. Biol. 170, 723–764.

Lewis, P. N., Momany, F. A., & Scheraga, H. A. (1973). Chain reversal in proteins. Biochim. Biophys. Acta 303, 211–229.

Martin, A. C. R., Cheetham, J., & Rees, A. R. (1989). Modeling antibody hypervariable loops: A combined algorithm. Proc. Natl. Acad. Sci. USA 86, 9268–9272.

Martin, F., Toniatti, C., Salvati, A. L., Venturini, S., Ciliberto, G., Cortese, R., & Sollazzo, M. (1994). The affinity selection of a minibody polypeptide inhibitor of human interleukin-6. EMBO J. 13, 5303–5309.

Milner-White, E. J., Ross, B. M., Ismail, R., Belhadi-Mostefa, K., & Poet, R. (1988). One type of gamma turn rather than the other gives raise to chain reversal in proteins. J. Mol. Biol. 204, 772–782.

Milner-White, E. J. (1990). Situation of gamma turns in proteins. J. Mol. Biol. 216, 385–397.

Montelione, G. T. & Scheraga, H. A. (1989). Formation of local structures in protein folding. Acc. Chem. Res. 22, 70–76.

Moult, J. & James, M. N. G. (1986). An algorithm for determining the conformation of polypeptide segments in proteins by systematic search. Proteins 1, 146–163.

Moult, J. & Unger, R. (1991). An analysis of protein folding pathways. Biochemistry 30, 3816–3824.

Ouzonis, C., Sander, C., Scharf, M., & Schneider, R. (1993). Prediction of protein structures by evaluation of sequence-structure fitness. Aligning sequences to contact profiles derived from three-dimensional structures. J. Mol. Biol. 232, 805–825.

Pascarella, S. & Argos, A. (1992). Analysis of insertions/deletions in protein structures. J. Mol. Biol. 224, 461–471.

Pessi, A., Bianchi, A., Crameri, A., Venturini, S., Tramontano, A., & Sollazzo, M. (1993). A designed metal binding protein with a novel fold. Nature 362, 367–369.

Reichmann, L., Clark, M., Waldmann, H., & Winter, G. (1988). Reshaping human antibodies for therapy. Nature 332, 323–327.

Richardson, J. S. (1981). The anatomy and taxonomy of protein structure. Adv. Protein Chem. 34, 167–339.

Rose, G. D. (1978). Prediction of chain turns in globular proteins on a hydrophobic basis. Nature 272, 586–590.

Rose, G. D., Gierasch, L. M., & Smith, J. A. (1985). Turns in peptides and proteins. Adv. Protein Chem. 37, 1–109.

Rost, B. & Sander, C. (1993). Prediction of protein secondary structure at better than 70% accuracy. J. Mol. Biol. 232, 584–599.

Rost, B., Schneider, R., & Sander, C. (1993). Progress in protein structure prediction? Trends Biochem. Sci. 18, 120–123.

Saragovi, H. U., Fitzpatrick, D., Raktabutr, A., Nakanishi, H., Kahn, M., & Green, M. J. (1991). Design and synthesis of a mimetic from an antibody complementarity-determining region. Science 253, 792–795.

Saragovi, H. U., Green, M. I., Chrusciel, R. A., & Kahn, M. (1992). Loops and secondary structure mimetics: development and applications in basic science and rational drug design. BioTech 10, 773–778.

Sibanda, B. L. & Thornton, J. M. (1985). Accommodating sequence changes in beta hairpins in proteins. Nature 317, 170–174.

Sippl, M. J. & Weitckus, S. (1992). Detection of nativelike models for amino acid sequences of unknown three-dimensional structure in a data base of known conformation. Proteins 13, 258–271.

Skolnick, J. & Kolinski, A. (1989). Computer simulations of globular protein folding and tertiary structure. Annu. Rev. Phys. Chem. 40, 207–235.

Skolnick, J., Kolinski, A., Brooks, C. L., III, Godzik, A., & Rey, A. (1993). A method for predicting protein structure from sequence. Curr. Biol. 3, 414–423.

Smith, G. P. (1991). Surface presentation of protein epitopes using bacteriophage expression systems. Curr. Opinion Biotech. 2, 668–673.

Smith, G. P. & Scott, J. K. (1993). Libraries of peptides and proteins displayed on filamentous phage. Methods Enzymol. 217, 228–257.

Tramontano, A., Chothia, C., & Lesk, A. M. (1989). Structural determinants of the conformations of medium-sized loops in proteins. Proteins 6, 382–394.

Tramontano, A., Chothia, C., & Lesk, A. M. (1990). Framework residue 71 is a major determinant of the second hypervariable region in V_H domains of immunoglobulins. J. Mol. Biol. 215, 175–182.

Tramontano, A. & Lesk, A. M. (1992). Common features of the conformations of antigen-binding loops in immunoglobulins and application to modeling loop conformations. Proteins 13, 231–245.

Vasquez, M. & Scheraga, H. A. (1985). Use of buildup and energy minimization procedures to compute low-energy structures of the backbone of enkephalin. Biopolymers 24, 1437–1447.

Venkatachalam, C. (1968). Stereochemical criteria for polypeptides and proteins. V. Conformation of three linked peptide units. Biopolymers 6, 1425–1436.

Venturini, S., Martin, F., & Sollazzo, M. (1994). Phage display of the minibody: a beta scaffold for the selection of conformationally constrained peptides. Protein Pept. Lett. 1, 70–75.

Vriend, G. (1990). WHAT IF: a molecular modelling and drug design program. J. Mol. Graphics 8, 52–56.

Wetlaufer, D. B. (1973). Nucleation, rapid folding, and globular intrachain regions in proteins. Proc. Natl. Acad. Sci. USA 70, 687–701.

Wilmot, C. & Thornton, M. J. (1988). Analysis and prediction of the different types of beta turns in proteins. J. Mol. Biol. 203, 221–232.

INDEX

Ab initio calculation, 241, 252-253
Accepted Point Mutation, 139
Acceptor site, 224
"Address labels", 2
Alignment, application to, 144-150
    optimal score, 144
Altplets, 128
Alu database, 172, 177, 186
    Alu-like sequences, 172, 178-179, 187
Amino acid usage, 124-125
    biased amino acid compositions, 170
    correlations, 125-126
    run and periodic patterns, 126-128
    substitution scoring matrices, 138-139
Amphiphilic molecules, 18
Ancient conserved regions (ACRs), 213
Anisotropic interaction, 29
Annealing technique, 69
Antigen binding sites, 244
Archetype hypothesis, 68-69
Artifactual matches, 172
Associative memory procedures, 67
Autoimmune diseases, 127
Averaged distances, 79

B1/B2 repeats, 172
B-factors, 106
B-hairpin/turns, 243

Baccatin core, 89
Back modeling, 113
Bacteriorhodopsin, 54-55
Basal gene biochemistry, 221-222
Best score, 194
Bioactive conformer, 89
Biological membranes and integral proteins, 16-17
Biological significance, measurement of, 167
BLAST, 176-179, 187-189, 199
    algorithm, 168
    scoring system, 169-170
BLASTN, 164, 178
BLASTP, 137-138, 143, 146, 176-178, 185-186, 200-202
BLASTX, 168, 178
BLOCKS, 164
BLOSUM matrices, 140, 143, 157
Bootstrapping, 115

C+G content, 217
"Canonical structures", 244
Cap-signal, 221
Carrier molecule, 92
CCAAT-box, 221
CDNA sequence, 213
Cell adhesion proteins, 75
Cell membrane, 17
Cellular recognition, 75
Channel, formation, 26-30
    transmembrane channel, 58-60

Chaperonins, 110
Charge clusters, 126
Chloroplast transit peptides, 6
Circulin A, 91
Coding regions, measures, 214
　statistical regularities, 214-215
Codon usage table, 214
Coexisting phases, 32
Coherence length, 38, 44-46
Computer-simulation techniques, 30-34
　finite-size scaling, 30, 32-34
　Monte Carlo techniques, 21, 30, 78
　reweighting techniques, 30-32
　see also Distribution functions
Configuration space, 71
Consensus sequence, 218
Cooperative phenomen, 22
Coordinate binding, 223
Critical mixing, 41-44
Critical point, 34, 38
Crystal packing, 107
Crystallography, 94

Database, 114, 163-172
　homology, 188
　"look up", 184-187
　matches, 172
　predictive powers, 184
　problems, 164-165
　searches, 143-144, 242
　similarity searches, 212
dbEST, 165, 213, *see also* ESTs
DBSITE, 197, 202
DDBJ, 163, 168, 172
Defensins, 90-92
Definite clause grammar, 225
Degenerate distances, 85
Deletions, 242
Density of states, 31
Dimer, formation of, 27
　gramicidin-A dimer, 26-28, 56-58

Dipalmitoyl phosphatidylcholine (DPPC) bilayer, 24, 36
Discriminant, 214-215
Displacements, 104
Distance measure, 150
Distance restrained, 69
Disteaoyl phosphatidylcholine (DSPC), 38, 54
Distribution functions, 30-32
　extreme value distribution, 196
　quantile distributions of amino acid usage, 124-125
　random scores, 194
Disulfur bonding scheme, 91
Dominant structures, 85
Donor site, 224
Dynamic membrane heterogeneity, 38-40
Dynamic programming, 144, 151, 227

Effective force fields, 67
Electron density map, 95
EMBL, 163, 168, 172
Energy barriers, 72
Entropy, 50, 57, 142
Equivalent protons, 85
Error estimates, 109
ESTs (expressed sequences tags), 162-165, 185-187
Excursion plot, 134
Exons, 180, 184, 188, 223
　detection of exons, 188
　prediction, 182
Expectation maximization (EM) algorithm, 152-157
Extremal spacings, 130

FASTA, 189
Feature table, 224
Feedback-restrained molecular dynamics (FRMD), 68-71
　archetype hypothesis, 68-69
　calculation, sensitivity, 114

Index

methodology, 69-75
RGD-containing peptides, 75-80
SA simulation of GRGDS, 78-82
taxol protocol, 84-86
trajectory, 80
validation, 85-89
Fibronectin, 76
Filamentous phage, 254
Filter option, 167, 176
Final geometries, 80
Finite-size scaling, 30, 32-34
First-order transitions, 33
Fisher-Tippet (extreme value distribution), 196
Fluctuations, 19, 37-40, 50-51
Fluidity, 17
Folding homology, 68, 109
Force field, 69, 114
Fosljun DNA-binding proteins, 156
Fragmentary results, integration of, 223-228
Free energylike functions, 32
Functional motif, 200

G-site, 105
Gap scores, 70, 144
GC-box, 221
GenBank, 163-164, 168, 172, 180
Genes, gene expression, regulation of, 222-223
  identification, 210-211
Genetic algorithm, 68
Genome sequencing, 68, 162
Gibbs sampling, 152, 155-157
Global comparison values, 146
Global minimum, 72
Global sequence features, 124-126
Glucanohydrolase motif, 201
Glucose-6-phosphate-isomerase sequences, 147-150
Glutathione S-transferases (GSTs), 93-94
Grail, 181-184, 187-188

Gramicidin-A dimer, 26-28, 56-58
GRGDS peptides, 76
Gumbel distribution, 196

H-site, 105
Harmonic restraints, 69, 93
Heat Shock protein, 110-111
  modeling, 108-114
Helix-turn-helix-DNA binding proteins, 156-157
Heterogeneity, 17
Hexamers, 181
Hidden Markov Model (HMM), 152, 225
High Scoring Segment Pairs (HSSPs), 132-134, 144-145
  identification, 145-146, 150
  maximal scoring order, 136
High temperature dynamics, 72, 80-82
Histogram, 31
HLA motifs, 108, 110
Homology, 95, 110, 241
  folding, 68, 109
  modeling, 69, 98, 102, 109
  protein, 68
Homooligo peptide, 128
Horizontal transfer, 150
HOVERGEN, 164
Human genomic contig, 183
"Humanizing" antibodies, 253-256
Hydrogen bonding, 57
Hydrophobic effect, 24-26
  hydrophobic matching, 20-21, 25, 40, 45, 51
  hydrophobic segments, 132
  hydrophobic thickness, 19, 56

Immunoglobulin hypervariable loops, 242, 244
Integral proteins, 16-17, 40-41, 53-55
Integration, 223-228
  kinds of, 224

logical, 225-227
syntactical, 224-225
Integrin family, 75
Interconverting conformations, 90
Interdiffusion, 52
Interfaces, 51, 230
Interfacial tension, 32, 38
Interleukin 6, 255
Interproton distance, 70, 80, 84
Intron, 223
Inverse protein folding, 69
Ion permeability, 39
Ionic channel, 28, 56
Irrelevant matches, 180

Jellyroll topologies, 68
"Junk" sequences, 177

Kendall Tau correlation coefficient, 125
Knowledge-based potentials (KBP), 67

Large molecule aggregates modeling methods, 93-98
   crystallographic restraint, 94-95
   feedback restrained molecular dynamics, 95-98
   Glutathione S-transferases, 93-94
Large-scale sequence analysis, 166-168
Lateral pressure, 24
"Leucine zippers", 129, 157
Line-1, 177
Linear motifs, 2
Lipid bilayers, compositional profiles, 51-53
   critical mixing and phase diagram, 41-44
   domains, 38-40
   Gramicidin-A dimer, 26-28, 56-57
   lateral distribution, 45-51

lipid-acyl chain order-parameter, 44-45
lipid-polypeptide mixtures, 41
one-component bilayers, 34-37
polypeptide aggregation, 58-61
specificity, 51-52
structure and thermodynamics, 53-55
transmembrane channel formation, 58-61
two-component bilayers, 37-38
Lipid-protein interactions, 19-20, 25-26
   computer simulations, 21-22
   Gramicidin-A dimerization, 26-28, 56-57
   polypeptide aggregation and channel formation, 28-30
   pure lipid bilayers, 23-25
Liposomes, 18
Local comparison values, 149
Local minimum, 75
Local sequence features, 126-128
Logical integration, 225-227
Loops, 240
   conformations, 251-252
   immunoglobulin hypervariable loops, 242, 244
   medium-sized loops, 248-250
   short loops/sequence patterns, 243-248
   structure, 242
   transplant, 247
   WS-loops, 244
Low temperature protocol, 75
Low-entropy matches, 187
   overwhelming output problem, 168-169
   simple sequences, 168, 171

Markov Model (Hidden), 152, 225
Matching score, 189
Mattress model, 25, 41, 54

MD self-consistent analysis, 95, 109
Membrane proteins, B-Barrel proteins, 9
  Helix-bundle proteins, 8-9
  membrane heterogeneity, 38-40
  topology and structure, 7-9, 18
Memory weight, 75
MER, 177
Methyl hydrogens, 85
MHC class II peptide, 110
Microscopic models of lipid-protein interactions, 23-30
Minibody, 254
Minimization, 67
Missing data problem, 191
Mitochondrial targeting peptides, 5-6
MKSITE program, 201-202
Molecular dynamics (MD) techniques, 22, 69-70
Molecular replacement, 94
Monte Carlo methods, 21, 30, 78
Motifs, 189
  identification, 152-157
  motif scores, 195, 199
  as profiles, 151-152
  search of, 199
Multiple alignment, 151
Multiple conformations, 79, 83
Multiple sequence comparisons, 123, 150-157
Multiplets, 128

Neural network, 3, 67, 181
NMR analysis, 69
Noninformative matches, 174
Nontransitivity, 150
Nuclear localization signals, 6-7
Nuclear Overhausen effect (NOE), 69-70, 80
  NOE-derived distances, 78, 89

Open reading frame, 217
Order-parameter profiles, 44-45

Pairwise sequence comparisons, 136-137
  amino acid substitution scoring matrices, 138-139
    BLOSUM matrices, 140, 143, 157
    others, 141-142
    Pam matrices, 139-140, 143, 192
  statistical theory, 137-138
PAM matrices, 139-140, 143, 192
Pattern recognition, functional sites, 218-221
Periodic patterns, 127, 129
Permeability, 26
Pero-166, 168, 174-175, 179-180

R-scans, 129-130
  statistics, 130
Ramachandran plots, 80-82, 99
Random mutation modeling, 113-114
Random scores, 194
  best score, 196
  distribution of, 194
Random sequence models, 126
"Relative mutability", 139
Repetitive structures, 128-129, 172
Residues, clustering of, 126
Restrained molecular dynamics, 69-70
Restraint force, 104
Retention signals, 7
Reverse modeling, 107
Reweighting techniques, 30-32
RGD-containing peptides, 75-80, 75-82
Rotamer library, 109
RPS4 gene, 184

SAPS program, 128-129
Scale factors, 93
Score threshold values, 152, 194, 198, 202
Score-based sequence analysis, 131-136
SEG program, 175
Self-modeling, 111-113
Semiempirical calculations, 252
Sequences, 162-163
    alignment, 66
    anonymous, 188
    data banks, 68
    genomes, 162
    homogeneity, 129
    motifs, 188-189, 213, 242
    partial cDNA sequences, 162
Short period repeats, 175
Signal peptides, 3-5
Significant repeats, 128
Similarity searches, 212-214
Simulated annealing (SA) protocol, 72, 98
Single sequence statistics, 123-136
    analysis of spacings, 129-131
    global sequence features, 124-126
        composition, 124-126
        length, 124
    local sequence features, 126-128
    repetitive structures, 128-129
    score-based sequence analysis, 131-136
Smart search strategies, 67, 70
Sorting signals in proteins, 2-7 *see also* Proteins
Spacings, analysis of, 129-131
    minimal and maximal spacings, 130
    R-scan statistics, 130
Specific heat, 33, 36, 54
Spectroscopic order parameters, 40
Splice sites, 186, 189, 220-221, 225

SSPA (Significant Segment Pair Alignment), 144-145, 150
    determination of SSPA values, 146-150
Statistical models, 108, 122, 137-138
    phylogenetic reconstructions, 123
    structure prediction, 123
Statistical significance, 167, 188
    motif scores, 195, 199
Steric interaction, 89
Stochastics orderings, 125
Substitution scores, 137-139, 142-144
Success run, 127
Swiss-Prot sequences, 164, 200, 202
Synchronism, 104
Syntactical integration, 224
Systemic lupus erythematosus, 127

"Target scores", 132, 134
TATA-box, 221
Taxol, 83-90
    FRMD protocol for taxol, 84-85
TBLASTN, 178, 187
TBLASTX, 185, 187
Temperature ramp, 74
Template-forcing technique, 69, 95
Tethered MD calculations, 95
Thermalization, 75, 97
Threading, 241
Transcription factors, 222
Transit peptides, 6
Transition enthalpy, 24, 43
Transition matrix, 192-193
Transition temperature, 24
Translation initiation, 222

Validation protocol, back modeling, 113
    experiments, 107-108
    random mutation modeling, 113-114
    self-modeling, 111-113
Vesicles, 18

W-loops, 244
Water channel, 28
Weight matrices, 151, 198, 202
Wetting phenomena, 53

XBLAST, 177-180, 186
XNU, 179-180

"XNU + SEG" option, 186
XNU program, 175-176
Xplor, 78

Y-turns, 244

J A I P R E S S

Advances in Computational Biology

Edited by **Hugo O. Villar,** *Terrapin Technologies Inc., South San Francisco, California*

Volume 1, 1994, 262 pp. $97.50
ISBN 1-55938-633-9

CONTENTS: Fluctuations in the Shape of Flexible Macromolecules, *Gustavo A. Arteca.* Modeling Nucleic Acids: Fine Structure, Flexibility and Conformational Transition, *Richard Lavery.* Molecular Modeling: An Essential Component in the Structural Determination of Oligo- and Polysaccharides, *Serge Perez,* Anne Imberty and *Jeremy P. Carver.* Hydration of Carbohydrates as Seen by Computer Simulation, *J. Raul Grigera.* Studies of Salt-Peptide Solutions: Theoretical and Experimental Approaches, *Gail E. Marlow and B. Montgomery Pettitt.*

> **FACULTY/PROFESSIONAL** discounts are available in the U.S. and Canada at a rate of 40% off the list price when prepaid by personal check or credit card and ordered directly from the publisher.

JAI PRESS INC.
55 Old Post Road # 2 - P.O. Box 1678
Greenwich, Connecticut 06836-1678
Tel: (203) 661- 7602   Fax: (203) 661-0792

Advances in Organ Biology

Edited by **E. Edward Bittar,** *Department of Physiology, University of Wisconsin Medical School*

Volume 1, Pregnancy and Parturition
In preparation, Summer 1996
ISBN 1-55938-639-8   Approx. $97.50

Edited by **Tamas Zakar,** *Perinatal Research Centre, University of Alberta*

CONTENTS: Preface, *Tamas Zakar.* Late Pregnancy and Parturition in the Sheep, *Wendy J. McLaren, Ian R. Young, and Gregory E. Rice.* Regulation of Oviposition, *Frank Hertelendy, Kiyoshi Shimada, Miklós Tóth, Mikós Molnár, and Kousaku Tanaka.* Initiation of Parturition in Non-human Primates, *Jonathan J. Hirst and Geoffrey D. Thorburn.* The Physiology of Human Parturition, *Jane E. Mijovic and David M. Olson.* The Trophoblast as an Active Regulator of the Pregnancy Environment in Health and Disease: An Emerging Concept, *Donald W. Morrish, Jamal Dakour, and Hongshi Li.* The Endocrinology of Late Pregnancy and Parturition, *Tamas Zakar and Brian F. Mitchell.* Steroid Modulation of Myometrial Structure and Function During Pregnancy, *Charles A. Ducsay and Joon W. Rhee.* Determinants of Reproductive Mortality and Preterm Childbirth, *John A. McCoshen, P. Audrey Fernandes, Michael L. Boroditsky, and James G. Allardice.*

> **FACULTY/PROFESSIONAL** discounts are available in the U.S. and Canada at a rate of 40% off the list price when prepaid by personal check or credit card and ordered directly from the publisher.

JAI PRESS INC.
55 Old Post Road # 2 - P.O. Box 1678
Greenwich, Connecticut 06836-1678
Tel: (203) 661- 7602   Fax: (203) 661-0792

Biomembranes
A Multi-Volume Treatise

Edited by **A.G. Lee,** *Department of Biochemistry, University of Southampton*

"Progress in understanding the nature of the biological membrane has been very rapid over a broad front, but still pockets of ignorance remain. Application of the techniques of molecular biology has provided the sequences of a very large number of membrane proteins, and led to the discovery of superfamilies of membrane proteins of related structure. In turn, the identification of these superfamilies has led to new ways of thinking about membrane processes. Many of these processes can now be discussed in molecular terms, and unexpected relationships between apparently unrelated phenomena are bringing a new unity to the study of biological membranes.

The quantity of information available about membrane proteins is now too large for any one person to be familiar with anything but a small part of the primary literature. A series of volumes concentrating on molecular aspects of biological membranes therefore seems timely. The hope is that, when complete, these volumes will provide a convenient introduction to the study of a wide range of membrane functions."

— *From the Preface*

Volume 3, Receptors of Cell Adhesion and Cellular Recognition
In preparation, Summer 1996
ISBN 1-55938-660-6   Approx. $97.50

CONTENTS: Molecules of Cell Adhesion and Recognition: An Overview, *R. Marsh and R. Brackenbury.* Cell Recognition Molecules of the Immunoglobulin Superfamily in the Nervous System, *G. Gegelashvili and E. Bock.* T-cell Antogen Receptors, *C. Horgan and J.D. Fraser.* The Major Presentation and Histocompatibility Complex, *J. Colombani.* Cadherins: A Review of Structure and Function, *J. Wallis, R. Moore, P. Smith and F.S. Walsh.* The Integrin Family, *R.D. Bowditch and R.J. Faull.* The Selectin Family, *M.A. Jutila.* The CD44 Family of Cell Adhesion Molecules: Functional Aspects, *C. B. Underhill.* Membrane-Associated Mucins, *H.L. Vos, J. Wesseling and J. Hilkens.* Platelet Membrane Glycoproteins, *K. J. Clemetson.* Immunoglobulin Fc Receptors: Diversity, Structure and Function, *P.M. Hogarth and M.D. Hulett.*

Also Available:
Volume 1 (1995) $97.50
   Volume 2 (2 Part Set) $195.00